Green Building Design and Life Cycle Management

Green Building Design and Life Cycle Management

Edited by Jackson Owen

Clanrye International
www.clanryeinternational.com

Clanrye International,
750 Third Avenue, 9th Floor,
New York, NY 10017, USA

ISBN: 978-1-64726-683-7

Cataloging-in-Publication Data

Green building design and life cycle management / edited by Jackson Owen.
p. cm.
Includes bibliographical references and index.
ISBN 978-1-64726-683-7
1. Sustainable buildings. 2. Buildings--Energy conservation. 3. Architecture--Environmental aspects.
4. Architectural design. 5. Sustainable buildings--Design and construction. I. Owen, Jackson.
TH880 .G74 2023
720.47--dc23

For information on all Clanrye International publications
visit our website at www.clanryeinternational.com

CLANRYE INTERNATIONAL

Contents

Preface VII

Chapter 1 **Techno-Economic Analysis of Green Building Codes in United Arab Emirates Based on a Case Study Office Building** 1
Hiba Najini, Mutasim Nour, Sulaiman Al-Zuhair and Fadi Ghaith

Chapter 2 **Review on Energy and Fire Performance of Water Wall Systems as a Green Building Façade** 23
Uthpala Rathnayake, Denvid Lau and Cheuk Lun Chow

Chapter 3 **Risk Factors Impacting the Project Value Created by Green Buildings** 50
Wael Alattyih, Husnain Haider and Halim Boussabaine

Chapter 4 **Impact of Heat Pump Flexibility in a French Residential Eco-District** 82
Camille Pajot, Benoit Delinchant, Yves Maréchal and Damien Frésier

Chapter 5 **Impact of Service Life on the Environmental Performance of Buildings** 96
Shahana Y. Janjua, Prabir K. Sarker and Wahidul K. Biswas

Chapter 6 **Mechanical and Durability Properties of Green Star Concretes** 119
Faiz Shaikh

Chapter 7 **Comprehensive Evaluation of Carbon Emissions for the Development of High-Rise Residential Building** 131
Stephen Y. C. Yim, S. Thomas Ng, M. U. Hossain and James M.W. Wong

Chapter 8 **Industrial Data-Based Life Cycle Assessment of Architecturally Integrated Glass-Glass Photovoltaics** 150
Jeeyoung Park, Dirk Hengevoss and Stephen Wittkopf

Chapter 9 **Life-Cycle Asset Management in Residential Developments Building on Transport System Critical Attributes via a Data-Mining Algorithm** 169
Umair Hasan, Andrew Whyte and Hamad Al Jassmi

Chapter 10 **Heat Stress Pattern in Conditioned Office Buildings with Shallow Plan Forms** 206
Upendra Rajapaksha

Chapter 11 **Photocatalytic Functionalized Aggregate: Enhanced Concrete Performance in Environmental Remediation**......227
Amer Hakki, Lu Yang, Fazhou Wang, Ammar Elhoweris, Yousef Alhorr and Donald E. Macphee

Permissions

List of Contributors

Index

Preface

This book was inspired by the evolution of our times; to answer the curiosity of inquisitive minds. Many developments have occurred across the globe in the recent past which has transformed the progress in the field.

A green building is defined as an integrated process that takes into consideration the impact of built environment on the natural environment. It aims to minimize harmful environmental effects and increase the value throughout the life cycle of a building or structure including its design, construction, operation, maintenance, renovation and demolition. The concept of green building emphasizes on making buildings more resource-efficient in terms of their usage of materials, water and energy while minimizing their negative effects on the natural environment and people's health. Green building technology focuses on low consumption, high efficiency, environmental protection, integration and optimization. This book includes some of the vital pieces of works being conducted across the world, on various topics related to green building design and their life cycle management. It will serve as a reference to a broad spectrum of readers.

This book was developed from a mere concept to drafts to chapters and finally compiled together as a complete text to benefit the readers across all nations. To ensure the quality of the content we instilled two significant steps in our procedure. The first was to appoint an editorial team that would verify the data and statistics provided in the book and also select the most appropriate and valuable contributions from the plentiful contributions we received from authors worldwide. The next step was to appoint an expert of the topic as the Editor-in-Chief, who would head the project and finally make the necessary amendments and modifications to make the text reader-friendly. I was then commissioned to examine all the material to present the topics in the most comprehensible and productive format.

I would like to take this opportunity to thank all the contributing authors who were supportive enough to contribute their time and knowledge to this project. I also wish to convey my regards to my family who have been extremely supportive during the entire project.

Editor

Techno-Economic Analysis of Green Building Codes in United Arab Emirates Based on a Case Study Office Building

Hiba Najini [1], Mutasim Nour [1,*], Sulaiman Al-Zuhair [2,*] and Fadi Ghaith [1]

[1] School of Engineering and Physical Sciences, Heriot-Watt University, Edinburgh EH14 4AS, UK; hn57@hw.ac.uk (H.N.); F.Ghaith@hw.ac.uk (F.G.)

[2] Chemical Engineering Department, UAE University, Al-Ain 15551, UAE

* Correspondence: mutasim.nour@hw.ac.uk (M.N.); s.alzuhair@uaeu.ac.ae (S.A.-Z.)

Abstract: Green building regulations in the United Arab Emirates are required to obtain building permits so that future construction projects can create a sustainable living environment. Emirates such as Abu Dhabi, Dubai, and Sharjah have specific green building regulations, whereas other emirates follow Abu Dhabi's regulatory criteria. Previous work fails to present a techno-economic cross-code analysis for various green building regulations in the UAE by evaluating energy and water performance. A case study using an existing high-rise green office building was formulated using the Integrated Environmental Solution: Virtual Environment (IES-VE) platform and the U.S. Leadership in Energy and Environmental Design (U.S. LEED) water consumption evaluation tool to study its energy and water performance, respectively. The archived results were used to devise an economic study based on the discounted cash flow technique. The principal findings of this research allowed us to determine a cross-code analysis and propose cost-effective trade-offs. These will aid the consultants and contractors in choosing appropriate green building regulations in the UAE by highlighting the potential of each parameter within green building regulations in terms of energy, water, and economic performance.

Keywords: business appraisal; energy modeling; IES-VE; energy performance; water performance

---

1. Introduction

Energy efficiency measures in the built environment sector are considered the low-hanging fruit because prior research indicates that the building sector consumes 80% of overall energy demand in the United Arab Emirates (UAE) and 40% across the globe [1,2]. These energy efficiency standards are promulgated within green building regulations, but constructing green buildings requires higher investment capital than ordinary buildings. The UAE has set a high bar by requiring all new construction projects to comply with green building regulations specific to each emirate to obtain a building permit in that region. Meanwhile, prescriptive guidelines in the UAE are becoming more stringent for sustainable living. Evaluating the technical performance of green building regulations in compliance with economic viability explores a balanced option for introducing (or revising) the benchmarking criteria that may be adopted by various stakeholders [3]. In other terms, this work proposes cost-effective trade-offs by performing a cross-code analysis based on technical and economic performance.

To evaluate the technical performance of regulations, relevant studies on Gulf Cooperation Council (GCC) countries are reviewed in this work. Green buildings and energy conservative measures were first recognized in the UAE in 2003 upon the issuing of Degree 66 under the legal affairs department, restricting envelope measures and suggesting heating, ventilation, and air conditioning (HVAC) designs. Later in 2008, the Government of Dubai tailored and calibrated a set of prescriptive and descriptive regulations confined to Dubai ports and the free-zone area based on the U.S. Green Building Council's Leadership in Energy and Environmental Design (USGBC LEED) evaluation system and liberalized to the UAE's environmental requirements known as Trakhees. These mainly focused on warehouses and factories, before Trakhees was revised to incorporate residential, commercial, recreational, industrial, and institutional development within the defined free-zone perimeter initiated in 2011.

After that, the Dubai municipality issued the prescriptive Dubai Municipality Green Building regulations (DMGB) [4,5] regulation endorsing sustainable living for all public sector buildings in 2012. These regulations govern the grantsmanship of green building permits based on design, construction, and operation compliance. Since March 2014, these codes have been revised to comply with the private building sector. Most importantly, these mandated regulations focus on constructional design specifications. Recently, the Dubai Municipality (DM) has started issuing Al'Safat regulations [6,7]. It is a rating system based on DMGB regulations, LEED, and ASHRAE to enforce a minimum of a bronze rating system for all new and existing building construction upon building permit application and renewal [8,9]. This enforcement technique is similar to Estidama, as all buildings within Dubai are required to meet green building standards as per Dubai's 2021 vision proclaimed in "Al'Safat".

On the other hand, in Abu Dhabi, Estidama was created in 2007 to help reach Abu Dhabi's 2030 sustainable living goal under the Urban Planning Council (UPC) [10]. Similar to Dubai's current (2016–2017) regulations, Abu Dhabi had issued prescriptive regulations since 2010 on Estidama's official website [11]. It requires that there be a minimum of a pearl 1 rating for all new communities, villas, and buildings within the private sector and a minimum of pearl 2 for all governmental projects for schools, mosques, and government buildings as of September 2010 [12]. Conversely, other rating systems cannot be transferred to the pearl rating system. Moreover, building permit renewal for existing buildings has to be re-indited to comply with Estidama guidelines [11,13]. Furthermore, all of these esteemed guidelines and regulatory standards comply with ASHRAE's HVAC requirements due to their renowned maturity within the sustainable environment field.

The parameters that impact the building energy intensity by targeting the cooling load demand were determined because space cooling is a major energy consumer in GCC countries classified under architectural, electrical, and mechanical design criteria [14–16]. Friess et al. [17] provided research-based contextual evidence that the variations in architectural design parameters corresponding to the thermal insulation structure of the building envelope yielded around 23.3% energy savings for skin-dominant buildings when the thermal transmittance value was changed from 2.4 to 0.6 W/m^2 K. The regulations may include the environmental impact based on the thickness of insulation as present in [18] for consultants to make an informed decision since a slight variation in thickness or R-Value for a high-rise building will be substantial [19]. Afshari et al. [20] and Radhi [21] identified 1–3% cooling load savings for non-skin-dominant buildings based on their simulation results. These results showed that energy-saving potential is dependent on the building's envelope feature, specifically, the window-to-wall aspect ratio. This conceptional idea was verified by Friess et al. [22] and Aboul-Naga et al. [23], who identified an energy-saving potential of 55% for high-rise non-skin-dominant commercial/office buildings, but electrical and mechanical measures were used in compliance with the architectural measures. Potential savings of 20–28% in cooling load were identified by modulating the thermal transmittance value of the fenestration from 2.4 to 1.47 W/m^2 K for high-rise buildings [17,21,24]. However, low-rise buildings with a similar wall-to-window ratio had energy savings of 12% [25] and a cooling load of 5% [20]. These identified potential margins indicate that the non-skin dominant high-rise buildings are sensitive to the slightest changes in glazing design parameters due to the reduction in solar gain through curtains with a large surface area [26,27]. Previous studies on the regulations in the UAE lack economical optimization techniques in terms of a cost-effective trade-off that considers energy efficiency.

Al-Absi et al. [28] reviewed various work on using phase change materials (PCMs) within the building walls to identify appropriate insulation materials. These PCMs predominately absorb energy, and the heat transfer from these materials can be controlled to potentially reduce HVAC energy consumption by maintaining the indoor temperature for longer if they are placed in close proximity to the indoor area. Li et al. [29] investigated the performance of PCMs for buildings in the United States, where 30%, 55%, and 60% reductions in HVAC requirements were identified for three sets of occupancy density. Thus, the applicability of this technique must be investigated with the GCC region alongside the use of geopolymetric coating for passive cooling [30]. However, Katanbafnasab [31] assessed Electrochromic Glazing in office buildings in the UAE that had up to 33% savings with the integration of photovoltaics (BIPV). Rehman et al. [32,33] identified solar insulation materials and summarized the performance of dry insulations through an experiment in Ras Al Khaimah. Akbari et al. [34,35] identified that a reflective roof yields 700 kWh/year of energy savings for a house and around 1000 kWh/year for an office building in very hot climate conditions. Ezzilden et al. [36] and Taleb [37] present energy-saving potential for multi-mode passive ventilation for low-energy and residential villas. The proposed technique is not suitable for high-rise buildings, but their observations of improvements in ventilation during nighttime may suggest that this could be useful in green building codes for residential typology.

Studies on the energy performance of a design that minimizes the internal gains through modifications of the architectural design measures showed energy savings of 28–46% [15,38,39]. The highest energy saving potential was achieved when envelope, glazing, heating, ventilation, and air conditioning (HVAC) set points were used with lighting [40,41], and the lowest was achieved using thermal insulation parameters [22,42–44]. The impact of these complex relationships between the parameters has not been reported in GCC-based research articles; however, Johnson et al. [45] and Wang et al. [46,47] identified the influence of orientation, window area, and glazing properties, such as the U-value, shading coefficient, and visible transmittance, on energy intensity for the office buildings in the United States using parametric sensitivity analysis techniques. Bande et al. [48] carried out a sensitivity analysis for a retrofit building simulation model in Abu Dhabi, confirming changes in design parameters produce different energy requirements. Banihashemi et al. [49] used a similar approach to verify the fenestration properties and reported different energy performance values depending on the location, climatic conditions, and physical properties in office buildings in the cities of Rasht, Ardabil, Yazd, and Bandar-e-Abbas. Ragab et al. [50] and Maria et al. [51] identified the impact of green roofs, and Callejas et al. [52] and Al-Sallal et al. [53] studied the potential of green roofs for passive cooling in tropical climates. These techniques showed potential, but high-rise buildings do not have enough area on which to install green roofs [54–56]. However, green roofs may be employed for villas, as carried out in [17,23,57], which constitute the other 25% of building typologies [52,58] in the UAE, with a potential saving of 12–18% [39]. Brien et al. [59] investigated the improvement of occupant working environments in building codes and standards. They analyzed the behavior pattern to identify the load profile to configure the HVAC settings in the Building Management System (BMS) [60,61]. Auer et al. [62] identified the critical settings needed for a school environment to enhance occupant well-being. Similar studies [60,63] within the region show similar results. The collated control algorithms that define load profiles for various office environments from international codes may be included as a suggestive code in the UAE, as there is a lack of detailed descriptive parameters. The primary goal of this paper is to identify cost-effective trade-offs while preparing an appraisal based on the technical performance and economic viability of various green building codes practiced in the UAE since new construction projects are required to comply with green building regulations. This research project used an existing green building as a case study by acquiring relevant building and unit price cost data from a reputable construction firm based in the UAE. According to market research [58,64,65], 75% of newly completed buildings in Dubai in 2019 were high-rise buildings. These high-rise buildings tend to have similar floor plans and are either of office building or residential building typology [66,67]. The HVAC system design in high-rise buildings is relatively common [40,44], and the occupancy density for an office building per unit area is much higher than that for residential buildings [68,69]. Additionally, the load profiles indicate the peak energy demand is during office hours. Thus, the selected case study building is a high-rise office building. The reference building was modeled using the building energy modeling (BEM) software (Integrated Environmental Solution: Virtual Environment (IES-VE) platform) to evaluate the building energy performance for each code [70]. The water consumption pattern was computed using the appropriate estimation tool provided by a renowned regulatory society (U.S. Green Building Council's Leadership in Energy and Environmental Design (U.S. LEED)) [71]. Then, a cash-flow plan for each regulatory standard was devised to determine its cost-effective benefits.

2. Methodology

Previous work has failed to present a cross-code analysis for evaluating the UAE's green building regulations in terms of technical and economic performance to determine a cost-effective trade-off. Figure 1 represents a five-phase framework methodology used in this research. Building energy modeling (BEM) under the IES-VE platform was used as the primary modeling tool to evaluate the building's energy performance.

Figure 1. Five-phase methodology block diagram implemented in this project.

The accuracy of the simulated results was determined through an uncertainty analysis. This analysis estimates the uncertainty of the simulation by quantifying possible error. The error margin is a relative percentage of the simulated value and measured value. The water consumption analysis was based on a computational approach prescribed within the U.S. LEED water consumption evaluation procedure, and the techno-economic evaluation was based on the discounted cash-flow financial evaluation technique. This technique provides a detailed financial analysis in terms of net present values and establishes a common base for comparing investment alternatives. The U.S. LEED water consumption evaluation uses flow rates, timing, occupant density, and frequency to estimate the water consumption precisely for various uses. This approach is superior to a water metering system, as it does not measure microscopic consumption patterns, such as the flow rates, to attain control over usage. These statistical results were combined with the simulated results to identify the potential of each green building standard.

2.1. Building Energy Performance Analysis

The energy demand associated with a building is broadly classified into the following categories:

- External Factors: yearly distributed climatic conditions such as the temperature, humidity, solar radiation, wind direction, and intensity.
- Prescriptive design parameters:
    - Geometrical conditions: the building orientation, shape, window-to-wall ratio, and exposed surface area created by shade from adjacent buildings.
    - Thermo-physical properties: the thermal transmittance of the external walls, roof, and external floors; thermal mass; and factor of solar heat gain through glazed surfaces.
- Hedonic parameters: set-point temperatures, the occupation density, the ventilation rate, and internal gains through lighting and equipment.

It must be noted that this research paper does not consider the use of renewable energy, as it is not required to attain green building status. However, this factor is required to attain net zero building status [72]. This research only focuses on evaluating the design parameter that contributes to energy demand, whereas renewable energy systems generate electricity to support the demand.

The quantitative design and hedonic parameters defined in Table 1 represent the synthesized fragment of each green building standard in the UAE, specifically, the Dubai Municipality Green Building regulations (DMGB); Trakhees; Al'Safat; the American Society of Heating, Refrigerating and Air-Conditioning Engineers (ASHRAE); and Estidama for office building typology. This research uses the high-rise office building with a high glazing ratio as its case study building. The use of this typology provides control over energy usage patterns from equipment and defined operational hours, whereas residential [73], hotel [42], or hospital [74] energy patterns are hedonic in nature due to varying occupancy and mentality. The variation in typology is predominately confined to the building usage that is controlled by equipment and lighting density within the regulations to an extent, as it is variable in terms of frequency of usage. Additionally, high-rise buildings with high glazing ratios are prominent in the UAE, where the market research indicates a 1.5-fold increase in the number of new buildings, with 90% of them being high-rise buildings, by 2020 [65,75,76]. Thus, the case study building includes the most critical envelope structure that cannot be controlled with communication systems, whereas the equipment and lighting power density may be controlled using smart communication systems for other typologies.

The parameters of the case study building are tabulated in the first column of Table 1 as a base reference for comparison. The three standards DMGB, Trakhees, and Al'Safat [77] were issued and are followed by the Emirate of Dubai, while Estidama is under the emirate of Abu Dhabi, and ASHRAE is an international standard used in the UAE. The quantitative design and hedonic parameters defined in Table 1 were predominately fed into the BEM software using the ModelIT package within the IES-VE platform. The hedonic parameter was defined within the existing case study building in terms of office use, which was then considered for the standards in this study. Figure 2 portrays a sample instance of solar shading analysis and an overview of the reference building's external structure. The building has four different floor plans to accommodate the ground podium, parking podium, office, mechanical floors, and terrace, with 15 defined zones for the entire building.

These variable categories govern the changes in energy intensity, while other categories were assumed to be constant. This is a significant limitation, and the numbers obtained may not be regarded as benchmarked numbers but simply a source for understanding the parametric behavior [78]. Equation (1) provides a computational formula for benchmarking the building energy performance commonly used by energy engineers in the building sector. This equation represents an index referred to as the energy utility index (*EUI*) for the ease of comparing building performance. It must be noted that this study normalizes electricity consumption per unit of conditioned floor area; this criterion is the most influential among other parameters, such as occupancy density, for the selected case study building typology.

$$EUI = \frac{Total\ Electricity\ Consumption\ (\mathrm{kWh})}{Total\ Conditioned\ Floor\ Area\ (\mathrm{m}^2)} \tag{1}$$

The simulation results were rationalized by incorporating the sensitivity analysis as represented by Equation (2) [79]. This sensitivity index is the percentage of output differences and is calculated by changing an input parameter from its minimum value ($E_{i,pi,min}$) to its maximum value ($E_{i,pi,max}$), normalized to the energy demand in the initial configuration of the building ($E_{d,i}$). The index indicates the parameter and model variability while confirming the true range of sensitivity of the independent input variable with respect to dependent output variable, i.e., the energy demand. A positive index confirms that the input has an independent nature and a lower probability contribution to the energy demand. However, a negative index indicates an indirect impact that increases the complexity of identifying the contributing factor affectivity and the certainty of the energy distribution profile. The strength of the dependency may be estimated using Pearson's correlation coefficient [80].

$$Sensitivity\ Index\ (SI) = \frac{E_i,\ p_{i,\ max} - E_i,\ p_{i,\ min}}{E_{d,\ i}} \tag{2}$$

Table 1. Summary of prescriptive regulatory specifications for analyzing energy performance.

Parameter	Reference Case Study Building	DMGB	Trakhees	Al' Safat	ASHRAE	Estidama
External Wall/Envelope (W/m^2 K)	0.57	0.57	0.28	0.57	0.7	0.32
Roof (W/m^2 K)	0.44	0.44	0.26	0.3	0.36	0.14
Wall and Roof (W/m^2 K)	0.25	0.25	0.13	0.2	0.24	0.1
Glazing (W/m^2 K)	3.18 (SC = 0.40)	3.28	2.1	1.9	6.9	2.2
Internal Light Power Density (W/m^2)	14	10	7.53	10	11.84	11.84
Equipment Power Density (W/m^2)	(200 W/desk + 10 W/m^2)	–	–	–	12	12
Occupancy (m^2/Person)	8.4	ASHRAE 62-2007	=	≥25 per 100 m^2 ≈ 4	20	20
HVAC Type	Chillers	Air conditioning equipment placed in Table 502.1 and efficiency requirement as per ASHRAE 90.1	Air conditioning equipment placed in Table 502.1 and efficiency requirement as per ASHRAE 90.1	Chillers with 95% of the year free cooling	Efficiency requirement as per ASHRAE 90.1	Air and water-based chillers
Heating Design Temperature	22 °C (±1)	25.5–22.5 °C	–	As per DM regulations	20 °C	20 °C
Cooling Design Temperature	23 °C (±1)	24 °C	24 °C	24 °C	23.8 °C	23.8 °C
Ventilation Rate (m^3/h/Person)	30.06	5% of the design volume	ASHRAE 62.2-2007, Ventilation	ASHRAE 62.2-2007, Ventilation	25.5	25.5
Infiltration Rate (ac/h) @ Normal Pressure	0.5	0.5	0.5	0.5	0.25	0.6552

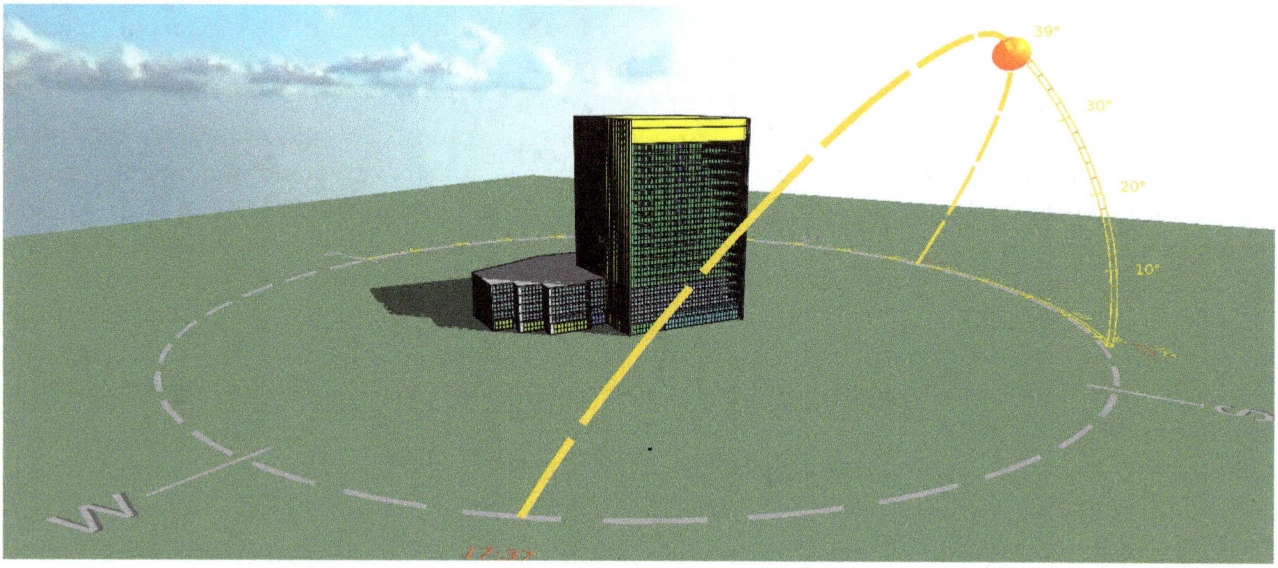

Figure 2. Reference building design using ModelIT tool in Integrated Environmental Solution: Virtual Environment (IES-VE) platform.

2.2. *Water Consumption Analysis*

This analysis was performed by employing the computation methods prescribed in the U.S. LEED water consumption evaluation procedure and using an Excel spreadsheet with the parameters relevant to office buildings presented in Table 2. This technique is recognized by official authorities and widely used at corporate levels within energy saving companies across the UAE. It measures the flow rates, timing per use, occupant density, and frequency of usage to perceive the consumption pattern. Utilizing a metered system does not provide a breakdown of usage patterns for controlling and conserving water.

Table 2. Summary of prescriptive regulatory specifications for estimating water consumption patterns.

Fixture/Appliances	Units	Maximum Flow Rate					
		Reference Case Study Building	DMGB	Trakhees	Al'Safat	ASHRAE	Estidama
Kitchen Taps	l/min	6	7		7		6
Bathroom Washbasin Tap	l/min	1.9 per use	6		6		6
Toilet Dual Flush	l/flush	6 or 4.8	6 or 3	Not Given	6 or 3	Not Applicable	6 or 4
Bidet spray	l/min	6	–		–		6
Shower head	l/min	9.5	8		8		9.5
Ablution Fixtures	l/min	6	–		–		6
Dishwasher	L/place setting	–	–		–		1.3
Washing Machines	L/kg of dry load	–	–		–		8.5
Urinal	L/flush or waterless	0.5	1		1		–

ASHRAE and Trakhees do not have a prescribed set of parameters and were considered unknown in the analysis (Table 2). These unknown parameters were replaced with the reference building's defined parameters and behavior patterns, which were assumed constant because they influence the water usage per person. The number of working days and occupancy density were used to normalize the consumption patterns using the simulated occupancy results from the previous BEM analysis. The net annual water consumption (Equation (3)) was calculated using the wastewater amount and condensate water consumption, which is directly dependent on the chiller flow supply, specific humidity in the air, and specific volume of dry air. The quality of the storm water collection system depends on net water consumption, which in turn depends on the annual rainfall and volumetric run-off coefficient for impervious covers.

$$Net\ Annual\ Consumption = Annual\ Consumption - Storm\ Water + Condensate\ water \quad (3)$$

Moreover, the accuracy of water consumption measurements is highly dependent on occupancy density, as this parameter is unpredictable and variable in nature. Therefore, the value of the occupancy density and usage pattern was assumed to be constant. This value corresponds to the specified occupant density in the standards (Table 1) and the assumed usage pattern to match the water utility consumption of the case study building.

2.3. *Techno-Economic Analysis*

The analysis was performed as a feasibility study to understand the costs and benefits of complying with the green building standards, to investigate the cash flow pattern by integrating the net present value technique under the discounted cash flow of the financial analysis. To calculate the cash flow, the cost of the entire project was estimated by considering a list of requirements as presented in Appendix A. A summary is presented below:

- Pre-requisite expenditures (permit, consultation, Environmental Impact Assessment (EIA), and land acquisition);
- On-site infrastructure expenditure;
- Operational expenditure;
- Structural construction expenditure;
- Payroll;
- Internal structure (cabling and soundproofing);
- Internal services such as mechanical, electrical, and plumbing (MEP) expenditure.

The net income was estimated from the rental charges (AED 115/sq ft) [81], and energy and water savings were computed at the end of the second and third stages.

3. Building Energy Performance Analysis Results

The overall energy performance efficiency evaluated using the energy utility index (EUI) approach is depicted in Figure 3 using the simulation results obtained from the BEM within the IES-VE platform by accommodating all the influential design parameters for each standard (Table 1) for high-rise office buildings. The building characteristic details are presented in the appendices.

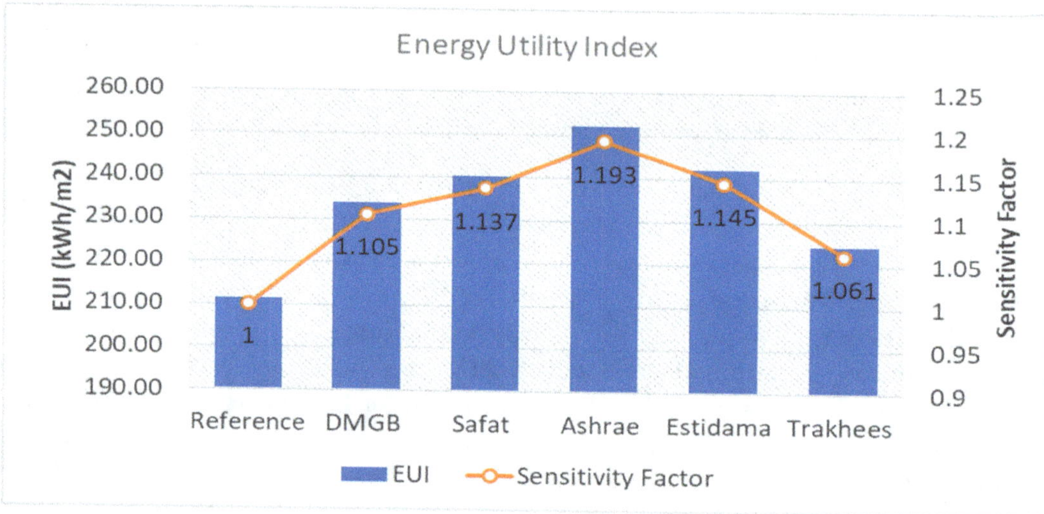

Figure 3. Cross-code analysis comparing green building standards in the UAE using the energy utility index approach, whose characteristic data were simulated using the IES-VE platform.

The cross-code analysis is illustrated in Figure 3 using the EUI approach and highlights the energy performance of each regulation, while the sensitivity factor is also the relativity factor, i.e., a ratio of actual and reference values [82]. Trakhees showed the best overall energy performance since its design parameters were developed to obtain a yield high performance. The intricate design conditions that enhanced the potential of each design criterion are related to its regulatory nature, confined to Dubai ports and free zones or being the first mandated regulation in the UAE in 2008 [83,84]. This finding is consistent with the findings of Jagannathan [85] while studying constructing low-carbon buildings using Trakhees' guidelines; however, the research of Jagannathan did not identify the potential of a few design criteria using all design conditions.

DMGB was regarded as a discrepancy in Figure 3, as Al'Safat and Estidama's design parameters had the potential to yield higher performance, indicating that the performance of each parameter was overshadowed by others. The effect of other variables on electric and chiller loads was computed using the sensitivity index (Equation (2)) [86] and is depicted in Figure 4. The EUI indicates the total electricity consumption pattern per unit of gross floor area. It does not specify the constituent distribution of the amount of energy lost through external or internal gains or the HVAC unit's energy requirements. Therefore, the simulation-based energy performance analysis does not necessarily paint a clear picture of the influence of each parametric design on the energy or cooling load outcome.

The simulated variables representing the total electricity and cooling load demand are not in agreement (Figure 4), implying that the variables that affect the electricity demand are different from those of the cooling load. Figure 4 shows that the chiller load is influenced by ventilation gain, internal gain (people, lighting, and equipment gain), and solar gain (convective and radiative heat gain through fenestration) [70]. This observation can be explained by the theory that these gains are responsible for the higher temperatures in a room. Thus, ventilation and glazing design parameters can control the chiller load.

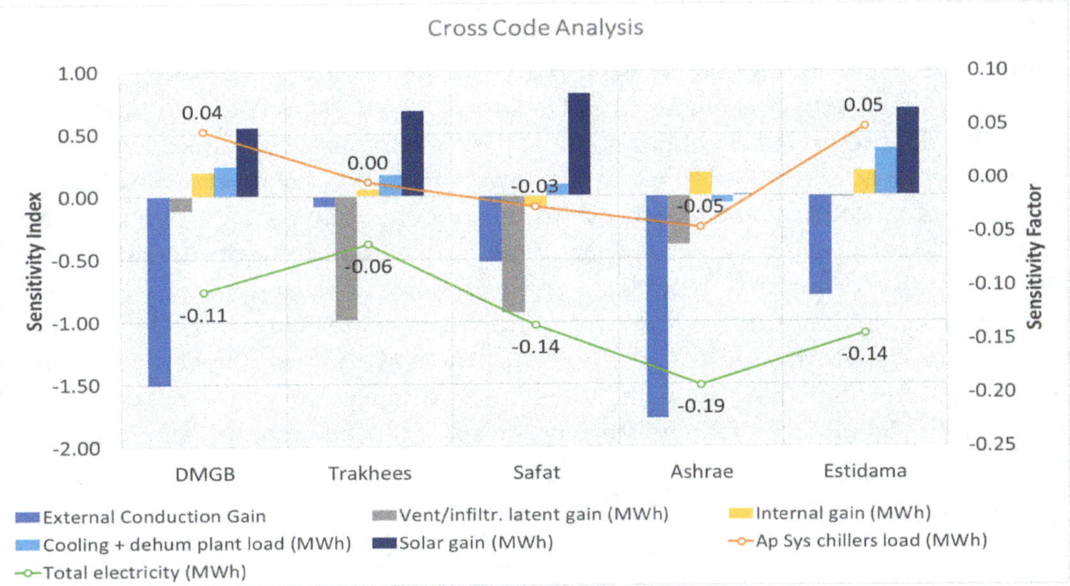

Figure 4. Impact on electric and chiller loads due to other variables in detailed cross-code analysis comparing green building standards in the UAE, where detailed cross-code analysis includes all other parameters to obtain the relationship pattern.

Electric load is directly dependent on lighting and equipment power density, but its direct dependence on conduction and convection gains indirectly indicates its reliance on the chiller load because internal gains contribute to the elevated heat in a room. The chiller load and electric load dependency affect the performance of DMGB in relation to Al'Safat because the Dubai municipal authority enhanced the individual design criteria for Al'Safat, which were based on DMGB's regulations. This was not an ideal modification because it did not balance the influence of other variables, and the slightest change in the design value greatly impacted the overall building performance. Moreover, the effect of the ventilation gain and envelope measure has the potential for reducing energy intensity to enhance building performance (Figure 4).

Uncertainty Analysis

The simulated results were used to perform energy, water, and economic analysis in the IES-VE platform. An uncertainty analysis was conducted for one variable, and it was assumed to have the same impact on other variables. The parameter chosen was the chiller load because this is an essential parameter in the Middle East, where the cooling load contribution to the energy distribution profile is around 60–70% in energy-efficient buildings [13,48]. The reference case study building is set up with four 1500 MW chiller plants with an estimated 5.47 MW load. The actual value was provided in the documentation collected during the data acquisition phase. The analysis presented a marginal error of 7.4% (Equation (4)).

$$Error_{Cooling\ load} = 1 - \frac{5873.46}{5469} = 7.4\% \quad (4)$$

The error margin of 7.4% is consistent with that in another study that determined the energy simulation accuracy of IES-VE, which varied by 0.4% from the reported range (6–7%) [87]. This variation was caused by the difference in the environmental conditions such as the external dry bulb temperature, wind intensity, or energy usage pattern defined by the user. Since the accuracy of the platform is in a reasonable range (under 10%), the simulated results were left unaltered throughout the project. The simulated results were not calibrated to project the real-time results, as this error margin confirms how close the simulation results are to real-life evaluation.

4. Water Consumption Analysis Results

The parameters listed in Table 2 for each regulatory standard were used to perform a water consumption analysis based on the U.S. LEED water estimation procedure. Since ASHRAE and Trakhees do not have a prescribed set of parameters, they were considered unknown. These unknown parameters were replaced with the reference building's defined parameters and behavior pattern, which were assumed to be constant, as they influenced the water usage per person. However, the number of office working days and occupancy density were used to normalize the consumption pattern using simulated occupancy results from a previous BEM analysis. The net annual water consumption (Figure 5) constitutes waste and condensate water consumption, which is directly dependent on the chiller flow supply, specific humidity in the air, and specific volume of dry air. It must be noted that the quality storm water collection is deducted from the net water consumption, which depends on the annual rain fall and volumetric run-off coefficient for impervious cover under coverage [88].

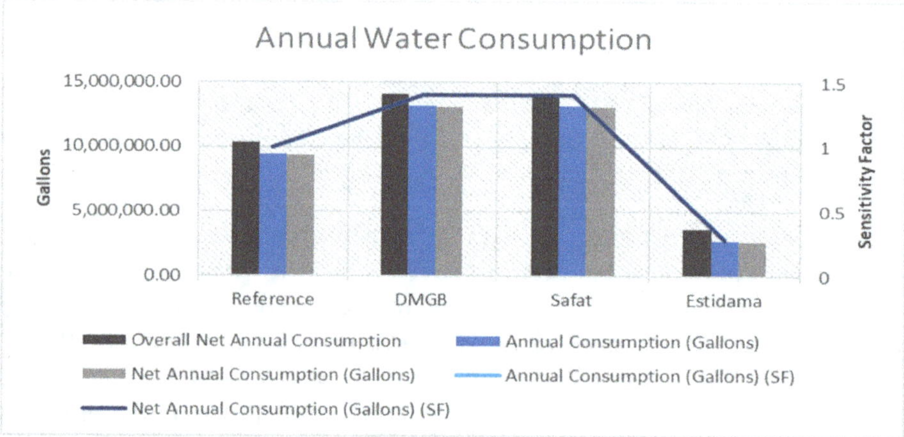

Figure 5. Cross-code annual water consumption pattern for green building standards in the UAE.

DMGB and Al'Safat have the same consumption scheme (about 140% of the reference buildings), whereas Estidama consumes only 29%, as shown in Figure 5. This competitive cross-code water consumption performance study showed that Estidama had the most effective regulative design parameters.

Figure 6 demonstrates the simulated variation in the supply air flow in cubic feet per minute to contemplate the end condensate water consumption, with a variation under 1.29% across each standard.

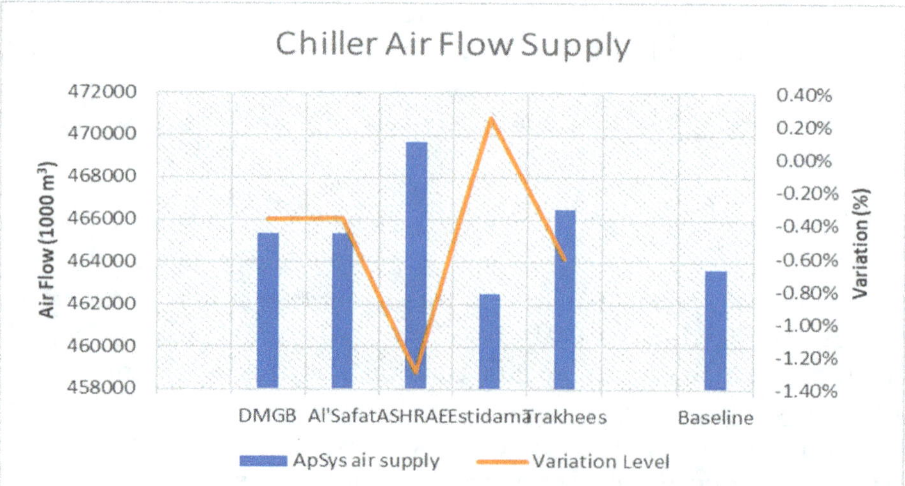

Figure 6. Chiller system's air flow supply for each green building standard obtained using simulation results on IES-VE platform.

Figure 7 demonstrates the cross-code analysis in the normalized form (index value), where the water consumption level of ASHRAE and Trakhees was estimated using the reference building's pattern. The number of occupants and its respective index values are presented numerically above the blue and orange bar graphs, respectively.

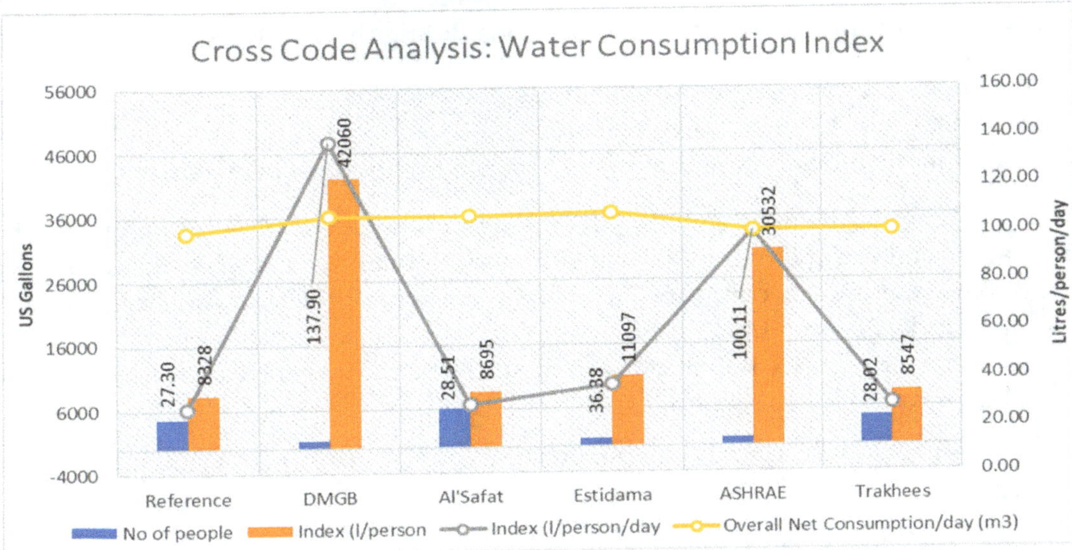

Figure 7. Cross-code analysis: water consumption index depicted considering all parameters using IES-VE platform.

Figure 7 contradicts Figure 5 because the water consumption per person index (gray line) shows that Al'Safat outperforms the other standards, followed by Trakhees and Estidama; ASHRAE and DMGB have the least significant standards in terms of water consumption patterns. Moreover, there are differences among the indices of ASHRAE, Trakhees, and the reference building owing to the variations in occupancy density, even though the total water consumption was assumed to be the same. The number of people in the building was incorporated in the simulation while performing the energy analysis modeling in the IES-VE platform using the occupant density accordingly.

Consequently, the parameter occupant density is identified as an opportunity for enhancing water consumption standards. For instance, the trade of Estidama's water consumption standard can be optimized by considering the value of the occupant density from Al'Safat. Therefore, it can be concluded that the water consumption analysis depends not only on allocated design parameters but also on occupancy density. The HVAC load has an influence on the variations in condensate water. In this regard, the design parameters influencing the chiller load will either increase or decrease the amount of condensate water required, affecting the water consumption pattern. ASHRAE and Trakhees were disregarded because their standards do not comply with the strategic water analysis structure. This is because ASHRAE complies with the requirements to reduce and maintain the heating and cooling load, and Trakhees follows the free-zone standard, which was typically prescribed for warehouse buildings during the early 2000 s. Since then, the UAE has become a commercial hub with rapid development in free zones. For instance, Jumeriah, Dubai, is a free zone and currently considered a suitable base for office buildings [77,83]. Trakhees requires revisions to incorporate the water consumption strategy because it was one of the first regulatory standards in the UAE [17]. Figure 7 shows that Estidama has a water consumption index closer to Al'Safat's and the best water consumption design standards. The water consumption analysis was performed by strictly defining the behavior pattern and is considered a major limiting factor that can significantly affect the outcome.

5. Techno-Economic Analysis Results

The building energy performance and water consumption were evaluated to compare the energy saving potential of each set of green building regulations and to verify whether they were economically feasible. It is necessary to ensure the feasibility and to understand the economic benefits of energy conservation by considering the investment details presented this section. The net cash flow using the discounted cash flow technique was computed by considering the following:

- The two-year construction period for which there is no income;
- The loan is taken at a 7% annual interest rate;
- The loan amount includes 7% annual interest for the first two years of construction.

The net cash flow for the first scenarios depicted in Figure 8 was determined using the income from rental charges and energy savings. These were the simulated energy results in terms of the EUI from Figure 3 that were used to compute the profit margin for energy savings as shown in Table 3. The net cash flow indicates the return-on-investment path, i.e., a percentile of income with respect to the initial investment. This cash flow pattern includes the interest rates applied at 7%, as per the assumptions stated earlier in this section.

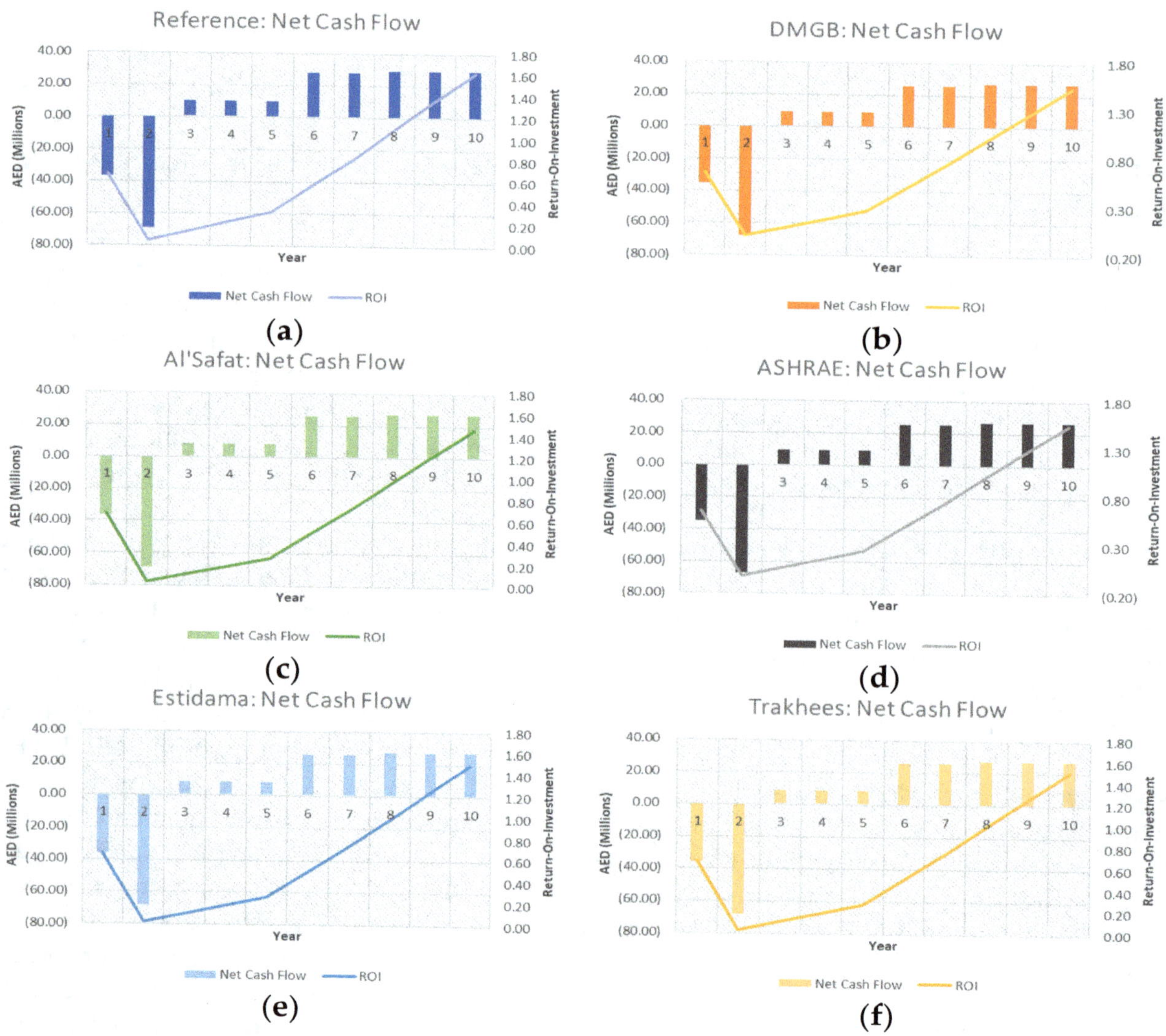

Figure 8. Net cash flow and return-on-investment (ROI) curve for each set of green building standards in the UAE for a 10-year period: (**a**) reference case study building; (**b**) DMGB; (**c**) Al'Safat; (**d**) ASHRAE; (**e**) Estidama; (**f**) Trakhees.

Table 3. Summary for techno-economic analysis with only electricity utility pay out.

	Investment (Millions)	Loan At 7% (Millions)	Payback Period (y)	ROI	IRR for 10 y Plan	Excess Money PBP (Millions)
DMGB	AED 70.19	AED 103.55	7	1.009	13.62%	AED 0.92
Al'Safat	AED 73.90	AED 109.03	8	1.261	12.68%	AED 28.41
Trakhees	AED 72.87	AED 107.51	8	1.293	13.20%	AED 31.48
Estidama	AED 72.20	AED 106.51	7	1.013	13.68%	AED 1.30
ASHRAE	AED 68.37	AED 100.86	8	1.284	13.02%	AED 30.25
Reference	AED 75.14	AED 110.84	8	1.273	12.95%	AED 30.28

Figure 8 represents the net cash flow diagrams for each green building regulation for the first scenario considering the income from energy-saving and rental charges, where each clustered column represents a one-year period. The primary vertical axis represents the annual net cash available in millions at the end of year, whereas the secondary vertical axis represents the return on investment (ROI). The payback period is determined when the line crosses "1.00", where it indicates a 100% return on investment.

Figure 8 demonstrates that regulatory standards have a payback period of seven or eight years, but they do not necessarily provide the investment details along with the return-on-investment (ROI) factor at the end of the payback period and the internal rate of return (IRR) after a ten-year investment scheme. A detailed summary of the net cash flow is tabulated in Table 3.

The results show that ASHRAE has the smallest investment, and Trakhees outperforms the others in terms of the IRR value at the end of 10 years, as this code has a positive effect on the EUI index compared to the rest. The 10-year mark was utilized to verify the IRR and identify it as a good investment. The significance of using the IRR evaluation method rather than the whole life-cycle cost, which considers the resale of the building, is prevalent due to the real estate market in the UAE. Moreover, the life-cycle cost was not considered since the design selection for the parameters must have a minimum lifetime of 25 years or more to qualify as a green building as per UAE regulations.

Table 3 illustrates the effect of energy performance on the economic assessment; the results for each regulation were approximately the same due to the minute variations in the cost per service pricing obtained from a reputable developer in the building sector in the UAE. Trakhees was considered a better choice, even though ASHRAE had the lowest investment and Estidama had the highest return-on-investment rates for the scenario when considering the income from the office space rent and profits from energy savings. Trakhees has a lower impact on the environment in the long run, due to higher energy savings and potentially lower carbon emissions. It has an IRR value close to that of Estidama because it requires a lower building maintenance investment due to its high building energy performance. Note that the income for this case study was from the energy-saving profits and rental charges, but the rental charges were not consistent with those for green buildings, even though green buildings receive 5 to 10% higher rental income. This was primarily done to estimate the values for the worst-case scenario.

A detailed techno-economic analysis considering water utility savings, metering, lighting, and furnishing fixtures is presented in Table 4 and assumes that the desalination plant does not change the unit price (4.6 fills per imperial gallon) in Dubai for the next 10 years.

Table 4. Summary for detailed techno-economic analysis including water and electricity savings.

	Investment (Millions)	Loan At 7% (Millions)	Payback Period (y)	ROI	IRR for 10 y Plan	Excess Money PBP (Millions)
DMGB	AED 89.01	AED 131.31	8	1.061	9.22%	(AED 7.98)
Al'Safat	AED 92.72	AED 136.79	8	1.021	8.41%	(AED 2.81)
Trakhees	AED 91.69	AED 135.27	8	1.050	9.03%	(AED 6.74)
ASHRAE	AED 87.19	AED 128.62	8	1.069	9.37%	(AED 8.87)
Estidama	AED 91.01	AED 134.27	8	1.052	9.07%	(AED 7.04)
Reference	AED 94.80	AED 139.85	8	1.032	8.69%	(AED 4.53)

Water consumption utility charges and interior furnishing charges brought the payback period to eight years for all the green building standards. These characteristics showed that the investment plan does not depend on the regulatory standard. This analysis considered Estidama to be a better choice, even though ASHRAE had the lowest investment plan and the highest IRR value at the end of 10 years. The water index for ASHRAE and Trakhees was estimated while considering the reference building's consumption, as these regulations did not have a prescribed set of standards for water performance. This is the reason the two scenarios for economic analysis were formulated to provide equal opportunities for each regulation to showcase its conservation potential. Moreover, the water consumption analysis was performed by strictly defining the behavior pattern and is considered a major limiting factor that can significantly affect the outcome. However, Al'Safat was not considered the best option for economic analysis based on the water utility scenario because its internal rate of return dropped significantly. The excess money at the end of the seven-year period observed in the previous scenario was spent, and it increased the return-on-investment period to eight years. This indicates that the impact of water consumption utility charges is quite significant, leaving Al'Safat with the lowest return-on-investment rates. Considering both the energy and water consumption structure, Estidama will have the upper hand in energy and water performance in terms of economic viability and environmental protection.

6. Conclusions

This research investigated the technical and economic performance of existing green building regulations in the UAE to propose a technical and cost-effective trade-off. This work identified Trakhees as an energy-effective standard using the EUI benchmarking technique. To propose an energy-effective trade-off, a strategy to control energy performance was determined by evaluating a variable influence diagram. Previous studies showed that the standards in the UAE enhanced the individual performance of design parameters, but those studies did not perform a feasibility analysis. Previous studies also failed to identify the parametric pairing trade-off that produced similar overall energy performance when all influential design criteria were considered. This gap in the literature will be considered in a future study to quantify the influence of each parametric pairing to serve as a guideline for policymakers or consultants for devising a strategic energy and cost-effective building plan or revising an existing green building standard. This work showed that the chiller loads were mostly influenced by the ventilation rate, infiltration rate, and glazing thermal transmittance values, as they are responsible for heat gain. The electric load was directly dependent on power density and indirectly dependent on envelope design due to its dependence on the cooling load.

The water consumption analysis used a straightforward computational approach using the U.S. LEED's water consumption calculation principle. The results showed that in new green building regulations, Al'Safat outperformed Estidama by a margin due to its occupancy density design parameter.

In the techno-economic analysis, employing the ASHRAE's prescriptive conditions was the cheapest considering energy utility, water utility, and internal furnishing expenditure. Complying with Estidama's regulations was economical considering its energy and water performance, which ensures a sustainable living environment. Even though Al'Safat outperformed the others in terms of energy and water performance, the cost of implementing those measures yielded 2.5 times lower profits at the end of the eight-year period compared to Estidama.

A competitive standpoint was deduced based on office buildings with high window-to-wall ratios, where the reference case study building had an 18% wall-to-window ratio. The DMGB's glazing condition paired with the Trakhees' HVAC set point, ventilation, and infiltration rate would be a plausible cost-effective trade-off to improve the chiller load performance with minimal cost because envelope measures have a significant effect on the convective gain, which impacts chiller load performance. For instance, office buildings with wall-to-window ratios below 20% contribute to the convection and conduction gain through the wall (around 20%). Hence, a large investment to maximize the wall area is not an economical approach, but coating the exposed envelope area with reflective paint will block the heat gain through the external surface. The use of blinds or internal tinting of glass may be used to further maximize energy performance.

Author Contributions: The conceptualization and methodology were developed by H.N. and M.N. The main parts of the research, including the system modeling and analysis of the obtained results, were developed by H.N. and M.N. All authors contributed to the writing and editing of the paper. S.A.-Z. and F.G. performed the visualization, validation, and revision and contributed to the final version of the manuscript along with M.N. All authors have read and agreed to the published version of the manuscript.

Acknowledgments: This research did not receive any specific grant from funding agencies in the public, commercial, or not-for-profit sectors. However, it was supported by Heriot-Watt and United Arab Emirates University academically and by the Director of Strategic Planning Corporate Facilities Management at Emaar in Dubai, UAE, providing insight and expertise that greatly assisted this research.

Appendix A

Table A1 presents building characteristic details of the reference case study building.

Table A1. Summary of reference building's estimated surface area.

	Podium Parking Floor	Office Floor
No. of Floors	5-zone	13-zone
Floor Area/Floor (m^2)	7164.71	1874
Window Area/floor (m^2)	941.4	571.2
Wall–Window Area	18.40%	18.40%
Total Floor Area (m^2)	35,823.55	24,362
Total Wall Area (m^2)	1061.38	1674.4
Total Window Area (m^2)	4706.97	7425.6
Building Wall Area (m^2)		2735.78
Building Window Area (m^2)		12,132.57
Building Floor Area (m^2)		60,185.55

Table A2 presents the extracted data from the BEM simulation based on energy consumption.

Table A2. Simulated energy consumption results.

Variable Consumption (MWh)	DMGB	Al' Safat	ASHRAE	Estidama	Trakhees	Reference
Total Electricity	14,066	14,464	15,177	14,565	13,502	12,729
Total Energy	14,201	14,599	15,355	14,688	13,614	12,828
Total Lights Energy	3258	3258	3877	3815	2434	1682
Lights Electricity	3249	3249	3866	3805	2427	1668
Lights Misc. A	8.6	8.6	10.19	10.19	6.48	13.88
Total Equip. Energy	5191	5191	5175	5158	5182	5187
Equip. Electricity	5187	5187	5171	5154	5178	5182
Equip. Misc. H	4.62	4.62	4.62	4.62	4.62	4.62

The investment description to compute the techno-economic analysis is listed; however, the unit cost was obtained through market research and the help of various consultants in the field [61]. The maintenance cost is a part of internal service expenditure, where each of the pieces of equipment or services was allocated a certain percentage of degradation value along with inflation relative to its initial cost annually until a decommissioning cost was estimated at the end of 10 years. The degradation value included the maintenance expenditure for annual maintenance or quarterly, which lowered the asset value.

1. Pre-Requisite Expenditure Description:
    - Land Acquisition Expenditure;
    - Feasibility Study;
    - Environmental Impact Analysis Study (EIA);
    - Professional Consultation Fee;
    - Land Acquisition and Commissioning Expenditure;
    - Permits, Licenses, and Fees (Before Construction);
    - Permits, Licenses, and Fees (After Construction);
    - Other Miscellaneous Expenditure.
2. Site and Infrastructure Expenditure Description:
    - Utilities (Electrical and Mechanical);
    - Transportation;
    - Waterway and Marine Construction;
    - Other Miscellaneous Expenditure.
3. Construction Expenditure Description:
    - Constructional Equipment Rent;
    - Constructional Equipment Maintenance (Non-Rental);
    - Other Miscellaneous Expenditure.
4. Payroll Description:
    - Constructional Laborers (On-Site);
    - Specialist and Technicians (Engineers, Supervisors, etc.) (On-Site);
    - Specialist and Technicians (Engineers, Managers, etc.) (Off-Site);
    - Consultants;

- Other Miscellaneous Expenditure.

5. Structural Construction Expenditure Description:

- External Wall (Concrete) (Light Weight) (Actual Building);
- External Wall (U = 0.7 W/m^2 K);
- External Wall (U = 0.57 W/m^2 K);
- External Wall (U = 0.28 W/m^2 K);
- External Wall (U = 0.32 W/m^2 K);
- Masonry Wall (Ordinary);
- Spandrel Panels (U = 0.31 W/m^2 K);
- External Window (Glass) (U = 1.5 W/m^2 K) (Actual Bldg) (SC= 0.34);
- External Window (Glass) (U = 6 W/m^2 K);
- External Window (Glass) (U = 3.3 W/m^2 K);
- External Window (Glass) (U = 2.2 W/m^2 K);
- External Window (Glass) (U = 2.1 W/m^2 K);
- External Window (Glass) (U = 2.1 W/m^2 K);
- External Roof (Actual Building) (U = 0.44 W/m^2 K);
- External Roof (U = 0.44 W/m^2 K);
- External Roof (U = 0.36 W/m^2 K);
- External Roof (U = 0.3 W/m^2 K);
- External Roof (U = 0.26 W/m^2 K);
- External Roof (U = 0.14 W/m^2 K);
- Internal Partition Wall (Approx.) U = ?;
- Internal Partition Door (Approx.) U = ?;
- Ceramic Tile Flooring;
- Other Miscellaneous Expenditure (Stairs and Other Rooms).

6. Internal Structure Expenditure:

- Painting (Cool Reflective Coating) (SHGC=) (External);
- Painting (Cool Reflective Coating) (SHGC=) (Internal);
- Thermal and Sound Insulation;
- Moisture-Proofing Insulant;
- Other Miscellaneous Expenditure (Electrical Wiring, Air Flow Duct, and Water Flow Pipes).

7. Internal Service Expenditure Description:

- Plumbing Service;
- AHU (Equipment);
- HVAC (Equipment);
- Water Supply Pumps for Fire and Domestic Water;
- Fire Protection Service;
- Electrical (Transformer, Substation, etc.);
- Furnishings and Fixed Equipment (Wash Room Taps and Desks for Offices);
- Gray Water Filtration System and Pumps;
- Solar Water Heater System;
- Elevators (Lifts);
- FAHU;
- Extract Fan;

- Pressurization Fan;
- BMS System (Infrastructure Set-Up);
- BMS System (Equipment);
- Other Miscellaneous Building Specialties.

Additionally, Tables A3 and A4 present relevant data obtained from market research and simulations to compute the net income, respectively.

Table A3. Relevant data for computing net income.

Cost/sq ft	Plots (sq m)	Plot (sq ft)	Revenue
115	24,362	262,230.39	AED 30.16 Million
105	24,362	262,230.39	AED 27.53 Million
135	24,362	262,230.39	AED 35.40 Million

Table A4. Simulated savings data for computing net income.

	DMGB	Al'Safat	ASHRAE	Estidama	Trakhees	Reference Bldg.
EUI	233.71	240.32	252.17	242.00	224.34	211.49
Energy Saving EUI				211.49		
Energy Saving per sq m	−22.30	−28.90	−40.76	−30.58	−12.93	−0.08
Energy Saving kWh (Total Area)	−1,341,981.27	−1,739,379.67	−2,452,881.40	−1,840,686.82	−778,096.08	−4644.08
Cost Savings (Millions)	−0.43	−0.56	−0.78	−0.59	−0.25	0.00

References

1. Clark, K. 80 Percent Energy Consumed by Buildings. *Khaleej Times Newspaper*, February 2016. Available online: http://www.khaleejtimes.com/nation/abu-dhabi/80-energy-consumed-by-buildings (accessed on 9 October 2019).
2. United-Nations Environment-Programme. *Buildings and Climate Change: Summary for Decision-Making UNEP SBCI*; Sustainable Buildings and Climate Initiative: New York, NY, USA, 2009.
3. Vyas, G.; Jha, K.N. Benchmarking green building attributes to achieve cost effectiveness using a data envelopment analysis. *Sustain. Cities Soc.* **2017**, *28*, 127–134. [CrossRef]
4. Water, Dubai, and Electricity Authority. *Manual of Green Building Materials, Products And their Testing Facilities*; Dubai Central Laboratory Department: Dubai, UAE, 2012.
5. Abu-Hijleh, B.A.K.; Jaheen, N. Energy and economic impact of the new Dubai municipality green building regulations and potential upgrades of the regulations. *Energy Strat. Rev.* **2019**, *24*, 51–67. [CrossRef]
6. Mokhtar, A. Challenges of retrofitting affordable housing to net-zero carbon in the United Arab Emirates. *IOP Conf. Ser. Earth Environ. Sci.* **2019**, *323*, 012011. [CrossRef]
7. Saseendran, S. Dubai Launched Green Building Rating System. Available online: http://gulfnews.com/news/uae/environment/dubai-launches-green-building-rating-system-1.1861227 (accessed on 10 September 2019).
8. Dubai Municipality. *Planning and Construction*; Dubai Municipality: Dubai, UAE, 2001. Available online: https://www.dm.gov.ae/municipality-business/planning-and-construction/al-safat-dubai-green-building-system/ (accessed on 16 September 2019).
9. Government of Dubai. *Dubai Green Building Guidelines UAE*; Government of Dubai: Dubai, UAE, 2009. Available online: https://www.dewa.gov.ae/~{}/media/Files/Consultants%20and%20Contractors/Green%20Building/Greenbuilding_Eng.ashx (accessed on 16 September 2019).
10. ABU DHABI Urban Planning Council. *The Pearl Rating System for Estidama*; ABU DHABI Urban Planning Council: Dubai, UAE, 2010. Available online: https://www.dmt.gov.ae/-/media/57346B04730942EEBAF60949CE1D94E7.ashx?newTab=1 (accessed on 16 September 2019).
11. ABU DHABI Urban Planning Council. Estidama FAQs. Available online: http://estidama.upc.gov.ae/faqs/general.aspx?lang=en-US#1 (accessed on 9 October 2019).

12. Zaman, S. Existing Buildings to be Made More Sustainable in Abu Dhabi. *Gulf Newspaper*, January 2012. Available online: http://gulfnews.com/news/uae/property/existing-buildings-to-be-made-more-sustainable-in-abu-dhabi-1.965778 (accessed on 3 November 2019).
13. Alobaidi, K.A.; Rahim, A.B.A.; Mohammed, A.; Baqutayan, S. Sustainability Achievement and Estidama Green Building Regulations in Abu Dhabi Vision 2030. *Mediterr. J. Soc. Sci.* **2015**, *6*, 509. [CrossRef]
14. John, I. GCC Needs Dollar 220b for New Cooling Power Capacities. *Khaleej Times*, July 2012. Available online: http://www.khaleejtimes.com/business/local/gcc-needs-220b-for-new-cooling-power-capacities (accessed on 3 November 2019).
15. Dakheel, J.A.; Aoul, K.T.; Hassan, A. Enhancing Green Building Rating of a School under the Hot Climate of UAE Renewable Energy Application and System Integration. *Energies* **2018**, *11*, 2465. [CrossRef]
16. Ntarangui, T.N. The Impact of Design Parameters on Energy Consumption in Office Buildings in Nairobi. Master's Thesis, University of Nairobi, Nairobi, Kenya, 2010. Available online: http://erepository.uonbi.ac.ke:8080/xmlui/handle/123456789/14634 (accessed on 19 October 2020).
17. Friess, W.A.; Rakhshan, K.; Hendawi, T.A.; Tajerzadeh, S. Wall insulation measures for residential villas in Dubai: A case study in energy efficiency. *Energy Build.* **2012**, *44*, 26–32. [CrossRef]
18. Raouf, A.M.; Al-Ghamdi, S.G. Effect of R-values changes in the baseline codes: Embodied energy and environmental life cycle impacts of building envelopes. *Energy Rep.* **2020**, *6*, 554–560. [CrossRef]
19. Radhi, H.; Assem, E.; Sharples, S. On the colours and properties of building surface materials to mitigate urban heat islands in highly productive solar regions. *Build. Environ.* **2014**, *72*, 162–172. [CrossRef]
20. Afshari, A.; Nikolopoulou, C.; Martin, M. Life-cycle analysis of building retrofits at the urban scale—A case study in United Arab Emirates. *Sustainability* **2014**, *6*, 453–473. [CrossRef]
21. Radhi, H. A systematic methodology for optimising the energy performance of buildings in Bahrain. *Energy Build.* **2008**, *40*, 1297–1303. [CrossRef]
22. Friess, W.A.; Rakhshan, K. A review of passive envelope measures for improved building energy efficiency in the UAE. *Renew. Sustain. Energy Rev.* **2017**, *72*, 485–496. [CrossRef]
23. Aboul-Naga, M.; Al-Sallal, K.A.; El Diasty, R. Impact of city urban patterns on building energy use: Al-Ain city as a case study for hot-arid climates. *Arch. Sci. Rev.* **2000**, *43*, 147–158. [CrossRef]
24. Radhi, H.; Sharples, S.; Fikiry, F. Will multi-facade systems reduce cooling energy in fully glazed buildings? A scoping study of UAE buildings. *Energy Build.* **2013**, *56*, 179–188. [CrossRef]
25. Al-Masri, N.; Abu-Hijleh, B.A.K. Courtyard housing in midrise buildings: An environmental assessment in hot-arid climate. *Renew. Sustain. Energy Rev.* **2012**, *16*, 1892–1898. [CrossRef]
26. Aboulnaga, M.M. A roof solar chimney assisted by cooling cavity for natural ventilation in buildings in hot arid climates: An energy conservation approach in Al-Ain city. *Renew. Energy* **1998**, *14*, 357–363. [CrossRef]
27. Selke, T.; Schubert, M. Leedplatinum awarded arabian green building with solar heat driven cooling technology. *Energy Procedia* **2016**, *91*, 824–831. [CrossRef]
28. Al-Absi, Z.A.; Isa, M.H.M.; Ismail, M. Phase change materials (PCMs) and their optimum position in building walls. *Sustainability* **2020**, *12*, 1294. [CrossRef]
29. Li, M.; Gui, G.; Lin, Z.; Jiang, L.; Pan, H.; Wang, X. Numerical thermal characterization and performance metrics of building envelopes containing phase change materials for energy-efficient buildings. *Sustainability* **2018**, *10*, 2657. [CrossRef]
30. Wu, C.-H.; Huang, C.-H.; Li, Y.-F.; Lee, W.-H.; Cheng, T. Utilization of basic oxygen furnace slag in geopolymeric coating for passive radiative cooling application. *Sustainability.* **2020**, *12*, 3967. [CrossRef]
31. Katanbafnasab, M. Assessment of the energy impact of using building integrated photovoltaic and electrochromic glazing in office building in UAE. *Engineering* **2013**, *5*, 56–61. [CrossRef]
32. Rehman, H.U. Steady state experimental analysis of various solar insulation materials and techniques for buildings in climatic condition of Ras Al Khaimah, UAE. *Energy Procedia* **2015**, *75*, 1419–1424. [CrossRef]
33. Rehman, H.U. Experimental performance evaluation of solid concrete and dry insulation materials for passive buildings in hot and humid climatic conditions. *Appl. Energy* **2017**, *185*, 1585–1594. [CrossRef]
34. Akbari, H.; Konopacki, S. Calculating energy-saving potentials of heat-island reduction strategies. *Energy Policy* **2005**, *33*, 721–756. [CrossRef]
35. Akbari, H.; Matthews, H.D. Global cooling updates: Reflective roofs and pavements. *Energy Build.* **2012**, *55*, 2–6. [CrossRef]

36. Ezzeldin, S.; Rees, S.J. The potential for office buildings with mixed-mode ventilation and low energy cooling systems in arid climates. *Energy Build.* **2013**, *65*, 368–381. [CrossRef]
37. Taleb, H.M. Natural ventilation as energy efficient solution for achieving low-energy houses in Dubai. *Energy Build.* **2015**, *99*, 284–291. [CrossRef]
38. Lin, B.; Yu, Q.; Li, Z.; Zhou, X. Research on parametric design method for energy efficiency of green building in architectural scheme phase. *Front. Arch. Res.* **2013**, *2*, 11–22. [CrossRef]
39. Haggag, M.; Hassan, A.; Elmasry, S. Experimental study on reduced heat gain through green façades in a high heat load climate. *Energy Build.* **2014**, *82*, 668–674. [CrossRef]
40. Kharseh, M.; Al-Khawaja, M.; Hassani, F. Comparison between different measures to reduce cooling requirements of residential building in cooling-dominated environment. *Energy Build.* **2015**, *88*, 409–412. [CrossRef]
41. Omrany, H.; GhaffarianHoseini, A.; GhaffarianHoseini, A.; Raahemifar, K.; Tookey, J. Application of passive wall systems for improving the energy efficiency in buildings: A comprehensive review. *Renew. Sustain. Energy Rev.* **2016**, *62*, 1252–1269. [CrossRef]
42. Ayoub, N.; Musharavati, F.; Pokharel, S.; Gabbar, H.A. Energy consumption and conservation practices in Qatar—A case study of a hotel building. *Energy Build.* **2014**, *84*, 55–69. [CrossRef]
43. Hammad, F.; Abu-Hijleh, B.A.K. The energy savings potential of using dynamic external louvers in an office building. *Energy Build.* **2010**, *42*, 1888–1895. [CrossRef]
44. Taleb, H.M.; Sharples, S. Developing sustainable residential buildings in Saudi Arabia: A case study. *Appl. Energy* **2011**, *88*, 383–391. [CrossRef]
45. Johnson, R.; Sullivan, R.; Selkowitz, S.; Nozaki, S.; Conner, C.; Arasteh, D. Glazing energy performance and design optimization with daylighting. *Energy Build.* **1984**, *6*, 305–317. [CrossRef]
46. Wang, Z.; Zhao, J. Optimization of passive envelop energy efficient measures for office buildings in different climate regions of China based on modified sensitivity analysis. *Sustainability* **2018**, *10*, 907. [CrossRef]
47. Wang, X.; Altan, H.; Kang, J. Parametric study on the performance of green residential buildings in China. *Front. Arch. Res.* **2015**, *4*, 56–67. [CrossRef]
48. Bande, L.; Cabrera, A.G.; Kim, Y.K.; Afshari, A.; Ragusini, M.F.; Cooke, M.G. A building retrofit and sensitivity analysis in an automatically calibrated model considering the urban heat island effect in Abu Dhabi, UAE. *Sustainability* **2019**, *11*, 6905. [CrossRef]
49. Banihashemi, S.; Golizadeh, H.; Hosseini, M.R.; Shakouri, M. Climatic, parametric and non-parametric analysis of energy performance of double-glazed windows in different climates. *Int. J. Sustain. Built Environ.* **2015**, *4*, 307–322. [CrossRef]
50. Ragab, A.; AbdelRady, A. Impact of green roofs on energy demand for cooling in Egyptian buildings. *Sustainability* **2020**, *12*, 5729. [CrossRef]
51. Kolokotroni, M.; Wines, C.; Babiker, R.M.; Da Silva, B.H. Cool and green roofs for storage buildings in various climates. *Procedia Eng.* **2016**, *169*, 350–358. [CrossRef]
52. Callejas, I.J.A.; Durante, L.C.; Diz-Mellado, E.; Galán-Marín, C. Thermal sensation in courtyards: Potentialities as a passive strategy in tropical climates. *Sustainability* **2020**, *12*, 6135. [CrossRef]
53. Al-Sallal, K.A.; Al-Rais, L.; Bin Dalmouk, M. Designing a sustainable house in the desert of Abu Dhabi. *Renew. Energy* **2013**, *49*, 80–84. [CrossRef]
54. Taleb, H.M. Using passive cooling strategies to improve thermal performance and reduce energy consumption of residential buildings in UAE buildings. *Front. Arch. Res.* **2014**, *3*, 154–165. [CrossRef]
55. Fahmy, M.; Radhi, H.; Sharples, S.; Taleb, H. Will cool roofs improve the thermal performance of our built environment? A study assessing roof systems in Bahrain. *Energy Build.* **2017**, *135*, 324–337. [CrossRef]
56. Pisello, A.L. State of the art on the development of cool coatings for buildings and cities. *Sol. Energy* **2017**, *144*, 660–680. [CrossRef]
57. AboulNaga, M.M.; Abdrabboh, S.N. Improving night ventilation into low-rise buildings in hot-arid climates exploring a combined wall–roof solar chimney. *Renewable Energy* **2000**, *19*, 47–54. [CrossRef]
58. Government of Dubai. *Multi Storey Building Permit 2019*; Government of Dubai: Dubai, UAE, 2000. Available online: https://dda.gov.ae/dda-services/zoning-services/development-control/building-permit-final-building-permit/ (accessed on 6 February 2020).

59. O'Brien, W.; Farhang, T.; Rune, K.A.; Elie, A.; Verena, B.; Zsofia, D.B.; Berger, C.; Dong, C.; De Simone, M.; Hong, T.; et al. An international review of occupant-related aspects of building energy codes and standards. *Build. Environ.* **2020**, *179*, 106906. [CrossRef]
60. Anbouhi, M.H.; Farahza, N.; Ayatollahi, S.M.H. Analysis of thermal behavior of materials in the building envelope using building information modeling (BIM)—A case study approach. *Open J. Energy Effic.* **2016**, *5*, 88–106. [CrossRef]
61. Iwaro, J.; Mwasha, A. The impact of sustainable building envelope design on building sustainability using Integrated Performance Model. *Int. J. Sustain. Built Environ.* **2013**, *2*, 153–171. [CrossRef]
62. Auer, T.; Vohlidka, P.; Zettelmeier, C. The right amount of technology in school buildings. *Sustainability* **2020**, *12*, 1134. [CrossRef]
63. Chaudhry, S.I.; Das, M. A stable energy saving adaptive control scheme for building heating and cooling systems. *J. Power Energy Eng.* **2014**, *2*, 14–25. [CrossRef]
64. Emporis, Building Information Directory. In 2017. Available online: https://www.emporis.com/statistics/new-high-rise-buildings (accessed on 6 February 2020).
65. Globaldata Construction in the UAE-Key Trends and Opportunities to 2023. Available online: https://store.globaldata.com/report/gdcn0454mr--construction-in-the-uae-key-trends-and-opportunities-to-2023/ (accessed on 6 February 2020).
66. Ameer, B.; Krarti, M. Impact of subsidization on high energy performance designs for Kuwaiti residential buildings. *Energy Build.* **2016**, *116*, 249–262. [CrossRef]
67. Al-Ajmi, F.F.; Hanby, V. Simulation of energy consumption for Kuwaiti domestic buildings. *Energy Build.* **2008**, *40*, 1101–1109. [CrossRef]
68. El-Fadl Abou, S. Cooling energy saving in residential buildings in UAE by shading and night ventilation. In Proceedings of the Fifth Annual UAE University Research Conference, Al Ain, UAE, 25–27 April 2006.
69. Al Horr, Y.; Arif, M.; Kaushik, A.; Mazroei, A.; Katafygiotou, M.; Elsarrag, E. Occupant productivity and office indoor environment quality: A review of the literature. *Build. Environ.* **2016**, *105*, 369–389. [CrossRef]
70. Reeves, T.; Olbina, S.; Issa, R.R.A. Guidelines for using building information modeling for energy analysis of buildings. *Buildings* **2015**, *5*, 1361–1388. [CrossRef]
71. Assaf, S.; Nour, M. Potential of energy and water efficiency improvement in Abu Dhabis building sector—Analysis of Estidama pearl rating system. *Renew. Energy* **2015**, *82*, 100–107. [CrossRef]
72. Alnaser, N. Building integrated renewable energy to achieve zero emission in Bahrain. *Energy Build.* **2015**, *93*, 32–39. [CrossRef]
73. Radhi, H. Evaluating the potential impact of global warming on the UAE residential buildings—A contribution to reduce the CO_2 Emissions. *Build Environ.* **2009**, *44*, 2451–2462. [CrossRef]
74. Taleb, H.M. Enhancing the skin performance of hospital buildings in the UAE. *J. Build. Eng.* **2016**, *7*, 300–311. [CrossRef]
75. Emporis Building Information Directory. Available online: https://www.emporis.com/city/100485/dubai-united-arab-emirates (accessed on 3 November 2019).
76. Bayomi, N.; Fernandez, J.E. Trends of energy demand in the Middle East: A sectoral level analysis. *Int. J. Energy Res.* **2017**, *42*, 731–753. [CrossRef]
77. Dubai Municipality. Al'safat Dubai Green Building Evalaution System. Available online: https://www.dm.gov.ae/wps/wcm/connect/336d1d11-ae11-4271-a614-07dd98d18520/english+alsafat+book.pdf?MOD=AJPERES&ContentCache=NONE&CACHEID=336d1d11-ae11-4271-a614-07dd98d18520 (accessed on 16 October 2019).
78. Ma, Y.; Saha, S.C.; Miller, W.; Guan, L. Parametric Analysis of Design Parameter Effects on the Performance of a Solar Desiccant Evaporative Cooling System in Brisbane, Australia. *Energies* **2017**, *10*, 849. [CrossRef]
79. Hamby, D.M. A review of techniques for parameter sensitivity analysis of environmental models. *Environ. Monit. Assess.* **1994**, *32*, 135–154. [CrossRef] [PubMed]
80. Benesty, J.; Chen, J.; Huang, Y.; Cohen, I. Pearson Correlation Coefficient. In *Noise Reduction in Speech Processing*; Springer: Berlin/Heidelberg, Germany, 2009; pp. 1–4. [CrossRef]
81. Barnard, L. Office rents in prime Dubai sites costly but rates fall elsewhere. *The National Newspaper*, 2016.
82. Lam, J.C.; Hui, S.C. Sensitivity analysis of energy performance of office buildings. *Build. Environ.* **1996**, *31*, 27–39. [CrossRef]

83. Dubai Municipality. EHS Trakhees about Us. 2014. Available online: http://trakhees.ae/en/Pages/About-Us.aspx (accessed on 16 October 2019).
84. Ports, Customs and Free Zone Corporation; Government of Dubai. Green Building Regulations for Dubai World Developments. Available online: https://www.trakhees.ae/en/ehs/gb/Pages/gb.aspx#:~{}:text=Green%20Buildings%20are%20also%20commonly,buildings%20or%20environmentally%20friendly%20buildings.&text=EHS%2DTRAKHEES%2C%20the%20Regulatory%20arm,to%20implement%20within%20its%20jurisdiction (accessed on 6 February 2020).
85. Jagannathan, P.R. Achieving low carbon buildings through policy instruments from 2014: Sharing the trakhees-ehs dubai experience. In Proceedings of the 2014 International Conference and Utility Exhibition on Green Energy for Sustainable Development (ICUE), Pattaya, Pattaya City, Thailand, 19–21 March 2014; pp. 1–10.
86. Heiselberg, P.K.; Brohus, H. Sensitivity analysis applied in design of low energy office building. In Proceedings of the 29th AIVC Conference in 2008: Advanced building ventilation and environment technology for addressing climate change issues, Kyoto, Japan, 14–16 October 2008; Volume 2, pp. 33–40.
87. Mostafavi, N.; Farzinmoghadam, M.; Hoque, S. Envelope retrofit analysis using eQUEST, IESVE Revit Plug-in and Green Building Studio: A university dormitory case study. *Int. J. Sustain. Energy* **2013**, *34*, 594–613. [CrossRef]
88. Akan, A.O.; Houghtalen, R.J. *Urban Hydrology, Hydraulics and Stormwater Quality: Engineering Applications and Computer Modeling*; John Wiley & Sons Inc.: Hoboken, NJ, USA, 2003.

2

Review on Energy and Fire Performance of Water Wall Systems as a Green Building Façade

Uthpala Rathnayake [1], Denvid Lau [1,2] and Cheuk Lun Chow [1,*]

[1] Department of Architecture and Civil Engineering, City University of Hong Kong, Hong Kong, China; rrathnaya2-c@my.cityu.edu.hk (U.R.); denvid.lau@cityu.edu.hk (D.L.)
[2] Department of Civil and Environmental Engineering, Massachusetts Institute of Technology, Cambridge, MA 02139, USA
* Correspondence: cheuchow@cityu.edu.hk

Abstract: Glass façades are widely utilized in green buildings. Ensuring fire safety while reducing the energy need without compromising occupants' comfort is a challenge in the modern-day green buildings with glass façades. One way of achieving both aspects is to construct a water wall system as a building façade. A water wall system has a water layer between two glass panes and can be considered as a glass façade system. The focus of this review, which builds on the published studies, is how water wall systems can help ensure fire safety and reduce energy demand in green buildings. The water layer within two glass panes of the water wall system store the solar radiation heat throughout the daytime, reducing the amount of heat transferred through the building facade. The reduced heat transfer effects lessen the need for air conditioning to sustain the thermal comfort of the building occupants. The stored energy is released during the nighttime. The transparency of the water wall system also allows daylight to enter the building, thus reducing artificial lighting needs. Furthermore, the water layer acts as a fire safety mechanism in case of a fire. However, the water wall systems are not much utilized in the modern-day green buildings due to their unpopularity and the unavailability of design guidelines. On the basis of the findings of the literature review, stakeholders and the public are encouraged to adopt water wall systems in green building projects as an energy-efficient strategy and a fire safety mechanism.

Keywords: green building; glass façade; water wall; energy; fire safety

---

1. Introduction

The façade, which is the skin of the building, separates the interior space from the external environment. There is an urgent need to design and construct the building façades to be more energy-efficient, as an approximate 20–60% of building heating and cooling energy is influenced by the façade design and construction [1,2]. Therefore, the construction industry is moving towards green façade constructions, putting much focus on the energy-efficiency of the façade [3], and glass is widely used in the façade constructions [4]. Apart from energy efficiency, green building façades have a less negative impact on the environment and provide a better indoor environment that might contribute to wellbeing, productivity, and performance of the occupants. For this purpose, a green building façade should satisfy several requirements, such as facilitating outside views, withstanding wind loads, air and water penetrations, sustaining its dead load weight, allowing daylight to the building interior, obstructing undesirable solar heat gain, blocking outside noise and reducing temperature fluctuations, and ultimately boosting the overall building performance [5–8]. The materials used in the façade construction have a considerable effect in performing these green façade functions [9,10]. Glass is considered an appropriate material to be used in green building façades. In consideration of the complete building life cycle, the negative environmental impact of glass is minimal [11]. The primary constituent of glass is sand, which is a non-polluting material, glass manufacturing processes are considered as energy-efficient as they require a low amount of water, and waste generation is less and recyclable at the end of the life cycle [12,13]. The glass façades are transparent, allowing outside views and substantially reducing the energy use for artificial lighting as they allow sufficient daylight to enter the building during the daytime. However, if not correctly designed, solar gain through heat transmittance of glass façades can be as high as 85% of the incident radiation [14]. This transferred heat raises the air temperature inside the building, creating a high demand for air conditioners to keep up the occupants' thermal comfort. It increases the building energy consumption, as air conditioners are high energy consumers and would result in high energy bills [15]. Furthermore, using more energy means more carbon emissions [16], and thereby is contrary to the primary aim of using a green building. Having a system to prevent heat transmittance through the façade lowers the building energy demand [17], which results in less carbon emissions [18,19]. Therefore, careful consideration is needed in designing the glass façades to get the delicate balance of energy over the building envelope together with adequate daylight to reduce the artificial lighting [2,20–22].

Moreover, careful consideration of the fire safety qualities of glazed façades is essential in designing, as a glazed façade can be the most critical element of building fire spread if it is not designed with adequate fire resistance [23]. Façades or windows made of glass are the most fragile and weakest elements of buildings [24,25] owing to the comparatively poor tensile strength and the brittle behavior of glass relative to other building materials [26,27]. In a fire situation, the breakage and fallout of glass would create a new vent for fresh air entrainment and fire spread, which may significantly accelerate fire development [20,26–31]. Falling parts of façades onto the ground might create secondary fires, which is another problem associated with glass façades. Furthermore, in the absence of adequate fire protection systems, fire would escalate carbon emissions by 30–40 kg of CO_2/m^2 over the life cycle of a standard building [32], and if exposed to extensive fire hazards, it can add up to an extra 14%. The effort to improve energy efficiency without considering fire safety has the potential to expand the fire risk by a factor of three [33]. Therefore, the integration of fire safety features is vital for a building to be considered green. Water wall systems (WWS) allow the integration of both of these aspects, making it a solution that is well suited for the green building façades.

WWS are traditionally considered as an excellent design to maintain the occupants' thermal comfort while reducing the building's energy use [34–43]. WWS store the thermal energy during the day and release it during the night, reducing the energy need for summer cooling and winter heating. Furthermore, it reduces the energy consumption for artificial lighting in the daytime, as a portion of the solar energy enters the building through the WWS [35]. These advantages of WWS have been discussed by a significant number of authors in the literature. These studies have emphasized that the correct design of the WWS depends on a variety of comprehensive construction, building, and weather details. These include orientation of the facade glass transparency and thermal resistance, space thermal balance, façade thermal capacity, the openness of the interior, and local climate. This very complex multi-parameter feature makes it very difficult to develop simple design rules based on mathematical models. Besides, the experimental results are strictly related to a unique morphology, thus generalizing the results is difficult. Advanced simulation studies are needed for this reason.

Apart from energy-saving, WWS offer an additional benefit for fire safety, as there is always a water layer between the two glass panes. The water layer absorbs the heat from interior fire or exterior fire. As water has a high specific heat capacity of 4.2 kJ/K, water absorbs a large amount of energy from the fire before the glass temperature escalates to its breakage temperature. The glass panes heat up more slowly with the water layer and delay the glass breakage. With this effect, the WWS protect the building from both inner and exterior fires compared to conventional glazed façades. Furthermore, the water layer helps in maintaining the glass panes at a uniform lower temperature and may not easily break due to the sudden cooling by the water sprayed by the firemen during a fire. It reduces property damages during the firefighting. However, knowledge on the fire performance of WWS is not available in the scientific literature. The assessment of the fire performance of a façade depends on the investigation of flame spread, smoke spread, and the potential damage. A hybrid fire simulation method with the fire performance of the WWS tested in the laboratory and the remaining building numerically simulated is beneficial [44], as it has the ability to study many aspects of fire propagation while avoiding the high costs of full-scale tests.

The objective of this paper is to review the existing literature on WWS and analyze the findings of the reviewed literature on the impact of WWS on the energy efficiency and fire safety of modern-day green buildings. Therefore, the central question of the review is formulated as "What is the state of knowledge on water wall system impacts on green buildings' energy efficiency and fire safety?" We formulated three sub-questions under the central question: 1. How do WWS affect energy efficiency? 2. How do WWS affect fire safety? 3. How suitable are WWS for green buildings? This paper discusses the applicability of WWS for both energy efficiency and fire safety, and in turn applicability for green buildings. This paper provides a useful reference for both industry practitioners and academics who are interested in WWS developments in order to enhance energy efficiency and fire safety in green buildings while maintaining occupant comfort. In addition, the paper provides the engineers in the construction field with an overview of WWS as a new concept which they can integrate into their building designs.

2. Materials and Methods

The authors selected the literature published up until August 2020 to conduct the review for this article. An overview of the literature search method is shown in Figure 1.

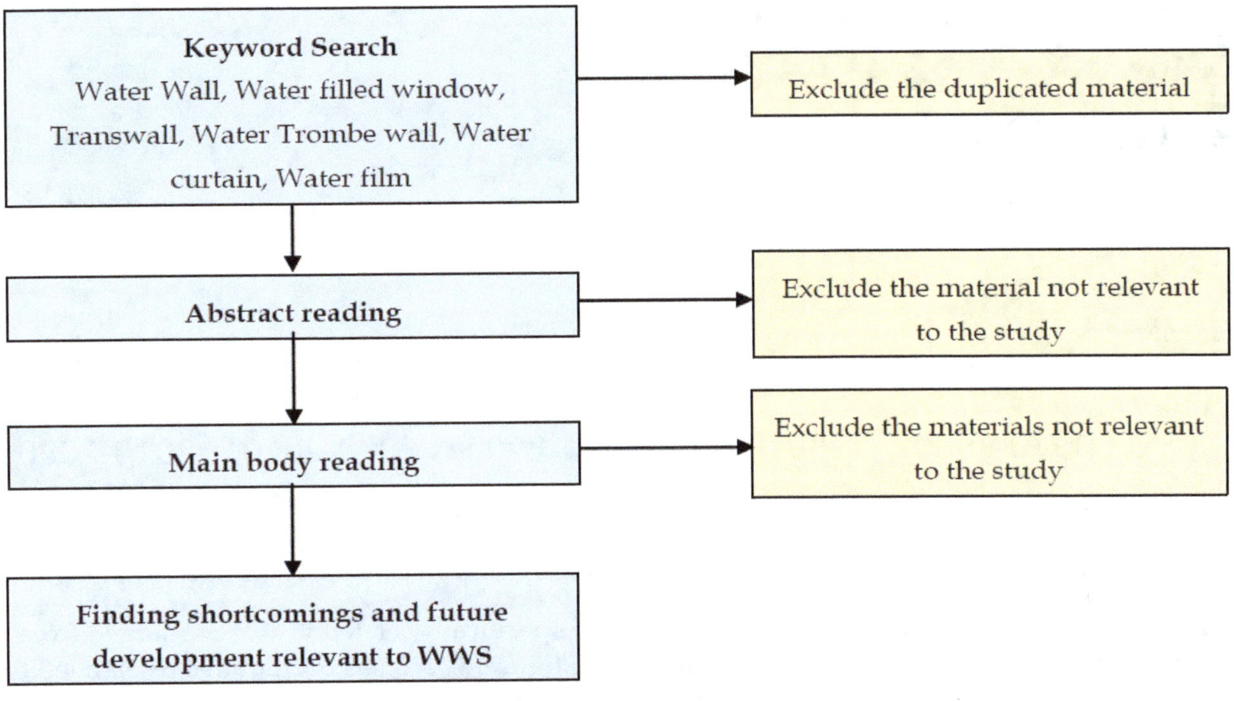

Figure 1. An overview of data collection and the selection procedure for the study.

The literature search process started with the defining of keywords. The keywords were defined after sample searches and sample paper reading. The keywords were Water Wall, Water filled window, Transwalls, and Water Trombe walls. All of these terms were synonyms or similar types of façade systems to WWS. Then the titles and abstracts of the search results were read to exclude the literature unrelated to the objectives of the research. The screened results were categorized according to energy-saving and fire safety. As there were no search results on the fire safety aspect of WWS; additionally, the key word Water curtain crossed with Fire safety was used to gain insight on fire safety from WWS. At last, the analysis of the contents of all the screened literature articles, books, chapters, proceeding papers, online publications, reports, standards, and theses revealed the up-to-date knowledge on the WWS performance for energy efficiency and fire safety, and uncovered the under-researched areas related to the application of WWS in green buildings.

The year-wise distribution of papers that resulted from a primary keyword search is shown in Figure 2. The literature search process started with the defining of keywords. The keywords were defined after sample searches and sample paper reading. The keywords were Water Wall, Water filled window, Transwalls, and Water Trombe walls. All of these terms were synonyms or similar types of façade systems to WWS. Then, the titles and abstracts of the search results were read to exclude the literature unrelated to the objectives of the research. The screened results were categorized according to energy-saving and fire safety. As there were no search results on the fire safety aspect of WWS; additionally, the key word Water curtain crossed with Fire safety was used to gain insight on fire safety from WWS. At last, the analysis of the contents of all the screened literature articles, books, chapters, proceeding papers, online publications, reports, standards, and theses revealed the up-to-date knowledge on the WWS performance for energy efficiency and fire safety and uncovered the under-researched areas related to the application of WWS in green buildings.

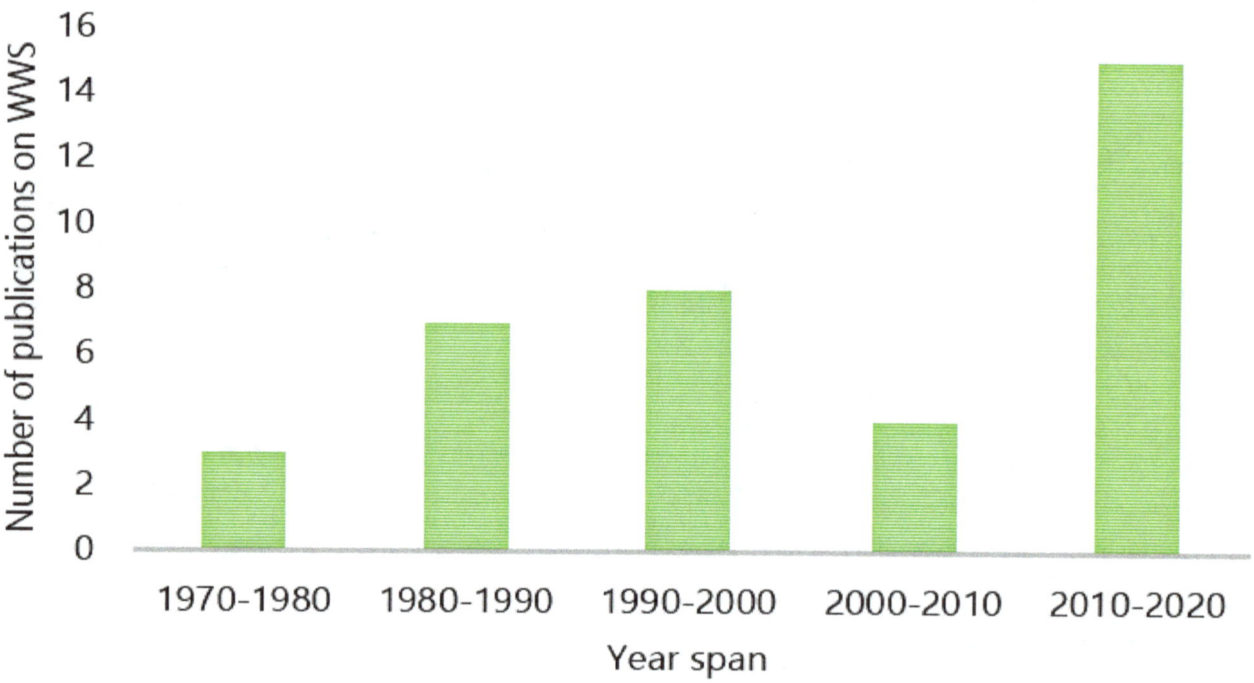

Figure 2. The number of publications under primary keyword Water Wall, Water filled window.

The year-wise distribution of publications that are specifically related to WWS emerged from the keyword search is shown in Figure 2. As shown in the figure, WWS has been under-researched for a long time, and there is an increasing trend on the WWS from the last decade.

3. Water Wall Systems

A Water Wall System (WWS) is "a façade that acts as a Thermal Energy Storage (TES) to store the heat energy for a specific period of time and release the stored energy later" [36]. It is capable of decreasing the energy consumption of buildings. There are different technologies used in TESs that can primarily be categorized as sensible heat storage and latent heat storage as per the heat storage medium. A latent TES retains or releases a significant amount of heat energy from the storage medium during the phase change cycle [45]. Sensible TES stores or releases heat energy with the storage medium's temperature change [46]. High thermal capacity materials like concrete, brick, or water are used in sensible TES to regulate the variations in temperature. Among these materials, water is more efficient in energy storage due to the high specific heat compared to other materials. Sensible TES is further classified into short-term and long-term. The short-term TESs store the heat energy throughout the day and release it at night, while long-term TES also operates for a seasonal period [46]. WWS falls under the short-term TES category and is useful in reducing the energy consumption of the building while preserving the occupants' thermal comfort.

3.1. *Evolution of Water Wall Systems*

The water wall built by Hoyt and his students (1947) in Massachusetts Institute of Technology in the United States using a series of 1.5 gallon cans, located behind a double-pane glass and painted in black, is claimed as the first recorded WWS in the world [47,48]. This WWS was able to provide 38–48% of the heating demand in the building. After some years, in Mexico, a school was designed with a large water tank for each building which serves as a water storage tank as well as a WWS. The buildings were stated to be very comfortable, but there are limited details available on this WWS. In 1972, WWS emerged back again in New Mexico in a space-age architecture designed by Steve Baer named Corrales house. This WWS was comprised of 55-gallon drums full of water, and as reported, this system worked very well. Steve Baer used water walls on both residential and commercial ventures. Many building designers were influenced by Steve Baer's work, including Jon Hammond, Tod Neubauer, Marshall Hunt, David Brianbridge, Marshall Hunt, and Denny Long, who did excellent water wall retrofit projects. In later stages, the water drums used for the energy storage have been replaced with thin water tubes made of different materials. At this stage, the commercial production of the WWS has been initiated. Further, the focus has been shifted for the designing of more aesthetical building elements, where the designers have recognized the glass as an attractive material for WWS. Figure 3 depicts the evolution of different types of water wall systems.

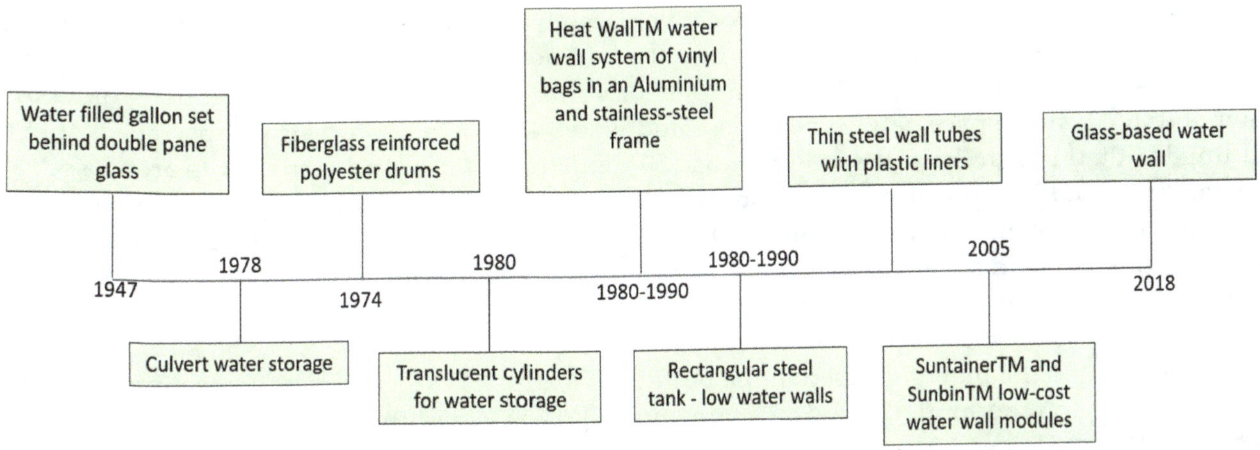

Figure 3. The evolution of water wall systems (WWS) [49,50].

The designing of water walls in early days has mainly dispersed over the United States. However, research on WWS is becoming a trend in recent days, with much focus on energy performance all over the world. A more descriptive discussion on the research on WWS is present in Section 4.

3.2. *Classification of Water Wall Systems*

The WWS, as a façade component, has a long history. The development of WWS started with the introduction of water drums or containers to the building façade to act as a thermal mass. These walls were known as solar barrel walls. Since then, WWS has undergone several technological and design enhancements to cater for the growing needs of sustainable development. With these enhancements at present, the designs of WWS are much focused on the inclusion of water as a layer between two panes made of different materials. This pane material should be chosen for the construction purpose. Based on the position of the water wall within the building, "Water Wall Systems" can be mainly categorized into two types, as direct gain systems and as collaborative systems, as shown in Figure 4. In both, water acts as a thermal mass for heating and cooling.

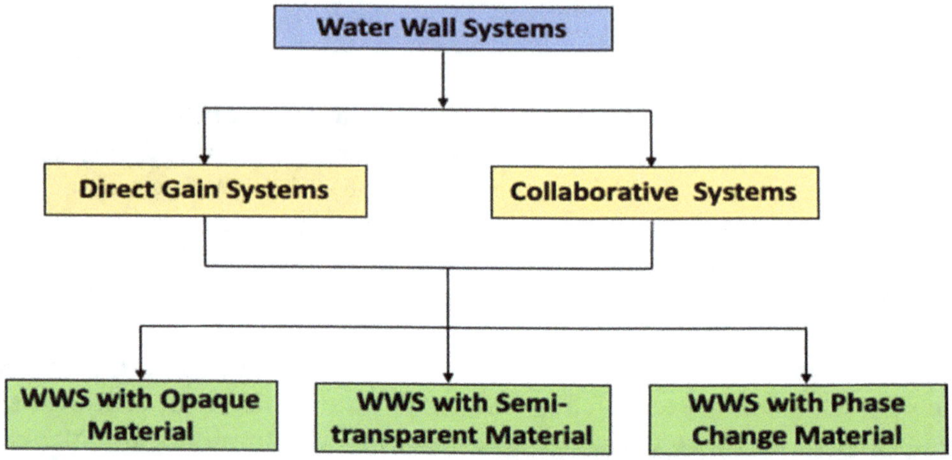

Figure 4. A categorization of WWS. Depending on the location within the building, the WWS are classified into direct gain systems and collaborative systems. This classification is further classified based on the material used in the WWS construction.

The basic categorization of WWS is direct gain systems and collaborative systems. In the direct gain system, the water wall directly faces the sun, and it is adjacent to the ambient. These systems receive solar energy without any intrusion, and heat collection, storage, and distribution all occur within the same space [51,52]. In the collaborative passive systems, the water wall is integrated with the other passive mechanisms for energy saving. It could be a sunspace, a solar chimney, a greenhouse, or a WWS with clerestory. The wall has dampers or vents at the base with the goal that colder air is drawn into the heated space between the glass and water wall. The heated air is drawn back into the building inside through the vents in at the top [53]. During the nighttime, the vents are closed, and the heat contained in the wall radiates into the building. WWS could be coupled with some other passive strategies, e.g., water wall combined with sunspace, and water wall combined with solar chimney. Sunspace and solar chimney are direct gain systems, while WWS is an indirect gain system. As revealed in previous studies, when a WWS and a sunspace or a solar chimney are combined, the thermal efficiency of the configuration is significantly improved.

Depending on the materials used in the building façade, a WWS is generally grouped into three types depending on the configurations: i.e., opaque WWS, WWS with a semi-transparent WWS, and WWS with Phase Change Materials (PCMs).

3.2.1. Opaque Water Wall

Opaque water walls are the basic configuration of water walls. In this configuration, a WWS is integrated into an opaque building façade that is a detached living space from the outer environment, for instance, concrete, metal plates, PVC pipes, and thermal insulating panes [35]. There are no outside viewing options with this configuration.

3.2.2. Semi-Transparent Water Wall

The WWS combined with a semi-transparent external façade, for example, light-porous plastic or glass, enables some portion of the daylight to infiltrate the WWS from outside. In this sort of system, the interior side of the WWS may be built with semi-transparent material or with opaque material

Depending on the material of the interior side, daylight can penetrate the room further. Transwall is a commonly used semi-transparent WWS which has a water layer in between two glass panes and a semi-transparent absorption plate at the middle In this kind of system, 80% of the incident solar radiation is absorbed by the semi-transparent plate and transmits the remaining 20% [54]. This form of wall thus incorporates both direct gain and indirect gain techniques, and is ideal for areas with high daytime temperatures [55]. In this type of system, baffles have often been used to lessen the flow of convective heat through the walls. To improve water viscosity and avoid micro-organisms from growing, bio-inhibiting agents and gelling need to be included in the water [55].

3.2.3. Water Wall with Phase Change Materials (PCMs)

WWS could be coupled with Phase Change Material (PCM). The combination of the PCMs with WWS offers the benefits of both the WWS and the PCMs, thereby enhancing thermal efficiency. The configuration of PCMs with the water wall can have different arrangements as at the outer side, at the inner side, or both sides of the WWS. In the daytime, PCMs melt by absorbing energy from the surrounding environment and then re-solidify at night, releasing the stored energy. This process reduces the need for cooling in the daytime and heating at night [56,57].

Depending on the above-discussed characteristics of different types of WWS, semi-transparent WWS allow more liberty for energy saving through thermal heat storage and daylighting options. Moreover, the semi-transparent WWS provide more design options for better aesthetic appearances compared to the other types. The following are the components of a semi-transparent WWS.

3.3. Components of a Semitransparent WWS

The essential components of a semi-transparent WWS are the two glazing panes and the water layer. The amount of heat transferred through the WWS mainly depends on the properties of these two components. Apart from these, the properties of objects used for fixing WWS into the façade such as frames, screws, nuts, and bolts, need to be considered in calculating the exact amount of heat transferred through the system. This section discusses the function of glass panes and a water column in a WWS.

3.3.1. Glazing

The glazing affects the efficiency of the WWS by controlling the transmission of solar radiation [58]. The material type, number of glass layers, and thickness of each layer are key attributes for designing a WWS [59]. Generally, the water temperature in a WWS is significantly reduced when the glass transmissivity is reduced from 0.9 to 0.45 [35]. Furthermore, the maximum daily temperature of the air in the attached room is significantly reduced with decreased glass transmittance, while there is not much change in the minimum daily temperature. In other research, reduced glass transmissions have been reported to prevent daytime indoor heating while preserving the required level of thermal comfort at night [60]. Thus, glazing has a direct impact on the performance of the WWS and is an essential consideration in designing the WWS.

3.3.2. Water Layer

The ideal water layer thickness of a WWS is correlated to the location of the building, latitude, and climate [52]. The thickness of the water layer has a direct impact on the performance of any kinds of WWS. With a dense layer of water, heat energy takes more time to penetrate the interior, creating thermal distress for the occupants in cold climate [61], but a high thick water column could result in smaller temperature fluctuations [37]. In contrast, a very thin water column can overheat the building's interior. Therefore, the optimum water column thickness must be determined before constructing the façade, as changing the water column thickness after constructing the building is technically difficult.

4. Energy Performance of Water Wall Systems

The energy efficiency or the energy loss of a glass façade depends on the design of the glass façade and the glass material used in the façade construction. The façade should be able to allow the maximum light gain while limiting the heat gain to the building interior. The optimum combination of these two requirements is the key to the glass façade energy performance. The appropriate selection of glazing types for green buildings can result in a reduced space cooling load due to external impact as well as the reduced power consumption for air-conditioning and artificial lighting [60,62].

WWS increase the energy efficiency in buildings by reducing energy usage for heating, cooling, and artificial lighting (only for transparent and semi-transparent WWS) by creating transparent envelopes as shown in Figure 5.

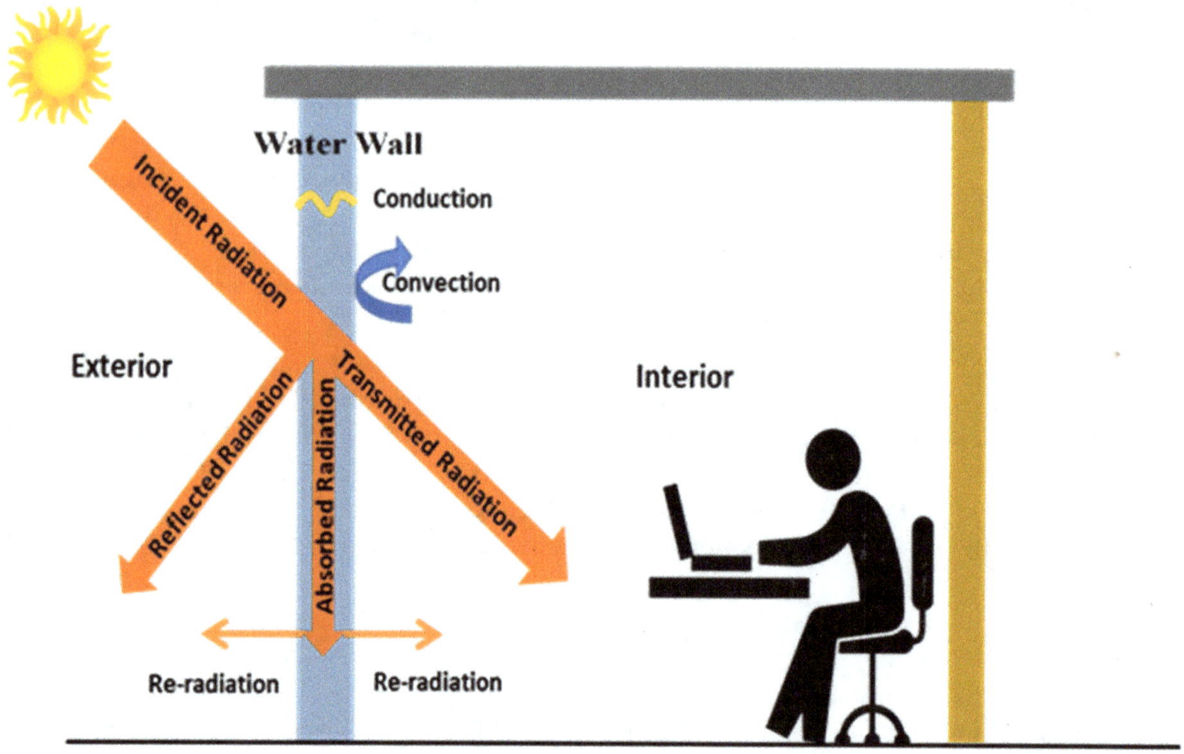

Figure 5. A Water Wall System as a building façade. A part of the solar radiation is transmitted through the glass in the daytime and heats the building interior. A solar radiation portion is reflected out of the glass. Part of the solar radiation is absorbed into the glass, conducted through the glass, and then radiated both inwards and outwards. This transmitted radiation lights up the room. Most of the heat transmitted into the wall reducing the heat transmission to the indoor environment.

The WWS's operational criteria is to increase the thermal comfort in the building interior against the annual environmental changes in the outdoor environment. This is accomplished mainly through the thermal mass gain of water, which has the ability to thermally stabilize envelopes in summer and reduce heat loss in winter through them. Besides, a transparent WWS helps in increased energy efficiency by maintaining an optical transparent stage all the time so that visible light can be transmitted directly through them during the day. In principle, the work of WWS depends primarily on the thermal and optical characteristics of the cavity filling materials. As shown in Figure 6, the solar energy incident on a semi-transparent water wall is partially absorbed and partially transmitted. The absorbed portion is eventually transported to the room while the transmitted fraction induces direct heating as well as daytime lighting in the room. When the outdoor temperature is declining in the nighttime, the WWS moderately heats the building interior [39] as shown in Figure 7.

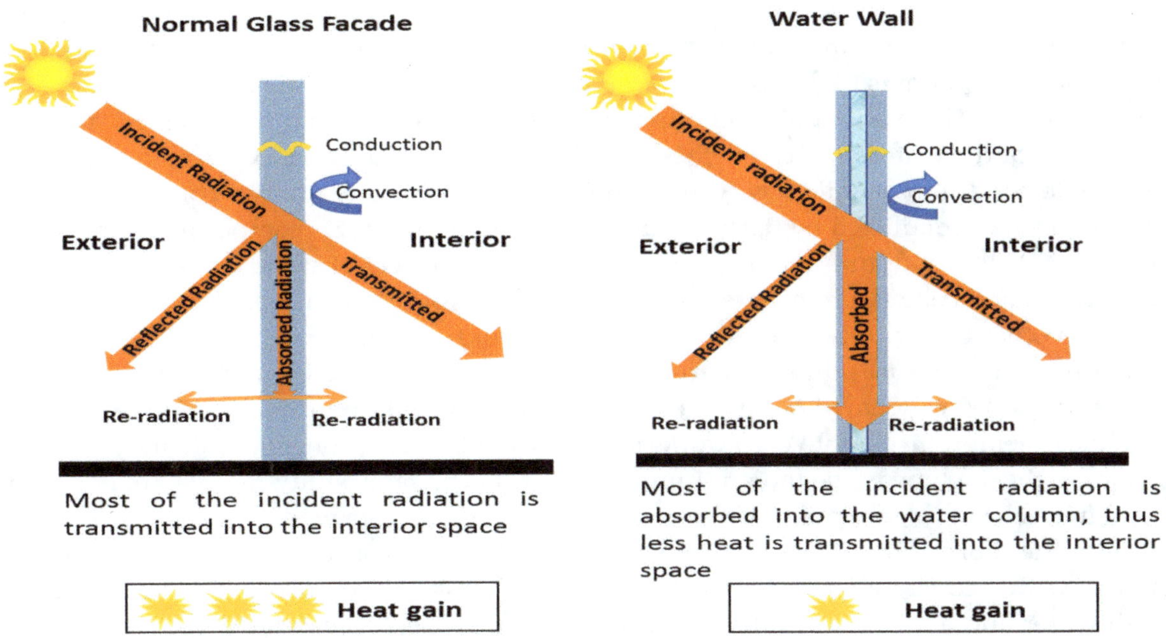

Figure 6. The heat gain through an ordinary glass façade vs. a water wall system.

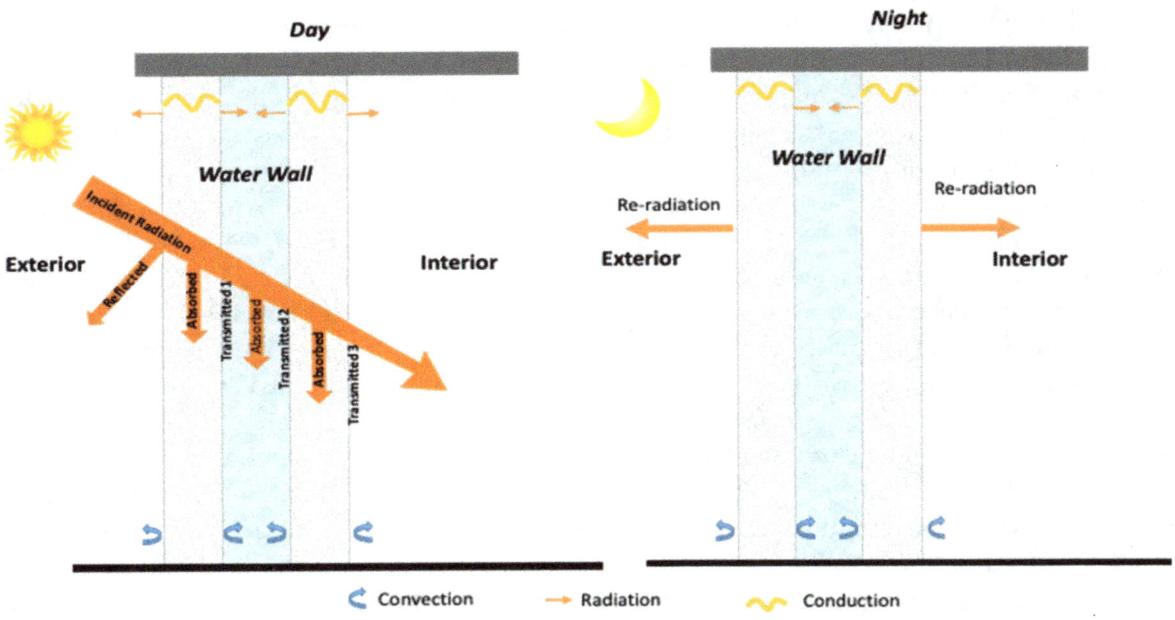

Figure 7. A water wall system stores the energy during the daytime and releases it during the nighttime.

4.1. Past Research on Assessing Thermal Performance of WWS

The review of the literature shows that researches on WWS were conducted in three different ways, namely, experimental modeling, an analytical method based on Heat Balance Models (HBM)s, and numerical modeling through Computational Fluid Dynamics (CFD) programs and Building Energy Simulation (BES) models. The following section discusses these methods in detail with examples from the literature.

4.1.1. Experimental Modeling of Thermal Performance of WWS

The research on WWS started with the experimental modeling, whereas the history of experimental modeling of thermal performances of WWS could be traced back to 1947, which was conducted by Hoyt Hottel and his students [35]. However, after that, there are no records for the research on water walls until the 1970s. There is a significant number of researches carried out from the 1970s to the 2000s on WWS. Much of the water wall researches have been conducted using experimental modeling in the real climate conditions, while few were carried out in laboratories under controlled conditions. The results of a study conducted by Balcomb and Mcfarland [63] in 1977, comparing the performances of five different WWS configurations, suggested that the WWS located at the rear side of a glass pane and the WWS located behind a transparent insulation panel worked with considerably higher thermal efficiency than other configurations in winter. The other configurations were the WWS located inside the room with a temperature equivalent to the room temperature, the internal WWS located inside the room, and the WWS located behind an opaque wall. Nayak, Bansal, and Sodha [64], in 1981, confirmed that a semi-transparent WWS works more efficiently than a concrete wall in terms of thermal performance in the daytime in the Indian summer climates. The same study revealed that a water wall with a 0.10 m thick water column in between a 0.22 m thick concrete wall gives nearly a continuous heat flux into the room at nighttime. In 1986, Sutton and McGregor [65] compared the efficiency of a water wall with a concrete wall. The water wall was constructed with 27 galvanized steel tubes filled with water and a concrete wall of 300 mm (12in) thickness. The study showed a significant heating energy saving in the house with a WWS relative to the concrete-walled house. The house with WWS consumed 70.8% of the concrete-walled house's energy consumption. Another study was conducted by Turner et al. [66] in 1994 to show that during the summer daytime, the building inside could be cooled by a WWS made-up of 7.6 cm diameter plastics tubes mounted on a traditional wall which was charged by cold ambient air at night. The performance of a WWS combined with PCMs was investigated in 1988 by Yadav and Tiwari [67]. The results revealed that the combined effect of a WWS with PCMs gave the best thermal performance over all other WWS arrangements considered. In 2003, Din et al. [68] conducted a study on a WWS attached to a greenhouse to record the plant temperature, the water temperature, and the room temperatures for a day in the winter season by changing the factors such as absorptivity of the black colored WWS surface, water wall thickness, and the transmitted solar energy fraction. The results showed that a north facing WWS with a thickness of 27.5 cm could increase the room temperature by up to 4–5 °C at night and 3–4 °C in the daytime. In the later stages, the performance of water walls was assessed when they were combined with other passive systems. For example, in 1996, Tiwari and Singh [69] revealed that a semi-transparent water wall in a sunspace could reduce temperature fluctuations significantly, and in 2018, Wang and Lei [39] revealed that incorporating a water wall system to a solar chimney could result in an increased thermal comfort level in the attached room. For the literature reviewed, most of the previously conducted researches on WWS focused on the heating ability of WWS for summer seasons, where less attention is given to the performance of WWS for the winter season.

Experimental modeling in real atmospheric conditions has several advantages and several disadvantages [35]. Experiments allow for the gathering of real-time performance data, for instance, solar irradiation and angle, atmospheric temperature, wind velocity and direction, and the like. At the same time, the experiments are usually inflexible. When an experimental model is built up, it is typically troublesome or costly to adjust the model for various designs. Conversely, reduced-scale models might be flexibly adapted for different purposes. Furthermore, full-scale experiments are costly compared with the reduced scale laboratory experiments. Full-scale design models demand a large land area for the installation and entail substantial development expenses. If the experiment includes the study of seasonal weather changes from winter to summer or the other way around, it might take as long as a large portion of a year to finish the testing. Most importantly, it is typically challenging to rehash full-scale experiments under real whether conditions. Even though experimental modeling is considered a compelling method of studying the thermal performance of WWS [35], the numerous constraints of the experimental approach have kept it from being broadly practiced by researchers in the recent past.

4.1.2. Analytical Modeling of Thermal Performance of WWS

The analytical modeling of WWS has been conducted mainly through the Heat Balance Model (HBM)s and has been used for analyzing the performances of WWS in different types and different locations from the 1970s. In 1979, a group of researchers studied the thermal performances of a semi-transparent WWS, a Trombe wall, and a direct gain system using HBM, which they called thermal network models, and concluded the solar heat gain in a Transwall is very close to the solar heat gain of a Trombe wall [70,71]. Bansal and Thomas [72] in 1991 compared the performances of semi-transparent WWS, mass wall, solarium, and a Trombe wall using steady-state heat transfer equations and concluded that a water wall performs better than the others if night losses are reduced through the use of movable insulation. A similar type of research conducted by Bhandari and Bansal in 1994 confirmed that water walls and Trombe walls are more effective for marginally higher solar radiation applications [73]. Nisbet and Mthembu [74] in 1992 compared the performances of a WWS with dyed water and water-gel solutions with an HBM based on the Fourier conduction equation and revealed that WWS with water-dye solution has a high heat release at evenings. The same study showed that a Transwall installed in a house in the west of Scotland can have an energy saving of 23%, while a fora house in the south of France is about 62%. This study further revealed that the optimum wall thickness should be about 150 mm for a maximum performance.

The HBMs were also used for the analysis of WWS attached to different other passive techniques. It was revealed in 1988 by Yadav and Tiwari [75] that an opaque WWS incorporated in a sunspace in a residential single-story house could significantly reduce the temperature changeability in the space linked with the WWS. Tiwari and Kumar [76] compared the performances of an opaque WWS, a Transwall, and an air-collector in a sunspace using energy balances of the different components. The results revealed that a Transwall as a linking wall increases the room temperature on winter nights and outperforms the other two systems.

The HBMs also can configure the optimum thickness of WWS. In 1991, Upadhya, Tiwari, and Rai [77], who conducted a study in Sri Nagar, India, revealed that for maximum heat gain in the winter daytime with less temperature fluctuations, the thickness of the trap material needs to be maximized, while the inner pane glass thickness needs to be minimized. Wang and Lei [38] in 2019 investigated the optimum configurations for a WWS including different glass pane thicknesses, different air gap widths, different water column thicknesses, surface tinting, and the relative solar chimney location using a Transient Heat Balance Model (THBM). The results showed that increased glass pane thickness or reduced water column thickness could result in better thermal performance of the combined system, while increased air gap results in better ventilation with decreased room temperature.

The water wall researches in the recent past involved with the BES programs too. The majority of the studies have utilized the TRNSYS whole building simulation model. Wang et al. (2012) utilized TRNSYS to examine the energy utilization and thermal performance of an opaque WWS in China [78]. The simulation results of a year-round period revealed that building architecture can have an effect in reducing 8.6% of annual energy usage and can raise the thermal comfort assessment index by 12.9%. Sánchez-Ostiz et al. [79] in 2014 simulated the energy performance of sunspace with an opaque WWS for both winter and summer climates using TRNSYS. The results of the study revealed that the energy requirement for the heating of a sunspace with a WWS was 12.9–16.7% less than a sunspace without a WWS. However, due to the limitations of the available BES programs, direct analysis is restricted to the opaque WWS. It requires external simulation programs such as CFD for the simulation of semi-transparent WWS.

In summary, much of the analytical research on WWS are based on HBM models, as it can foresee the essential elements of a building. However, these calculations are based on many assumptions, such as constant convective heat transfer coefficients, uniform surface temperatures, gray radiative surfaces, diffuse radiative surfaces, and one-dimensional heat transfer within the system. Furthermore, radiation interchanges between internal surfaces are often ignored in HBMs, while some models recognize the radiation released from external surfaces. However, Wu and Lei [36] showed that the internal surface-to-surface radiation exchange and time variations of heat transfer coefficients have a substantial impact on system efficiency. Due to these limitations of the models, the real performance of the system could be varied from the calculated performances.

4.1.3. CFD Modeling of Thermal Performance of WWS

The application of CFD modeling to WWS research is a new approach, and still a limited number of researches have been conducted by means of CFD modeling. Karabay, Arici, and Sandik [40], in 2013, researched the thermal efficiency of a concrete mass wall with built-in pipes filled with constant temperature water using CFD where turbulent water flow was presumed. The simulations were limited to steady-state conditions. In 2015, Moustafa and Aripin [41] assessed the steady-state thermal efficiency of a WWS combined with a porous ceramic pipes system, which is an advanced version of a potter wall. Coupling of BES tools with CFD is a growing research direction. In 2018, Sameti [80], with the use of TRNSYS, together with CFD, showed that the thermal efficiency of an opaque solar WWS with a water storage tank was higher in thermal performance than a conventional opaque WWS. Furthermore, a new thermal network model was developed by Liu, Guo, and Wang [43] and was used for the simulation of semi-transparent WWS through EnergyPlus together with a CFD study. The findings of these studies establish that the BES models together with CFD are useful in analyzing the energy performance of WWS in detail. Some recent studies have developed and used transient CFD models that considered the time effect of solar energy fluctuations and ambient temperature in evaluating the thermal efficiency of an opaque WWS and semi-transparent WWS [81]. However, many of the above researches have neglected the refractive radiation index and glass panes radiation reflection. These studies assumed a bulk value for the attenuation coefficient of solar radiation in water within the wavelength spectrum. However, in actuality, the attenuation coefficient of solar radiation in water depends on the wavelength. In this manner, a multiple-waveband attenuation model merits further research to get progressively precise outcomes. The evolution of WWS research is summarized in Table 1.

Table 1. A summary of past research on assessing thermal performance of WWS.

Year	Location	Type of WWS	Method	Results
1947 [35]	United States	WWS with opaque material	Experimental modeling	WWS has the capability to increase the thermal comfort level
1978 [63]	United States	WWS with semi-transparent material	Experimental modeling	Water drums behind glass panes improve the thermal comfort inside the room
1979 [70]	United States	WWS with semi-transparent material	Thermal Network Models	Solar heat gain in transwall is very close to the solar heat gain of Trombe wall
1981 [82]	India	WWS with semi-transparent material	Transient Heat Balance Model (THBM)	The thermal performance of a semi-transparent water wall is more efficient than a concrete mass wall
1986 [65]	Tasmania	WWS with opaque material	Experimental modeling	Energy consumption of a house with WWS is less (70%) than that of a concrete walled house
1987 [34]	India	WWS with semi-transparent material	Experimental modeling	Semi-transparent WWS is more efficient compared to a concrete mass wall in terms of thermal performance in daytime
1987 [83]	India	WWS with opaque material	Mathematical model	Large drums full of water kept inside a greenhouse can increase the thermal comfort
1987 [71]	-	WWS with opaque material	Mathematical model	Complex heat transfer coefficients for WWS are developed

Table 1. *Cont.*

Year	Location	Type of WWS	Method	Results
1988 [67]	India	WWS with Phase Change Materials (PCM)	Experimental modeling	Water wall combined with PCMs as a link wall can increase the thermal performance
1988 [75,84]	India	Collaborative-opaque water	Heat Balance Models (HBM)	The system reduces temperature fluctuations inside the room significantly
1991 [72]	Leh	WWS with semi-transparent material	THBM	WWS performs better than the other considered thermal energy storage walls
1991 [77]	India	WWS with semi-transparent material	THBM	WWS performs better than the other considered thermal energy storage walls
1991 [76]	India	Collaborative-opaque water	THBM	For less temperature fluctuations with a maximum heat gain, the inner pane thickness of WWS should be at a minimum
1992 [74]	Scotland and France	WWS with semi-transparent material	THBM	WWS can achieve energy saving of 23%—west of Scotland, 62%—south of France, compared to a concrete walled house
1994 [66]	-	WWS with opaque material	Experimental modeling	WWS are efficient for summer days
1994 [73]	India	WWS with semi-transparent material	THBM	WWS performs better than the other considered thermal energy storage walls
1996 [69]	India	WWS with passive mechanisms	Experimental modeling	WWS reduces temperature fluctuations inside the room significantly
2003 [68]	India	WWS with opaque material	Experimental modeling	A north-facing WWS with a thickness of 27.5 cm can increase the room temperature by up to 4–5 °C at night, and 3–4 °C in the daytime
2007 [42] 2008 [85]	China	WWS with semi-transparent material	THBM	Natural convection effects are more influential than radiation effects, and the effect of natural convection in the water layer can be regulated if suitable geometric shapes are adopted
2011 [81,86] 2018 [87]	Hong Kong	WWS with semi-transparent material	Experimental modeling, THBM	WWS can significantly reduce the indoor heat gain in both warm and cool climates
2012 [78]	China	WWS with semi-transparent material	TRNSYS	WWS can decrease the room maximum temperature by 4 °C and increase the minimum temperature by 3 °C
2013 [40]	Turkey	WWS with opaque material	Computational Fluid Dynamics (CFD) (FLUENT)	The thermal performance of wall heating systems is better than the floor heating systems in terms of thermal performance
2013 [88]	Spain	WWS with semi-transparent material	Experimental modeling	The energy performance of WWS is more effective than the traditional windows
2014 [79]	Spain	Collaborative (Sunspace) -Water wall in a sunspace	TRNSYS	A sunspace with WWS needs 12.9–16.7% less heating energy compared to a sunspace without WWS
2015 [41]	Luxor, Egypt	WWS with opaque material	CFD	A pottery–water wall can reduce cooling and heating demand by 88% at extreme climatic conditions
2015 [89] 2017 [90]	China	WWS with semi-transparent material	CFD	The effect of header design on the performance of WWS is insignificant
2016 [37]	Sydney, Australia	WWS with semi-transparent material	CFD	Greater energy savings of WWS in winter climate compared to a concrete wall with same thickness, and in summer they perform in a similar way
2018 [39]	Sidney, Australia	WWS with Passive mechanisms—with sunspace	Experimental modeling	The thermal comfort level is increased in the attached room by raising room temperature in winter as well as with a considerable ventilation
2018 [43]	United States, Chicago	WWS with semi-transparent material	EnergyPlus + CFD	Thermal mass windows have a greater thermal performance compared to the traditional windows

Table 1. *Cont.*

Year	Location	Type of WWS	Method	Results
2018 [80]	-	WWS with semi-transparent material	TRNSYS + CFD	The proposed system is high in thermal performance and aesthetic appearance
2019 [38]	Sidney, Australia	Collaborative (Solar Chimney)-Semi-transparent	THBM	A combined solar chimney and WWS can provide good ventilation and thermal comfort throughout the day and night
2019 [91]	United States, Chicago	WWS with semi-transparent wall	Integrated Energy (IE) and CFD	Transparent water storage envelopes can result in 70% of heating and cooling load reduction compared to conventional glass facades
2020 [92]	13 cities	WWS with semi-transparent wall	Window Program + TRNSYS	WWS can result in energy saving of 3–84%/US\$0.99–24.25/m^2 based on the climate

The WWS is proven to be a potential building element to increase thermal performance and energy efficiency on a limited scale. The location of the WWS has a significant effect on the performance level, as the heat transfer through the system heat storage depends on the incident solar flux on the external side of the façade. However, as shown in Table 1, the WWS researches have been performed only in several countries, including the United States, India, China, Australia, and several other European countries. These researches have been carried out in small scales with prototype models, and the research results for the thermal performance of WWS in real buildings are limitedly available in the literature. This may be a reason for the limited application of WWS in real building projects. The research methods on WWS has evolved from experimental modeling and analytical modeling to CFD and whole-building energy simulations. In the majority of these researches, the CFD calculated values for the temperature of the surfaces and water have used as input data for the BES packages. This is because many of the available BES are not equipped with the capability of direct simulation of WWS. Having a direct simulation program may increase the WWS in real building applications, as then the building designers can design WWS for the advancement of energy efficiency and thermal performance in buildings.

The energy efficiency considerations have overlooked the effect of embodied energy throughout the building life cycle. The embodied energy calculation considers the energy use for extraction/manufacturing of materials, transportation, construction, maintenance, replacements during the entire life cycle, and the disposal of the system. If more energy is consumed in these phases of the WWS lifetime, the embodied energy can overperform the reduced energy consumption for air conditioning of the building. Therefore, the comparison of embodied energy against the expected energy savings is very much needed for the real building applications.

The future research directions concerning the energy performance of WWS are discussed in Section 7.

5. Fire Performance of Water Wall Systems

A building façade should be planned to limit the spread of fire to the inside of the building built to restrict fire spread inside the structure, to the adjacent floors, and to the outside of the building, allowing the evacuation of building occupants and ensuring the safety of the rescue team [93]. Moreover, the façade needs to be constructed in a manner that stops parts from collapsing throughout the evacuation and rescue team activities. However, a standard glass façade is more vulnerable to fire spread than a conventional façade made with bricks, stone, or concrete, and can easily break within a matter of minutes [94–96] when exposed to fire. This collapse of glass panes introduces a new airflow for fresh air and fire spread outdoors, enabling the phenomenon of flashover or backdraft [97]. Compartment fire dynamics may be drastically modified, triggering more serious disasters [98,99]. The possible ways of fire spread through a glass façade are elaborated in Figure 8.

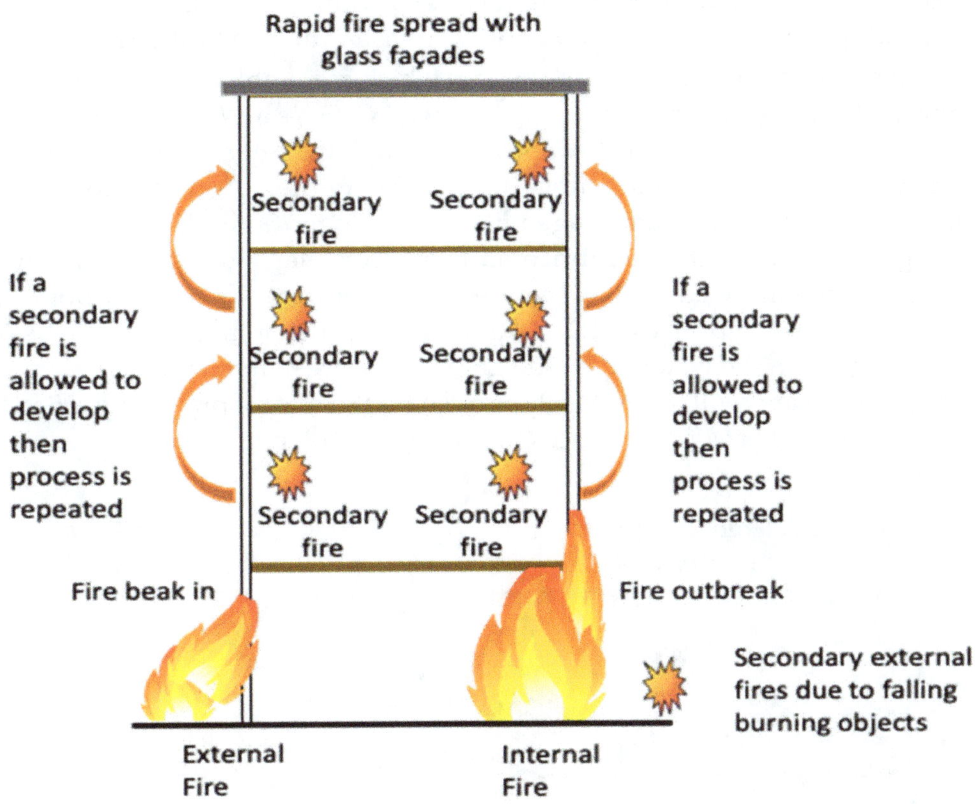

Figure 8. Rapid fire spread with glass façades. Potential fire risks for a building in fire accidents may come from foreign or local agents. For external causes, the fire may spread from nearby buildings or a firing item in the local environment. The radiation coming from the external fire can develop the thermal stress on the glass and can break the glass. Then the fire enters the building interior. Fire accidents from interior fire sources are more common in the real world, where the fire propagates from room to room/floor to floor. The internal fire can develop the thermal stress on the glass, and then the glass breaks. Then the flame will hit the upper floor façade, and the secondary fire will develop. The fallen burning objects can develop another secondary fire.

Glass façade fire performance is influenced by many factors, while thermal gradient is the primary cause of glass breakage [100]. Several researches have been carried out on the crack ignition of glass panes in a fire, and different results are established. For example, Keski-Rahkonen [101] revealed that an 80 °C temperature variance between the heated glass pane and the edge is needed for crack initiation. Pagni and Joshi [31,102] reported that only a 58 °C temperature difference is needed for crack initiation in a glass pane. Moreover, it is established that the fallout of a glass pane when exposed to fire also depends on the imposed heat flux. For example, a single-strength glass pane will crack with 4 to 5 kW m^{-2} of heat fluxes [103]. However, in a real fire situation, it is difficult to predict when the glass will fall out after breakage [100]. The glass pane thickness is another significant factor that affects the glass pane performance in a fire. For example, 3 mm window glass panes break at a temperature of around 360 ° For thicker, 4–6 mm glass pane, the glass pane will break at around 450 °C, and double-glazed glass panes with a 6 mm thickness are expected to break at about 600 °C [100]. It has also been proven with several experiments that the temperature differences required for glass pane cracking will increase with increasing thickness [28]. The type of glass pane selection, such as single pane/double pane, also has a significant effect on the glass fire performance [28,104,105]. Besides, the smoke movement [98], glass pane orientation [28,30,99], and location of the fire [30] are found to be of considerable significance for glass pane breakage [14].

The characteristics of WWS make itself potentially a fire safety construction element. Water is used to extinguish flames. Especially water film and water curtain are used on glass panes to prevent breakage [106–109] since glass breakage influences the fire spreading, as explained previously. However, glass cracking typically happens in the rapid cooling phase rather than the rapid heating process, owing to the thermal shock induced by sudden cooling, which is a problem associated with water films and water curtains [110]. This issue can be overcome by a WWS, as a WWS always has a continuous water layer between two glass panes. The heat passes through the interior or exterior glass pane and is absorbed by the water column from the fire initiation. Therefore, it is reasonable to expect that to heat up the glass pane until its breakage stage will take a longer time. The WWS offers abundant faster heat exchange than a typical wall fabricated from concrete or brick because the heat stored in water could also be decentralized through convection. Compared with other building materials, water has a high specific heat capacity of 4.2 kJ/kg.K, while concrete has a specific heat capacity of 0.880 kJ/kg.K only [111]. With the water layer, the glass panes are maintained at a lower and perhaps more uniform temperature and will not be easily broken by the water sprayed by a fireman during a fire. The mechanism by which a WWS enhances fire safety is shown in Figure 9.

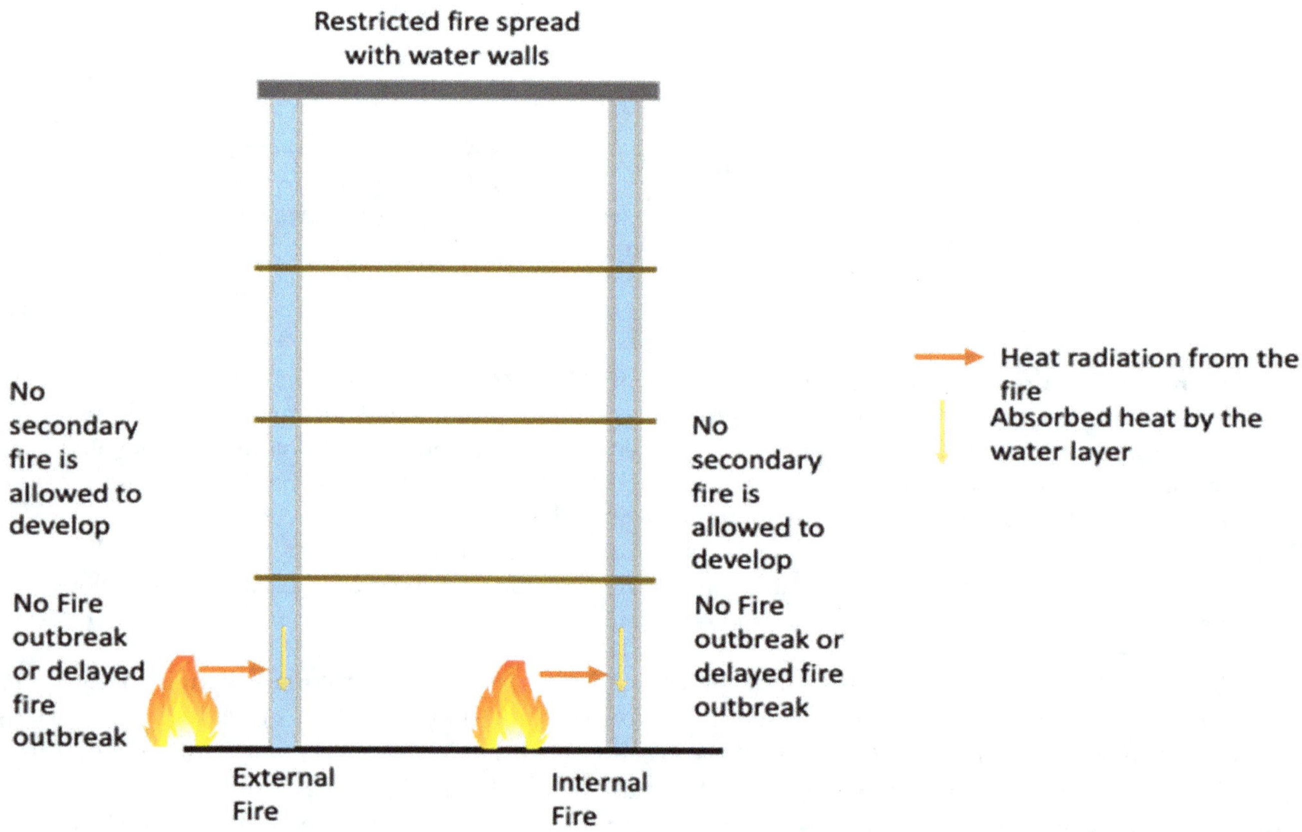

Figure 9. Restricted fire spread with water wall systems. The water layer traps the heat transmitted and conducted through the glass panes from the inner fire or the exterior fire. As the added heat from the fire is transferred to the water layer, the glass panes are heated up more slowly during a fire, delaying breakage of glass panes due to heat up. With this effect, the WWS protects the building from both interior and exterior fires.

6. Water Walls as a Green Building Façade

A green building needs to be more friendly with the environment and humans than conventional buildings [112]. For this, a green building must be designed, constructed, restored, managed, or recycled in an environmentally friendly, economical, and resource-efficient way [113,114]. It is expected to meet a range of particular targets: Energy efficiency, reduced CO_2 and Green House gas, pollution control, noise reduction, enhanced indoor air quality, and environmental harmonization. In this context, water walls are well suited for the green building façades [114].

6.1. Economic Cost of WWS

WWS are considered as simple and very economical for many reasons. WWS may also be built in new or renovated buildings with traditional construction techniques. WWS do not depend on auxiliary power, as no pumps or valves are incorporated. Therefore, WWS operate even though the electricity is off, which helps in reduced energy bills. Constructing and maintaining a WWS is less costly than other thermal storage systems, since WWS is constructed with standard materials and also has a long life [38]. For example, a water wall will cost about \$9612 for 100,000 BTU storage, while a concrete wall will cost about \$14,426 for the same heat storage in San Francisco, CA, USA [50]. Thus, WWS is more economical than that of concrete walls for constructions in San Francisco, CA, USA. As the economic cost of the building elements may vary for different locations, the economic cost needs to be calculated before the building design. However, depending on the lower material volume, ease of construction, lower maintenance needs, and the reduced energy consumption, it is expected that WWS are economical on the whole worldwide than that of high thermal mass façades such as concrete walls.

6.2. Embodied Energy of WWS

The embodied energy of WWS (which is the total energy required for extraction of raw materials, processing of WWS, delivery and construction of WWS, and the disposal after the building demolition) needs to be considered in assessing the suitability of WWS for green buildings. The embodied energy of a WWS primarily depends on the embodied energy of glass, as a WWS is built of two glass panes and a water layer. A material-wise comparison of the embodied energy of a WWS and a concrete wall is shown in Table 2. In all the selected countries, float glass has a much higher embodied energy than concrete for a unit mass. However, for effective comparison, embodied energies need to be compared on a per area basis for façade elements.

Table 2. A WWS embodied energy comparison against a concrete facade for different countries.

Location	Glass Embodied Energy per Unit Mass (MJ/kg)	Concrete-Embodied Energy per Unit Mass (MJ/kg)	WWS Embodied Energy per 1 m^2 of the Façade (MJ)	Concrete–Embodied Energy per 1 m^2 of the Façade (MJ)
Western Australia [115]	15	1.39	885	1737.5
New Zealand [116]	15	2	885	2500
USA [117]	12.7	1.5	749.3	1875

The thickness of façade elements is usually decided based on the structural requirements and the required U-value of the building envelope, which determines the amount of heat transfer through the building envelope. The U-value of a building element depends on the thermal conductivity of the material, convection coefficient from the material surface, and the thickness of the building element. For example, a WWS with 100 mm water layer and two 25 mm thickness glass panes has a U-value of 1.08 $Wm^{-2} K^{-1}$. In order to have the same U-value, a concrete wall needs to have about 400 mm thickness. The mass of glass required for the construction of a WWS facade is lower than the mass of concrete for a comparable concrete facade. A comparison of the embodied energy of WWS and a comparable concrete wall is presented in Table 2 and Figure 10.

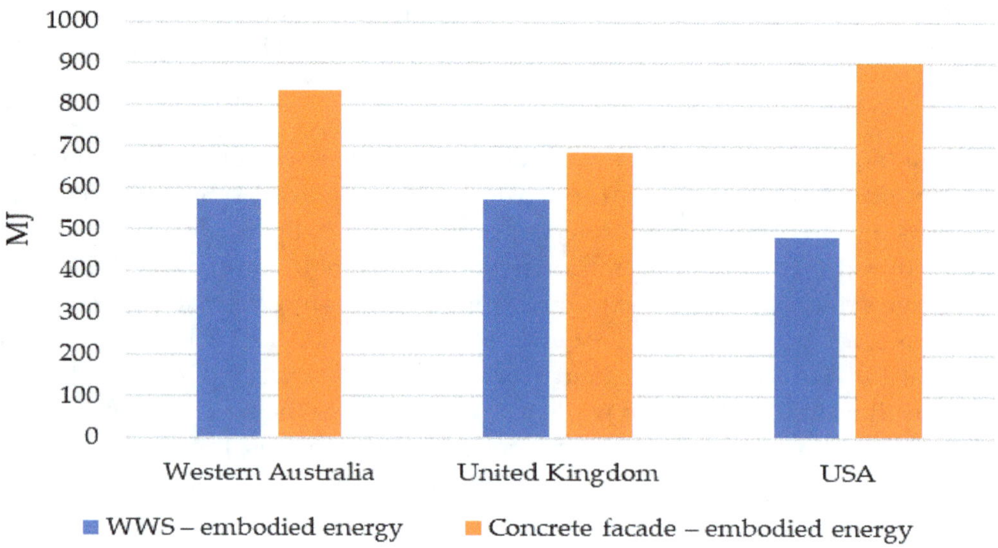

Figure 10. A WWS embodied energy comparison against a concrete facade for different countries (per 1 m^2 of the façade) (Sources: [115–117]).

As shown in Figure 10, even though the embodied energy value of WWS differs for different locations, it is always less than the embodied energy of concrete walls. Thus, WWS is a more sustainable feature than a concrete wall for a green building, considering the fact that the embodied energy of WWS is lower than that of the concrete facade on a unit area basis.

6.3. Energy Consumption and Greenhouse Gas Emissions

WWS decreases the heat loss and heat gain through the building facade, thus lowering energy usage for air conditioning. On the other hand, WWS get a large amount of light that aids in giving a high measure of natural daylighting as opposed to depending exclusively on artificial lighting [118], thereby reducing the energy consumption for artificial lighting [119,120]. The mechanism of energy saving through WWS has been explained in detail in Section 4.1.

The reduction of CO_2 and GHG emissions is one of the major purposes of green buildings to reduce the impact on the environment [17,19,121]. As energy consumption is reduced, the need for energy generation is also reduced. Consequently, fewer fossil fuels are burned for electricity generation reducing CO_2 and GHG emissions. Furthermore, during a fire, the WWS absorbs much of the heat radiation from the fire, thus reducing the thermal stress development on the glass pane. This helps in preventing more hazardous fire situations. The reduced fire risk also effects in reducing the possibility of CO_2 emissions.

6.4. Recyclability

The suitability of a material to match with the green concept needs to be evaluated by considering the entire life cycle, including construction and demolition. Glass panes are comprised of non-contaminating raw materials, has an energy-efficient manufacturing process requiring an insignificant amount of water, and creates a negligible amount of waste [122,123]. Its application in the structures chops down the use of concrete, which is again one of the primary sources of CO_2 generation. Additionally, glass panes are recyclable and can be reused, offering to ascend to a Cradle to Cradle approach [114], which makes it a clear choice that is causing a decrease in raw material needs.

6.5. Human Friendliness

A green building needs to be human friendly. Thermal comfort, acoustic comfort, and aesthetic appearance are some of the factors that affect human friendliness. As the temperature fluctuations within the buildings are reduced, the WWS deliver thermal comfort to the occupants. Furthermore, WWS provides a high level of acoustic solace by warding off sound from the outsides to the insides, thus guaranteeing a more quiet atmosphere inside [124]. WWS provides natural views to prevent claustrophobic experiences of complete masonry walls [125–127]. It is reported in the previous researches that better viewing options of the outside can lead to the improvement of the health and wellbeing of humans, which is possible with WWS [128,129]. Meanwhile, water is not toxic, contrary to other TES substances such as PCMs. Therefore, WWS are environmentally safe and human friendly.

Owing to the above-discussed design and working principle, it is fair to conclude that WWS is innovative, environmentally beneficial, economical, and socially approachable, and an excellent sustainable application to a green building façade. Figure 11 portrays the green benefits of water wall systems.

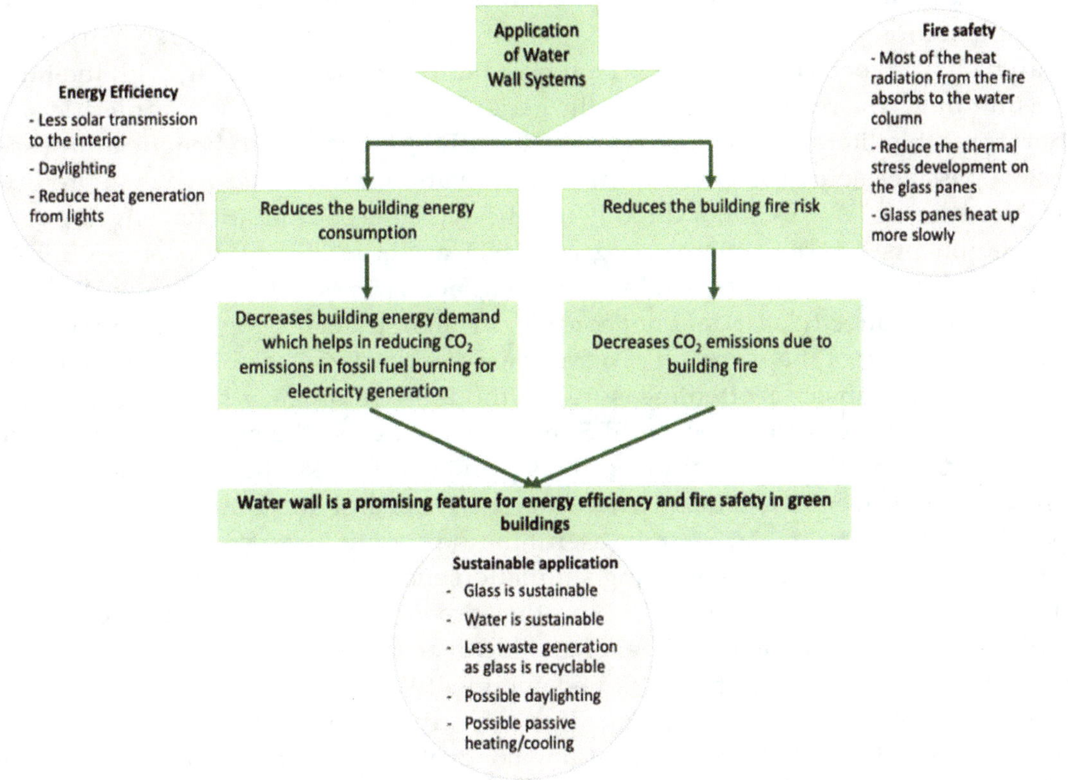

Figure 11. A water wall is well suited for a green building façade. The application of water wall systems helps in reducing building energy consumption and fire risk. Water wall systems help to reduce the building heat loss and heat gain, allow daylight within the daytime, and thereby reduce the energy consumption for air conditioning and artificial lighting. Consequently, fewer fossil fuels are burned for electricity generation. In a fire situation, the WWS absorbs much of the heat radiation from the fire, slowing down the glass pane heat up and reducing the thermal stress development on the glass pane. The reduced fire risk and reduced electricity generation effects in reducing the possibility of CO_2 emissions, which is one of the main aims of green building constructions. Further, the material used in WWS constructions are sustainable and recyclable.

7. Shortcomings of Past WWS Research and Future Directions

A considerable number of researches have been carried out on assessing the energy performance of different types of WWS. However, there is a gap in the literature assessing the economic cost of WWS. In fact, assessing the cost of WWS is important for before applying to the green buildings, as the economy is one pillar of sustainability, with the environment and humans being the remaining pillars. It is a common perception in the construction industry that "Green costs more". This is one of the reasons that WWS are limitedly applied in buildings worldwide. Therefore, to foster WWS developments in worldwide, studies on the economic aspect of WWS is a deliberate need.

The reviewed literature reveals that the application and the research on WWS are limited only to several countries. The results of those researches suggest that the efficiency of the system significantly varies with the climatic conditions, which cannot be generalized to the whole world [77,82,130,131]. Future studies should concentrate on the best WWS configuration concerning each climatic condition around the world. The results can increase the popularity of WWS in the global context.

Past research suggests some parameters (water wall type, the material used, location of the WWS in the building, size of the WWS) influence the effectiveness of WWS significantly. By modifying one or two of these parameters, WWS' overall efficiency could be greatly strengthened. To this end, future research is crucial to find the optimal design mix for the effective operation of the WWS.

The previous studies confirm that the incorporation of WWS into the buildings could result in reduced temperature fluctuations in most of the countries concerned. However, the impact of reduced temperature fluctuations on the energy efficiency of the building has been barely studied. Therefore, future research should focus on assessing the whole building energy saving with the integration of WWS to the building. The results will attract the general public for the WWS application.

Furthermore, this literature review reveals that BES programs are beneficial in assessing the performance of passive energy systems. Numerical simulation is more effective than prototype testing in research. Nonetheless, WWS cannot directly use the available computational envelope components or window components in current energy programs, such as Equest, EnergyPlus, IES-VE, TRNSYS, and TAS [132]. Since envelopes are predefined as opaque surfaces, it is unable to perform WWS simulations as a façade component directly, as it does not consider the light transmission to the indoors. In addition, window modules of most of these energy programs are standardized as objects without thermal mass parameters; thus, heat transfer is often measured without heat storage, which is crucial for the WWS operation. Therefore, the development of BES programs to assess the energy performance of WWS should be focused on by the future studies. This provides the likelihood of simulating advanced building environments over an extended period in merely several minutes. It is less time consuming, and inexpensive. Therefore, it is incredibly useful in finding out large-scale systems over extended periods of time, and having a whole building simulation enables proper building designs with WWS.

In addition, WWS could be connected to other green building elements such as green roofs. The integration could result in increased benefits. However, the evidence of such assessment lacks in the reviewed literature. Moreover, the embodied energy of WWS needs to be considered in assessing the suitability of WWS for green buildings. If more carbon emissions are associated with WWS applications, it is a detriment to the green building concept. Therefore, future research should assess the embodied energy of WWS and compare the efficiency of integrating WWS with other green building features.

As suggested in Section 5, depending on the characteristics of the WWS, it is fair to assume that the WWS can provide fire protection to the buildings which have not yet been investigated in the scientific literature. Therefore, more realistic studies are required to examine the WWS contribution to fire safety. The assessment of the fire performance of a façade depends on the determination of the flame spread, smoke spread and the potential damage. A hybrid fire simulation approach in which the WWS fire efficiency is measured in a fire testing facility while the remaining building is numerically simulated at each stage is advantageous [44], as it has the ability to study many aspects of the fire propagation, avoiding the high costs of full-scale tests.

The designing and construction of passive solar systems, such as WWS, are challenging [51]. The cost of installation and maintenance, lack of trained technical personnel, and no established building codes addressing the design and operation of WWS are some of the challenges associated with the limited use of WWS. However, the life cycle benefits of the system will outgrow the cost of installation and maintenance, and the proper design decisions for WWS will help to keep their installation and maintenance costs at a minimum. Thus, correct technical knowledge is essential in constructing a WWS [48]. When the systems are installed incorrectly, it creates additional risks. If a building system has established performance-based building codes, it helps in constructing more efficient systems. Therefore, future research should aim at the establishment of building codes concerning the WWS.

Provided their possible long-term advantages, WWS tend to be cost-effective. WWS will have long-term ecological, socioeconomic, and environmental advantages to make communities safer, healthier, and more prosperous. A systematic analysis of WWS long-term benefits worldwide is expected in future studies. Long-term evaluations might provide significant evidences to enable stakeholders to participate in large-scale WWS programs.

A summary of future research directions is shown in Figure 12.

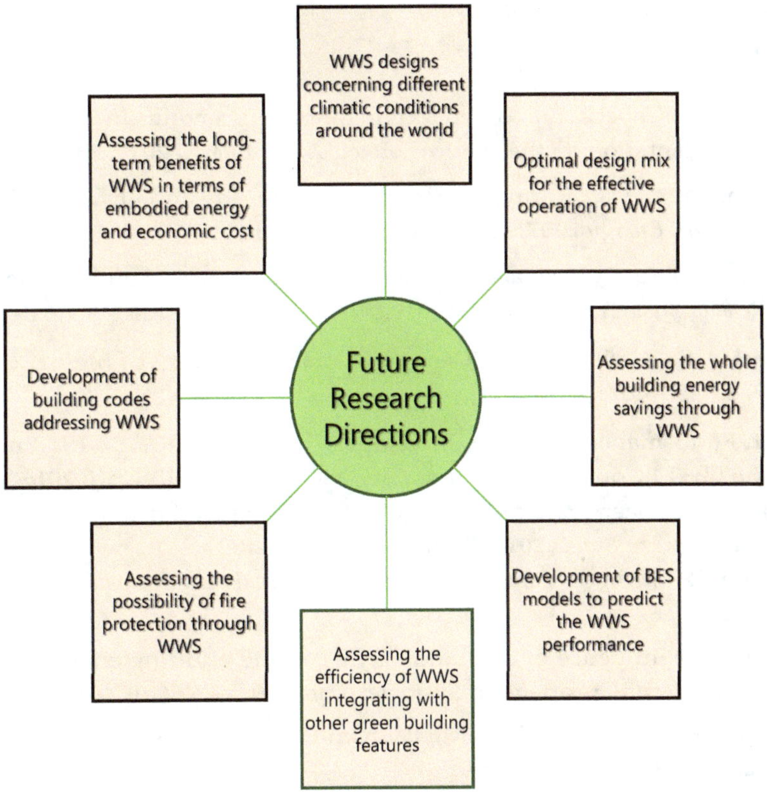

Figure 12. A summary of future research directions on WWS.

8. Conclusions

This paper reviews the research on WWS to assess WWS for energy-saving and fire safety when incorporated into green building façades. From the analysis, it is established that WWS has a significant impact on controlling energy efficiency and fire safety in green buildings. Nevertheless, careful attention is required in designing the façade in green buildings to minimize the solar gain to the building, allow sufficient daylight, and reduce the fire risk, so as to fully utilize the advantages of WWS. It is expected that water wall systems can provide an integrated solution for energy efficiency and fire safety of buildings attached with a water wall system as the façade. It is envisaged that this approach has a high potential to be applied in practice.

The possible energy efficiency of water wall systems has been studied comprehensively through different methods such as analytical modeling, experimental modeling, and numerical modeling (CFD). The studies suggest that the WWS are capable of reducing the temperature fluctuation, which can result in reduced energy cost when they are attached to the building façade. This system is economical and comes with a lot of green benefits.

Moreover, prospective studies must tackle various challenges by systemic examination. The identified research gaps and challenges demand a comprehensive analysis in the future. Future research should focus on assessing the WWS performance for different climatic conditions, find the optimal design mix, assess the performance of WWS when incorporated into other green building features, development of BES programs addressing the simulation of WWS, assessing the fire performance of WWS, WWS life cycle appraisal of embodied energy and economic cost, and development of performance-based building codes addressing the WWS. With the results of these researches, efficient design of WWS with numerous social, environmental, and economic advantages makes the application of WWS in green buildings more appealing.

Author Contributions: Conceptualization, D.L. and C.L.C.; Writing original draft preparation, U.R.; Writing review and editing, D.L. and C.L.C., Supervision, C.L.C.; Funding acquisition, C.L.C. All authors have read and agreed to the published version of the manuscript.

References

1. Pérez, G.; Coma, J.; Sol, S.; Cabeza, L.F. Green facade for energy savings in buildings: The influence of leaf area index and facade orientation on the shadow effect. *Appl. Energy* **2017**, *187*, 424–437. [CrossRef]
2. Tzempelikos, A.; Athienitis, A.K.; Karava, P. Simulation of façade and envelope design options for a new institutional building. *Sol. Energy* **2007**, *81*, 1088–1103. [CrossRef]
3. Chen, J.; Qiu, Q.; Han, Y.; Lau, D. Piezoelectric materials for sustainable building structures: Fundamentals and applications. *Renew. Sustain. Energy Rev.* **2019**, *101*, 14–25. [CrossRef]
4. Chew, M.Y.L.; Conejos, S.; Azril, F.H. Bin Design for maintainability of high-rise vertical green facades. *Build. Res. Inf.* **2019**, *47*, 453–467. [CrossRef]
5. Aksamija, A. High-Performance Building Envelopes: Design Methods for Energy Efficient Facades. In Proceedings of the BEST4 Conference, Kansas City, MO, USA, 13–15 April 2015.
6. Haggag, M.; Hassan, A.; Elmasry, S. Experimental study on reduced heat gain through green façades in a high heat load climate. *Energy Build.* **2014**, *82*, 668–674. [CrossRef]
7. Khanna, T.; Chhugani, R. *Intelligent Facades to Improve Sustainability & Comfort*; NürnbergMesse India Pvt. Ltd.: New Delhi, India, 2016.
8. Wang, Y.; Fukuda, H. The influence of insulation styles on the building energy consumption and indoor thermal comfort of multi-family residences. *Sustainability* **2019**, *11*, 266. [CrossRef]
9. Rakhshandehroo, M.; Yusof, M.J.M.; Najd, M.D. Green Façade (Vertical Greening): Benefits and Threats. *Appl. Mech. Mater.* **2015**, *747*, 12–15. [CrossRef]
10. Carletti, C.; Sciurpi, F.; Pierangioli, L. The energy upgrading of existing buildings: Window and shading device typologies for energy efficiency refurbishment. *Sustainability* **2014**, *6*, 5354–5377. [CrossRef]
11. Khasreen, M.M.; Banfill, P.F.G.; Menzies, G.F. Life-cycle assessment and the environmental impact of buildings: A review. *Sustainability* **2009**, *1*, 674–701. [CrossRef]
12. Joseph, P.; Tretsiakova-McNally, S. Sustainable non-metallic building materials. *Sustainability* **2010**, *2*, 400–427. [CrossRef]
13. Guo, P.; Meng, W.; Nassif, H.; Gou, H.; Bao, Y. New perspectives on recycling waste glass in manufacturing concrete for sustainable civil infrastructure. *Constr. Build. Mater.* **2020**, *257*. [CrossRef]
14. Bahadori-Jahromi, A.; Rotimi, A.; Mylona, A.; Godfrey, P.; Cook, D. Impact of window films on the overall energy consumption of existing UK hotel buildings. *Sustainability* **2017**, *9*, 731. [CrossRef]
15. Zhou, A.; Wong, K.W.; Lau, D. Thermal insulating concrete wall panel design for sustainable built environment. *Sci. World J.* **2014**, *2014*. [CrossRef] [PubMed]
16. Zhou, A.; Yu, Z.; Chow, C.L.; Lau, D. Enhanced solar spectral reflectance of thermal coatings through inorganic additives. *Energy Build.* **2017**, *138*, 641–647. [CrossRef]
17. Morakinyo, T.E.; Dahanayake, K.W.D.K.C.; Ng, E.; Chow, C.L. Temperature and cooling demand reduction by green-roof types in different climates and urban densities: A co-simulation parametric study. *Energy Build.* **2017**, *145*, 226–237. [CrossRef]

18. Wang, F.; Yang, W.J.; Sun, W.F. Heat transfer and energy consumption of passive house in a severely cold area: Simulation analyses. *Energies* **2020**, *13*, 626. [CrossRef]
19. Dahanayake, K.W.D.K.C.; Chow, C.L. Studying the potential of energy saving through vertical greenery systems: Using EnergyPlus simulation program. *Energy Build.* **2017**, *138*, 47–59. [CrossRef]
20. Chow, W.K.; Kong, H.; Gao, Y.; Dong, H.; Zou, G.W.; Han, S.S.; Kong, H.; Chow, C.L.; Kong, H. Experimental studies on fire response of glass façade systems. *IJEPBFC* **2006**, *8*, 54–68.
21. Khalaf, M.; Ashrafian, T.; Demirci, C. Energy Efficiency Evaluation of Different Glazing and Shading Systems in a School Building. *E3S Web Conf.* **2019**, *111*. [CrossRef]
22. Huang, Y.; Niu, J.L. Energy and visual performance of the silica aerogel glazing system in commercial buildings of Hong Kong. *Constr. Build. Mater.* **2015**, *94*, 57–72. [CrossRef]
23. Wang, Y.; Wu, Y.; Wang, Q.; Liew, K.M.; Chen, H.; Sun, J.; He, L. Numerical study on fire response of glass facades in different installation forms. *Constr. Build. Mater.* **2014**, *61*, 172–180. [CrossRef]
24. Stewart, M.G.; Netherton, M.D. Security risks and probabilistic risk assessment of glazing subject to explosive blast loading. *Reliab. Eng. Syst. Saf.* **2008**, *93*, 627–638. [CrossRef]
25. Vedrtnam, A.; Bedon, C.; Youssef, M.A.; Wamiq, M.; Sabsabi, A.; Chaturvedi, S. Experimental and numerical structural assessment of transparent and tinted glass during fire exposure. *Constr. Build. Mater.* **2020**, *250*. [CrossRef]
26. Bedon, C.; Zhang, X.; Santos, F.; Honfi, D.; Kozłowski, M.; Arrigoni, M.; Figuli, L.; Lange, D. Performance of structural glass facades under extreme loads—Design methods, existing research, current issues and trends. *Constr. Build. Mater.* **2018**, *163*, 921–937. [CrossRef]
27. Debuyser, M.; Sjöström, J.; Lange, D.; Honfi, D.; Sonck, D.; Belis, J. Behaviour of monolithic and laminated glass exposed to radiant heating. *Constr. Build. Mater.* **2017**, *130*, 212–229. [CrossRef]
28. Wang, Y.; Xie, Q.; Zhang, Y.; Wang, Q.; Sun, J. Sensitivity analysis of influencing factors on glass façade breakage in fire. *Fire Saf. J.* **2018**, *98*, 38–47. [CrossRef]
29. Cuzzillo, B.R.; Pagni, P.J. Thermal breakage of Double-Pane Glazing by Fire. *J. Fire Prot. Eng.* **1998**, *9*, 1–11. [CrossRef]
30. Wang, Y.; Wang, Q.; Sun, J.; He, L.; Liew, K.M. Influence of fire location on the thermal performance of glass façades. *Appl. Therm. Eng.* **2016**, *106*, 438–442. [CrossRef]
31. Pagni, P.J. Thermal glass breakage. *Fire Saf. Sci.* **2003**, 3–22. [CrossRef]
32. Cho, S.H.; Chae, C.U. A study on life cycle CO2 emissions of low-carbon building in South Korea. *Sustainability* **2016**, *8*, 579. [CrossRef]
33. Tidwell, J. *Bridging the Gap—Fire Safety and Green Buildings*; National Association of State Fire Marshals: Burns, OR, USA, 2010.
34. Nayak, J.K. Thermal performance of a water wall. *Build. Environ.* **1987**, *22*, 83–90. [CrossRef]
35. Wu, T.; Lei, C. A review of research and development on water wall for building applications. *Energy Build.* **2016**, *112*, 198–208. [CrossRef]
36. Wu, T.; Lei, C. Thermal modelling and experimental validation of a semi-transparent water wall system for Sydney climate. *Sol. Energy* **2016**, *136*, 533–546. [CrossRef]
37. Wu, T.; Lei, C. CFD simulation of the thermal performance of an opaque water wall system for Australian climate. *Sol. Energy* **2016**, *133*, 141–154. [CrossRef]
38. Wang, H.; Lei, C. Theoretical modeling of combined solar chimney and water wall for buildings. *Energy Build.* **2019**, *187*, 186–200. [CrossRef]
39. Wang, H.; Lei, C. Combined Solar Chimney and Water Wall for Ventilation and Thermal Comfort. In Proceedings of the 4th International Conference on Building Energy, Environment, Melbourne, Australia, 5–9 February 2018; pp. 705–710.
40. Karabay, H.; Arici, M.; Sandik, M. A numerical investigation of fluid flow and heat transfer inside a room for floor heating and wall heating systems. *Energy Build.* **2013**, *67*, 471–478. [CrossRef]
41. Moustafa, M.A.; Aripin, S. CFD Evaluation of the pottery water wall in a hot adrid climate of Luxor, Egypt. *J. Green Build.* **2015**. [CrossRef]
42. Xiangfeng, L.; Tianxing, S. Conceptual Development of Transparent Water Storage Envelopes. *Archit. Sci. Rev.* **2007**, *50*, 18–25. [CrossRef]
43. Liu, X.; Guo, J.; Wang, J. Numerical and experimental studies on the energy performance of thermal mass windows. *J. Build. Phys.* **2018**, *42*, 692–721. [CrossRef]

44. Wang, X.; Kim, R.E.; Kwon, O.S.; Yeo, I. Hybrid Simulation Method for a Structure Subjected to Fire and Its Application to a Steel Frame. *J. Struct. Eng.* **2018**, *144*, 1–11. [CrossRef]
45. Kuznik, F.; David, D.; Johannes, K.; Roux, J.J. A review on phase change materials integrated in building walls. *Renew. Sustain. Energy Rev.* **2011**, *15*, 379–391. [CrossRef]
46. Tamme, R.; Laing, D.; Steinmann, W.D.; Bauer, T. Thermal Energy Storage. *Sol. Energy* **2013**, 688–714. [CrossRef]
47. Perlin, J. *Let It Shine: The 6,000-Year Story of Solar Energy*; New World Library: Novato, CA, USA, 2013; ISBN 9781608681327.
48. Bainbridge, D.A. *A Water Wall Solar Design Manual*; San Diego, CA, USA, 2005; Available online: https://www.solaripedia.com/files/472.pdf (accessed on 13 October 2020).
49. Maloney, T.J.; Habib, V. *Design, Fabrication and Testing of a Marketable Waterwall Component*; One Design, Inc.: Winchester, VA, USA, 1979; Volume 84.
50. Bainbridge, B.D.A.; Haggard, K.; Cooper, P. Return of the Water Wall. *Sol. Today* **2007**, 38–41.
51. Saadatian, O.; Sopian, K.; Lim, C.H.; Asim, N.; Sulaiman, M.Y. Trombe walls: A review of opportunities and challenges in research and development. *Renew. Sustain. Energy Rev.* **2012**, *16*, 6340–6351. [CrossRef]
52. Saadatian, O.; Sopian, K.; Salleh, E.; Lim, C.H.; Riffat, S.; Saadatian, E.; Toudeshki, A.; Sulaiman, M.Y. A review of energy aspects of green roofs. *Renew. Sustain. Energy Rev.* **2013**, *23*, 155–168. [CrossRef]
53. Bevilacqua, P.; Benevento, F.; Bruno, R.; Arcuri, N. Are Trombe walls suitable passive systems for the reduction of the yearly building energy requirements? *Energy* **2019**, *185*, 554–566. [CrossRef]
54. Agrawal, B.; Tiwari, G.N. *Building Integrated Photovoltaic Thermal Systems: For Sustainable Development*; Royal Society of Chemistry: London, UK, 2011; ISBN 9781849732000.
55. Al-Karaghouli, A.; Kazmerski, L.L. Renewable Energy Opportunities in Water Desalination. *Desalin. Trends Technol.* **2011**. [CrossRef]
56. McLaggan, M.S.; Hadden, R.M.; Gillie, M. Fire Performance of Phase Change Material Enhanced Plasterboard. *Fire Technol.* **2018**, *54*, 117–134. [CrossRef]
57. Khan, R.J.; Bhuiyan, M.Z.H.; Ahmed, D.H. Investigation of heat transfer of a building wall in the presence of phase change material (PCM). *Energy Built Environ.* **2020**, *1*, 199–206. [CrossRef]
58. Richman, R.C.; Cianfrone, C.; Pressnail, K.D. More sustainable masonry façades: Preheating ventilation air using a dynamic buffer zone. *J. Build. Phys.* **2010**, *34*, 27–41. [CrossRef]
59. Stazi, F.; Mastrucci, A.; Munafò, P. Life cycle assessment approach for the optimization of sustainable building envelopes: An application on solar wall systems. *Build. Environ.* **2012**, *58*, 278–288. [CrossRef]
60. Kisilewicz, T. On the role of external walls in the reduction of energy demand and the mitigation of human thermal discomfort. *Sustainability* **2019**, *11*, 61. [CrossRef]
61. Yang, L.; Yan, H.; Lam, J.C. Thermal comfort and building energy consumption implications -A review. *Appl. Energy* **2014**, *115*, 164–173. [CrossRef]
62. Li, D.; Wu, Y.; Wang, B.; Liu, C.; Arıcı, M. Optical and thermal performance of glazing units containing PCM in buildings: A review. *Constr. Build. Mater.* **2020**, *233*, 117327. [CrossRef]
63. Balcomb, J.D.; Mcfarland, R.D. A simple empirical method for estimating the performance of a passive solar heated building of the thermal storage wall type. In Proceedings of the 2nd National Passive Solar, University of Pennsylvania, Philadelphia, PA, USA, 16–18 March 1978.
64. Nayak, J.K.; Bansal, N.K.; Sodha, M.S. Analysis of passive heating concepts. *Sol. Energy* **1983**, *30*, 51–69. [CrossRef]
65. Sutton, R.G.; McGregor, R.J. Solarwall project: Two demonstration houses with passive solar heating in Tasmania. *Archit. Sci. Rev.* **1986**, *29*, 2–11. [CrossRef]
66. Turner, R.H.; Liu, G.; Cengel, Y.A.; Harris, C.P. Thermal Storage in the Walls of a Solar House. *J. Sol. Eng.* **1994**, *116*, 183–193. [CrossRef]
67. Yadav, Y.P.; Tiwari, G.N. Analytical model of a solarium for cold climate-A new approach. *Energy Convers. Manag.* **1988**, *28*, 15–20. [CrossRef]
68. Din, M.; Tiwari, G.N.; Ghosal, M.K.; Srivastava, N.S.L.; Khan, M.I.; Sodha, M.S. Effect of thermal storage on the performance of greenhouse. *Int. J. Energy Res.* **2003**, *27*, 79–92. [CrossRef]
69. Tiwari, G.N.; Singh, A.K. Comparative Studies of Different Heating Techniques of a Non-air Conditioned Building. *Build. Environ.* **1996**, *31*, 215–224. [CrossRef]
70. Fuchs, R.; McClelland, J.F. Passive solar heating of buildings using a transwall structure. *Sol. Energy* **1979**, *23*, 123–128. [CrossRef]

71. Bansal, N.K.; Sodha, M.S.; Ram, S.; Singh, S.P. Evaluation of complex heat transfer coefficients for passive heating concepts. *Build. Environ.* **1987**, *22*, 259–268. [CrossRef]
72. Bansal, N.K.; Thomas, P.C. A Simple Procedure for Selection and Sizing of Indirect Passive Solar Heating Systems. *Build. Environ.* **1991**, *26*, 381–387. [CrossRef]
73. Bhandari, M.S.; Bansal, N.K. Solar heat gain factors and heat loss coefficients for passive heating concepts. *Sol. Energy* **1994**, *53*, 199–208. [CrossRef]
74. Nisbet, S.K.; Mthembu, N.S. Transwall modelling using effective conductivities. *Sol. Energy* **1992**, *49*, 127–138. [CrossRef]
75. Yadav, Y.P.; Tiwari, G.N. Large scale solar water heater: Experimental and theoretical studies. *Energy Convers. Manag.* **1991**, *31*, 337–351. [CrossRef]
76. Tiwari, G.N.; Kumar, S. Thermal evaluation of solarium-cum-passive solar house. *Energy Convers. Manag.* **1991**, *32*, 303–310. [CrossRef]
77. Upadhya, M.; Tiwari, G.N.; Rai, S.N. Optimum distribution of water-wall thickness in a transwall. *Energy Build.* **1991**, *17*, 97–102. [CrossRef]
78. Wang, W.L.; Tian, Z.; Niu, X.L.; Xu, X. Investigation on a Passive Solar House Equipped with Water Thermal Storage Wall. *Appl. Mech. Mater.* **2012**, *178–181*, 193–196. [CrossRef]
79. Sánchez-Ostiz, A.; Monge-Barrio, A.; Domingo-Irigoyen, S.; González-Martínez, P. Design and experimental study of an industrialized sunspace with solar heat storage. *Energy Build.* **2014**, *80*, 231–246. [CrossRef]
80. Sameti, M. A new design of a solar water storage wall: A system-level model and simulation. *Energy Syst.* **2018**, *9*, 361–383. [CrossRef]
81. Chow, T.T.; Li, C.; Lin, Z. Thermal characteristics of water-flow double-pane window. *Int. J. Therm. Sci.* **2011**, *50*, 140–148. [CrossRef]
82. Sodha, M.S.; Kaushik, S.C.; Nayak, J.K. Performance of trombe walls and roof pond systems. *Appl. Energy* **1981**, *8*, 175–191. [CrossRef]
83. Yadav, Y.P.; Tiwari, G.N. Transient analysis of a winter greenhouse integrated with solar still. *Energy Convers. Manag.* **1987**, *27*, 267–273. [CrossRef]
84. Tiwari, G.N.; Yadav, Y.P.; Lawrence, S.A. Performance of a solarium: An analytical study. *Build. Environ.* **1988**, *23*, 145–151. [CrossRef]
85. Xiangfeng, L.; Tianxing, S. The Development of Transparent Water Storage Envelopes (TWSE) through Theoretical Thermal and Optical Analyses. *Archit. Sci. Rev.* **2008**, *51*, 109–123. [CrossRef]
86. Chow, T.T.; Li, C. Water-filled double reflective window and its year-round performance. *Procedia Environ. Sci.* **2011**, *11*, 1039–1047. [CrossRef]
87. Lyu, Y.L.; Chow, T.T.; Wang, J.L. Numerical prediction of thermal performance of liquid-flow window in different climates with anti-freeze. *Energy* **2018**, *157*, 412–423. [CrossRef]
88. Gil-Lopez, T.; Gimenez-Molina, C. Influence of double glazing with a circulating water chamber on the thermal energy savings in buildings. *Energy Build.* **2013**, *56*, 56–65. [CrossRef]
89. Lyu, Y.L.; Chow, T.T. Evaluation of influence of header design on water flow characteristics in window cavity with CFD. *Energy Procedia* **2015**, *78*, 97–102. [CrossRef]
90. Chow, T.T.; Lyu, Y. Effect of design configurations on water flow window performance. *Sol. Energy* **2017**, *155*, 354–362. [CrossRef]
91. Liu, X.; Xu, M.; Guo, J.; Zhu, R. Numerical study on the energy performance of building zones with transparent water storage envelopes. *Sol. Energy* **2019**, *180*, 690–706. [CrossRef]
92. Gutai, M.; Kheybari, A.G. Energy consumption of water-filled glass (WFG) hybrid building envelope. *Energy Build.* **2020**, *218*, 110050. [CrossRef]
93. Sędłak, B.; Kinowski, J.; Sulik, P.; Kimbar, G. The risks associated with falling parts of glazed facades in case of fire. *Open Eng.* **2018**, *8*, 147–155. [CrossRef]
94. Hasheminasab, H.; Zolfani, S.H.; Bitarafan, M.; Chatterjee, P.; Ezabadi, A.A. The Role of Façade Materials in Blast-Resistant Buildings: An Evaluation Based on Fuzzy Delphi and Fuzzy EDAS. *Algorithms* **2019**, *12*, 119. [CrossRef]
95. Nguyen, K.; Weerasinghe, P.; Mendis, P.; Ngo, T.; Barnett, J. Performance of modern building facades in fire: A comprehensive review. *Electron. J. Struct. Eng.* **2016**, *16*, 69–86.
96. McKenna, S.T.; Jones, N.; Peck, G.; Dickens, K.; Pawelec, W.; Oradei, S.; Harris, S.; Stec, A.A.; Hull, T.R. Fire behaviour of modern façade materials – Understanding the Grenfell Tower fire. *J. Hazard. Mater.* **2019**, *368*, 115–123. [CrossRef]

97. Wang, Y.; Zhang, Y.; Wang, Q.; Yang, Y.; Sun, J. The effect of glass panel dimension on the fire response of glass façades. *Constr. Build. Mater.* **2018**, *181*, 588–597. [CrossRef]
98. Chow, W.K.; Hung, W.Y.; Gao, Y.; Zou, G.; Dong, H. Experimental study on smoke movement leading to glass damages in double-skinned façade. *Constr. Build. Mater.* **2007**, *21*, 556–566. [CrossRef]
99. Quinn, M.; Nadjai, A.; Ali, F.; Abu-Tair, A. Experimental and numerical investigation of localised fire on glazing facades having different orientations. *J. Struct. Fire Eng.* **2013**, *4*, 153–164. [CrossRef]
100. Babrauskas, V. Glass breakage in fires. *Fire Sci. Technol. Inc.* **2011**, 6–11. Available online: http://doctorfire.com/GlassBreak.pdf (accessed on 13 October 2020).
101. Keski-Rahkonen, O. Breaking of window glass close to fire. *Fire Mater.* **1988**, *12*, 61–69. [CrossRef]
102. Pagni, P.J.; Joshi, A.A. Glass breaking in fires. *Fire Saf. Sci. Proc. Third Int. Symp.* **2006**, 791–802. [CrossRef]
103. Mowrer, F. *Window Breakage Induced by Exterior Fires*; National Institute of Standards and Technology: Gaithersburg, MD, USA, 1997.
104. Chow, C.L. Numerical studies on smoke spread in the cavity of a double-skin façade. *J. Civ. Eng. Manag.* **2016**, *22*, 470–479. [CrossRef]
105. Manzello, S.L.; Gann, R.G.; Kukuck, S.R.; Prasad, K.R.; Jones, W.W. An experimental determination of a real fire performance of a non-load bearing glass wall assembly. *Fire Technol.* **2007**, *43*, 77–89. [CrossRef]
106. Wang, X.; Tan, Q.; Wang, Z.; Kong, X.; Cong, H. Preliminary study on fire protection of window glass by water mist curtain. *Int. J. Therm. Sci.* **2018**, *125*, 44–51. [CrossRef]
107. Turco, M.; Lhotsky, P.; Hadjisophocleous, G. Investigation into the use of a water curtain over openings to prevent fire spread. *MATEC Web Conf.* **2016**, *46*. [CrossRef]
108. Tseng, C.C.; Viskanta, R. Absorptance and transmittance of water spray/mist curtains. *Fire Saf. J.* **2007**, *42*, 106–114. [CrossRef]
109. Abdoh, D.A.; Ademiloye, A.S.; Liew, K.M. A meshfree analysis of the thermal behaviors of hot surface glass pane subjects to down-flowing water film via smoothed particle hydrodynamics. *Eng. Anal. Bound. Elem.* **2020**, *120*, 195–210. [CrossRef]
110. Shao, G.; Wang, Q.; Zhao, H.; Wang, Y.; Chen, H.; Su, Y.; Sun, J.; He, L. Maximum temperature to withstand water film for tempered glass exposed to fire. *Constr. Build. Mater.* **2014**, *57*, 15–23. [CrossRef]
111. Braun, J.E. Load control using building thermal mass. *J. Sol. Energy Eng. Trans. ASME* **2003**, *125*, 292–301. [CrossRef]
112. Shi, Y.; Liu, X. Research on the Literature of Green Building Based on the Web of Science: A Scientometric Analysis in Citespace (2002–2018). *Sustainability* **2019**, *11*, 3716. [CrossRef]
113. Cakmakli, A.B. Environmental Analysis of Construction Materials. In *Green Building Management and Smart Automation*; Solanki, A., Nayyar, A., Eds.; IGI Global: Pennsylvania, PA, USA, 2019; pp. 90–110. [CrossRef]
114. Akadiri, P.O.; Chinyio, E.A.; Olomolaiye, P.O. Design of A Sustainable Building: A Conceptual Framework for Implementing Sustainability in the Building Sector. *Buildings* **2012**, *2*, 126–152. [CrossRef]
115. Biswas, W.K. Carbon footprint and embodied energy consumption assessment of building construction works in Western Australia. *Int. J. Sustain. Built Environ.* **2014**, *3*, 179–186. [CrossRef]
116. Baird, G.; Alcorn, A.; Haslam, P. The Energy Embodied in Building Materials—Updated New Zealand Coefficients and Their Significance. *Trans. Inst. Prof. Eng. New Zeal. Civ. Eng. Sect.* **1997**, *24*, 46–54.
117. Lawson, B. *Building Materials, Energy and the Environment: Towards Ecologically Sustainable Development*; Royal Australian Institute of Architects: Victoria, Australia, 1996; ISBN 9781863180283.
118. Straube, J. Can Highly Glazed Building Façades Be Green. Available online: https://www.buildingscience.com/documents/insights/bsi-006-can-fully-glazed-curtainwalls-be-green (accessed on 19 October 2020).
119. Bunjongjit, S.; Ngaopitakkul, A. Feasibility study and impact of daylight on illumination control for Energy-saving lighting systems. *Sustainability* **2018**, *10*, 4075. [CrossRef]
120. Kaminska, A.; Ozadowicz, A. Lighting control including daylight and energy efficiency improvements analysis. *Energies* **2018**, *11*, 2166. [CrossRef]
121. Al-Mansour, A.; Chow, C.L.; Feo, L.; Penna, R.; Lau, D. Green concrete: By-products utilization and advanced approaches. *Sustainability* **2019**, *11*, 5145. [CrossRef]
122. Albayyaa, H.; Hagare, D.; Saha, S. Energy conservation in residential buildings by incorporating Passive Solar and Energy Efficiency Design Strategies and higher thermal mass. *Energy Build.* **2019**, *182*, 205–213. [CrossRef]
123. Aboulnaga, M.M. Towards green buildings: Glass as a building element—The use and misuse in the gulf region. *Renew. Energy* **2006**, *31*, 631–653. [CrossRef]

124. Altan, H.; Hajibandeh, M.; Aoul, T.K.A.; Deep, A. Passive design. In *Springer Tracts in Civil Engineering*; Springer: Cham, Switzerland, 2016; pp. 209–236. [CrossRef]
125. Burkett, K. The Hidden Architecture Behind 7 Deceiving Façades. Architizer, 2019. Available online: https://architizer.com/blog/inspiration/collections/deceiving-facades (accessed on 13 October 2020).
126. Nijsse, R.; Wenting, R. Designing and constructing corrugated glass facades. *J. Facade Des. Eng.* **2014**, *2*, 123–131. [CrossRef]
127. Allen, E. *How Buildings Work*; Oxford UniversityPress: New York, NY, USA, 2005; Volume 53, ISBN 9788578110796.
128. Maller, C.; Townsend, M.; Pryor, A.; Brown, P.; St Leger, L. Healthy nature healthy people: "contact with nature" as an upstream health promotion intervention for populations. *Health Promot. Int.* **2006**, *21*, 45–54. [CrossRef] [PubMed]
129. De Been, I.; Beijer, M. The influence of office type on satisfaction and perceived productivity support. *J. Facil. Manag.* **2014**, *12*, 142–157. [CrossRef]
130. Adams, S.; Becker, M.; Krauss, D.; Gilman, C. Not a Dry Subject: Optimizing Water Trombe Walls. *Sol. Energy Soc. Annu. Conf.* **2010**, *2010*, 1334–1340.
131. Tiwari, G.N.; Upadhya, M.; Rai, S.N. Relative thermal performances of south walls in winter. *Energy Build.* **1991**, *17*, 313–320. [CrossRef]
132. Crawley, D.B.; Hand, J.W.; Kummert, M.; Griffith, B.T. Contrasting the capabilities of building energy performance simulation programs. *Build. Environ.* **2008**, *43*, 661–673. [CrossRef]

3

Risk Factors Impacting the Project Value Created by Green Buildings in Saudi Arabia

Wael Alattyih [1,*], Husnain Haider [1] and Halim Boussabaine [2]

1 Department of Civil Engineering, College of Engineering, Qassim University, Buraydah, Qassim 51452, Saudi Arabia; husnain@qec.edu.sa
2 Faculty of Business and Law, The British University in Dubai, Dubai 345015, UAE; halimmay45@gmail.com
* Correspondence: alattyih@qec.edu.sa

Abstract: Green buildings are playing a pivotal role in sustainable urban development around the world, including Saudi Arabia. Green buildings subject to various sources of risk that influence the potential outcomes of the investments or services added in their design. The present study developed a structured framework to examine various risks that may lead to green buildings' value destruction in Saudi Arabia. The framework initiates with identification of 66 potential risk factors from reported literature. A questionnaire compiling a list of identified risk factors was hand-delivered to 300 practitioners (managers, engineers, and architects) having knowledge of value engineering in the construction industry, and an overall response rate of 29.7% was achieved. Subsequently, descriptive statistics ranked the risk factors based on scores given by the respondents. The principal component analysis extracted 16 components, based on the likelihood of risk factors impacting the value created by green building design. Finally, the factor analysis grouped the 35 most significant risk factors in 5 clusters—i.e., 8 in functional risk, 13 in financial risk, 3 in operational risk, 3 in environmental risk, and 8 in management risk cluster. The study enhances the understanding of the importance of the risk factors' impact on value creation. Based on the results, the value management (or engineering) teams and the top-level management can identify, manage, and control the risk factors that have a significant impact on the project value created by green building design.

Keywords: risk factors; value creation; value engineering; green building; construction; project management; project risk

1. Introduction

Green buildings possess a wide range of advantages over the conventional buildings over their life cycle, e.g., minimal environmental impacts and low maintenance costs [1]. In the recent past, rapidly growing urbanization trends have undergone the world to unusual climate change and environmental deterioration [2]. Due to global warming, the urban heat island (UHI) phenomenon has already raised, in comparison to the rural proximities, the temperature from 2 to 5 °C in many cities around the globe [3]. The potential impacts of global warming on temperature and rainfall patterns are expected to be highly significant in warm and arid regions like Saudi Arabia [4]. Decision-makers need to take proactive measures to minimize the production of greenhouse gasses from all types of development activities, particularly the growing building sector in the country [5].

As per Shin et al. [6], certified green buildings reduce the UHI intensity by around half a degree. To further mitigate the impact of buildings on climate change, He [2] proposed a concept of Zero UHI impact building. In addition to revising the sustainability assessment standards and economic regulations, the idea needs serious efforts towards technological interventions in building envelopes, materials, and construction equipment. Dwaikat and Ali [7] estimated the economic benefits of green buildings in Malaysia. They found that, over the life cycle with an annual increase of 1% in the energy price, a green building saves up to 5756 kWh/m^2 that corresponds to around 2.8 million US dollars. Using Harvard's co-benefits calculator, MacNaughton et al. [8] reported significant social cost reductions in terms of health benefits due to decreasing carbon emissions from Leadership in Energy and Environmental Design (LEED) certified green buildings. Considering the variations in life cycle costs and associated environmental impacts, there is a need for region (or country) specific studies for assessing the value created by green buildings as well as risks that might influence the created value.

Modern construction projects, such as green buildings, contain inherent complexities due to the engagement and coordination of multiple stakeholders—e.g., consultants, designers, contractors, and suppliers of materials. Value management helps to improve the communication between various projects' stakeholders, accommodate mutually agreed project objectives, provides a better quality project definition, builds innovation, and eliminates unnecessary costs [9]. Risk management improves construction projects in several ways by providing better control of uncertainties, helping the decision-maker to choose the right project during design phase, improving team communication and motivation, and protecting the project cost [10,11].

Value and risk are two of the main management concepts that improve both the quality and cost controls for the sustainability of the construction industry. Value management aims to achieve the desired value with lower prices without sacrificing the quality and functions, while risk management aims to identify, evaluate, and choose the strategies to maintain uncertainty and risk at a tolerable level for the organization to maximize its performance. The term 'value' can be understood in a multitude of ways including through economic, cultural, and social interpretations [12]. Gao et al. [13] stated that economic, technical, and education levels are the main factors influencing the development of green buildings in a country. The indicator for perceiving value in construction projects is obtaining success in those projects that involves tolerance levels of risk, affordability, timeliness, and safe completion within specific quality parameters while meeting normal requirements [14]. Past studies investigated how risk influences the perceived value [15]. Whole-life-cycle value exchange mechanisms are risky and complex, therefore, a deeper investigation of the risks associated with value exchange is needed to focus on developing value targets within the whole life-cycle agenda [16].

The construction industry in Saudi Arabia has not effectively incorporated the impact of risk factors on development projects [17], including green buildings. Value creation improves function longevity and operation and maintenance costs that enhance the performance reliability, quality, safety, and life-cycle cost of the green buildings. Based on the past work, further investigations are required to evaluate the impacts of perceived risk on the value exchange of green buildings [15,18].

Like other long-term investments, green buildings are also subject to various sources of risk that influence the potential outcome of investments or services provided throughout the lifecycle from inception to disposal. The low perceived risk adds high value from investments in green buildings. High risk is involved in investing capital value (both financial and human) in building assets during both the design and production operations; the primary perceived risk is the case when the project will not be completed and not all the anticipated value will be realized. Nevertheless, the potential for value destruction still exists even after the completion of the construction process. The risk to value returns from the green building operation stage is high [16].

In Saudi Arabia, organizations are also interested in the potential benefits of green buildings similarly to other high-income countries around the world. For improving value creation, there is a need of a framework that integrates both the value and risk in the development phase of green buildings projects in Saudi Arabia. In general, the practices towards risk management in construction projects are still at the earlier stages in the country [17]. Authors in their previous research developed the value creation drivers for the green buildings in Saudi Arabia [19]. The present study examines the risks that may lead to value destruction in green building development and operation. The purpose of identifying these risks is to evaluate the magnitude of their consequences on the value, should they occur, and the likelihood of their occurrence. The specific objectives are set to: (i) identify various risk factors that may influence the value creation, and (ii) analyze the influence of these risks on the project value created by the green buildings design.

The remainder of the paper is organized as follows. Section 2 defines 66 risk factors under five risk classifications based on an extensive review of literature. Section 3 describes the methodological framework and statistical methods used for selection of most important risk factors. Section 4 presents results and discussions, while Section 5 outlines conclusions of the research work and recommends future works for the continuation and practical implementation of the present research.

2. Risk Classification in Green Building Development

There is no clear consensus in the existing literature on how to classify the existing risks that may emerge from the development of green buildings. In an early study, Tiong [20] clustered risk into financial, political, and technical aspects of green buildings. Boussabaine and Kirkham [21] presented a systemic life cycle risk classification based on design, construction, operation, and disposal of building facilities. Medda [22] suggested additional classes to include, such as commercial, regulatory, and economic risks. Zou et al. [23] classified risks based on time, cost, quality, safety, and environmental aspects of performance. Yang et al. [24] classified previously defined risks into several categories and introduced the ethics/reputation risk cluster. Zou and Couani [25] classified a number of risks based on the perspective of the stakeholders involved in the process of green building supply chain.

Risk classification largely depends on the intended purpose of the investigation [26]. Based on the objectives of the present research, Table 1 groups the identified risks in, (i) functional risks related to how the building and its components will function in use; (ii) financial risks related to a project's financing parameters, capital and operational costs, and return on investments; (iii) operational risks related to safety and how easy and efficient the green assets are to operate; (iv) environmental risks related to protecting the environment from the impacts of the development; and (v) management risks related to stakeholders' interaction, knowledge, and contractual and organizational relationship. It should be noted that the purpose of our classification is to facilitate risk identification in value creation in green building development. These risks will aid the process of value engineering analysis. The sources of the identified risks in each cluster are described in the following sub-sections.

2.1. Functional Risks

Functionality in the green building's use plays a pivotal role in the optimization of the operational cost of core services and the productivity of the occupant. From a value point of view, the function of a building and its components are related to the purpose of their design and existence. The design of green buildings must be subject to risk analysis in order to assess and understand the uncertainties that are associated with the function and design parameters. Changes in the assumption of these parameters may lead to different levels of performance and reliability. Functional risks can be attributed to the state of the product or services provided. If the service or product fails to fulfil its functional requirements as expected, then all or most of the invested value will be lost.

Table 1 lists the functionality risks that may impact the value creation in green building development. Physical risks are related to the building asset's condition over the life span of the capital value investment. Typical risks include loss due to fire, corrosion, explosion, structural defect, war, etc. Technical risks are due to increased technology in manufacturing, communications, data handling, and interdependency of manufacturers, methods of storage, stock control, and distribution. These risks could also be associated with physical aspects of the green building's development. Building components and their function are subject to obsolescence in terms of service life, design life, and functional purpose over time, leading to both tangible and intangible value. The monetary value generation from a building's assets is directly related to quality and durability of the asset [27]. The building's systems should be easy to operate, robust, and efficient. It is essential that a green building's facilities should easily accommodate any changes in activity that are likely to occur throughout its life cycle, including when user requirements change in the future. This is essential to guard against risks of obsolesence.

Zou and Couani [25] compiled a list of key risk factors that may have a consequential impact throughout the green building's supply chain. Al-Yousefi [28] had previously highlighted risks due to the lack of quality, reliability, and performance in sustainable development. Non-complying products and materials and change in technologies due to green building are the two major risks that can impact the green building development [25]. Isa et al. [29] pointed out that physical risks in green buildings might result from events such as earthquake, flood, wear and tear, and user damage. The authors report that these risks will have an impact on the economic value of a property. The Green Building Council [30] stressed the importance of the following regulatory risks: (i) property value decrease due to changes in planning/transport policy; (ii) inability to compete with newer, greener properties; (iii) decrease in value due to low energy rating; and (iv) inability to lease due to new regulations.

The risk of obsolescence of the asset function or it is components will result in larger operating expenditure cost and may undermine the value of the asset that may lead to lower income, smaller capital receipts, higher costs, and the possibility of legal action. Furthermore, new technologies that can change demand-side behavior (e.g., wireless measurement of energy use at appliance level) could increase the risk of obsolescence and missed opportunities for reduced operational costs [30].

2.2. Financial Risks

Financial risks attribute to inadequate inflation forecasts, incorrect marketing decisions, and credit policies. Zurich [31] defines financial risks as "the additional costs of green buildings may affect completing projects on time and on budget, but must be weighed against the cost of not going green". Table 1 lists the financial risks that may found to have an impact on value creation in green building development. Haghnegahdar and Asgharizadeh [32] reported that 75% of the projects are not accomplished according to the allocated financial resources and time schedules. Zou and Couani [25] identified "higher investment costs to go green" and "costs of investment in skills development" as additional costs in the development of green buildings. Zurich [31] claimed that additional costs spent on design and construction of green buildings can be too costly for some companies and delay the completion of the projects within the specified budget. According to the NAO [33], buildings that consume large amounts of capital in their development and operation will end up having a negative impact on the user's business and performance.

Thus, if the budgets for both capital and whole-life costs are not estimated correctly and justified in the business case to be sustainable and affordable over the life of the green building facility, it leads to the risks of failure to recognize cost–value mismatches, failure to identify cost–value relationships, and losing potential revenue from the investment. It is also important that operational and maintenance costs are evaluated and kept within the budget. The investment appraisal must address various options for creating the required value from green buildings. Failure to consider the implications of economic conditions and to recognize the cost as resource expenditure will certainly lead to the risk of affordability and to the risk that whole-life cost estimates are not realistic but based on unreliable evidence.

Indirect factors—like the inflation rate, liquidity, and financing risks—will impact the capital and operational costs of developing green buildings. Lower economic activities may influence both the asset's economic value and the rental return [34]. Higher financing cost also results in value loss, leading to a longer period being required to recover the invested capital. Sustainability features in green buildings attract funding at competitive rates.

The investments in green buildings offer higher returns in the form of higher rent, capital appreciation, and cost savings [29]. It is becoming a standard procedure that real estate valuation take into consideration the value difference if environmental features are not incorporated into the construction and operation of the building estates [30]. The Council also suggest that the failure in meeting the benchmarking criteria of sustainability by the potential investment partners may result in the different potential market risks, including, (i) brown discounts (i.e., reduction in rent and asset value), (ii) increased speed of depreciation, (iii) lower occupancy rates, and (iv) shorter tenancies.

A marked link has been emerged between the market value and the associated performance of the green features of the building [35]. Ashuri and Durmus-Pedini [36] further compiled the following list of financial risks associated with green buildings:

- Due to the common belief that "green cost [is] much higher", lending amounts can be disproportionate to need resulting in needless debt;
- Return on Investment needs more of a historical perspective to become more predictable;
- Inexperienced teams might lack the skills to properly implement green-oriented technology, which could hinder its effectiveness;
- Company budgets are not usually structured to track life cycle cost (LCC) for a project, making longer-term gains harder to record;
- Costs associated with litigation between the architect/engineers and the owner if certification is not reached;
- Loss of tax incentives if certification is not reached;
- New green building materials might result in issues never encountered previously and be a source of litigation;
- Loss of possible financial gain if the building does not perform as it was intended to; and
- The possible unforeseen conditions of retrofitting existing buildings.

2.3. *Operational Risks*

Operational risks are concerned with maintaining, operating, and cleaning a green building facility once it is in use. Table 1 lists the operational risks that may found to have an impact on value creation in green building development. Operational performance of green buildings has a significant impact on their market value, both rental and capital. The project owners are starting to require additional contract provisions regarding the energy efficiency of green buildings; breaching of contract can increase the exposure to legal liabilities, such as tortuous, statutory, and contractual liabilities [31]. Lutkendorft and Lorenz [37] proposed that the value should be attributed to the quality of the indoor environment and its relationship with the efficiency in employees' productivity. Low energy cost reduces the potential occupants' operating cost that minimizes the vacancy risk and improves the rental value.

2.4. *Environmental Risks*

The whole endeavor of the green building ethos is to create facilities that must minimize waste and energy use during construction and operation stages. Investors are concerned with the inherent risk from the environmental perspective to the real estate portfolio [30]. The building should provide a comfortable and healthy working and living environment for people. The methods and materials used in construction should be selected based on their potential risk impacts on the environment [38]. The frequency by which building materials are replaced will have an influence/increase in carbon emissions over the life cycle of the green assets [39–41]. This stems from the fact that the replaced materials need to be disposed of, new materials have to be manufactured and transported, equipment must be utilized, and energy must be expended to rebuild or renovate the asset.

The waste from such activities increases the building's environmental impacts, such as global warming—from the building machinery and the operation of the transport and construction vehicles; acidification—caused by emissions from burning of diesel by the building machinery and the operation of the transport and construction vehicles; eutrophication—caused by indirect emissions from the source of electricity supply, and the burning of diesel from the use of building machinery and transport; winter smog—from waste transportation and the production of natural gas; heavy metals—due to toxic effects of heavy metals from disposing and recycling materials; and energy—from electricity and oil usage and production impacts [42]. These risks can be mitigated through the design and specification of robust structures and construction. Therefore, failure to consider maintainability and reparability increases the environmental risk.

The performance of new products and technologies that are being developed for green construction can also pose a risk [31]. This view is based on the fact that green materials are developed rapidly without robust testing of their performance and environmental credentials. This might lead to litigation over specifications or materials that are unfit for purpose or product failure. The legal liability risks—related to tortuous, statutory, and contractual liabilities—eventually reduce the client investment clave. Risk of obsolescence due to a green building's non-conformance with sustainability issues and consumption of resources may undermine the value of the green real estate [30].

2.5. Management Risks

Effective management by an integrated project team is inevitable to create value in the development and operation of green and traditional facilities [43]. Risk management is an important aspect of designing and operating green buildings [44]. There are opportunities to maximize the value and minimize waste in each stage of a building project, i.e., planning, design, procurement, construction, and operations [33]. If the process of development is not well managed, risks may emerge from the lack of integration, coordination, and communication between the project team. The project team should have the foresight to develop and communicate a clear brief and make a realistic budget and cost estimation from the outset. Also, the team should be given enough time, as per the need during the whole project cycle, to plan and complete the project.

If the project execution plan is poorly conceived, it may lead to risks related to poor definition of scope and output specification, poor communication, and poor lines of decision-making. Other management risks include psychological risks associated to the choice of service or product selection and procurement. If the wrong product or service is chosen, capital value may be lost and it might have other negative effects on the whole life cycle chain. In addition, lack of coordination is considered one of the most prevalent endemic risks in a construction project's development. Furthermore, stakeholder involvement and teamwork is essential for adding and creating value throughout the life cycle of green building development [45]. Table 1 lists 17 managerial risks that have an impact on value creation in green building development.

The identified risk factors in Table 1 were put forward for evaluation by industry professionals to test their impact on value creation in green buildings.

3. Methodology

3.1. Framework for Identification of Risk Factors

Figure 1 illustrates the methodological framework adopted in present research. Through a literature survey and expert judgment, 66 risk factors were found relevant to the context of Saudi Arabia and elsewhere (refer to Table 1). These risk factors were group under five main categories in Section 2, as described above. A questionnaire was developed to obtain the view of professionals in the country on the importance and likelihood of the identified risk factors. Subsequently, the significance of the selected risk factors was evaluated through a hand-delivered questionnaire survey. The responses were statistically analyzed with the help of Statistical Package for Social Sciences (SPSS). Finally, the risk factors with the highest importance were ranked and grouped into clusters to facilitate the shareholders and designers to enhance performance efficiency and obtain more value from investment in green building assets. All these steps are discussed in detail in the subsequent sub-sections.

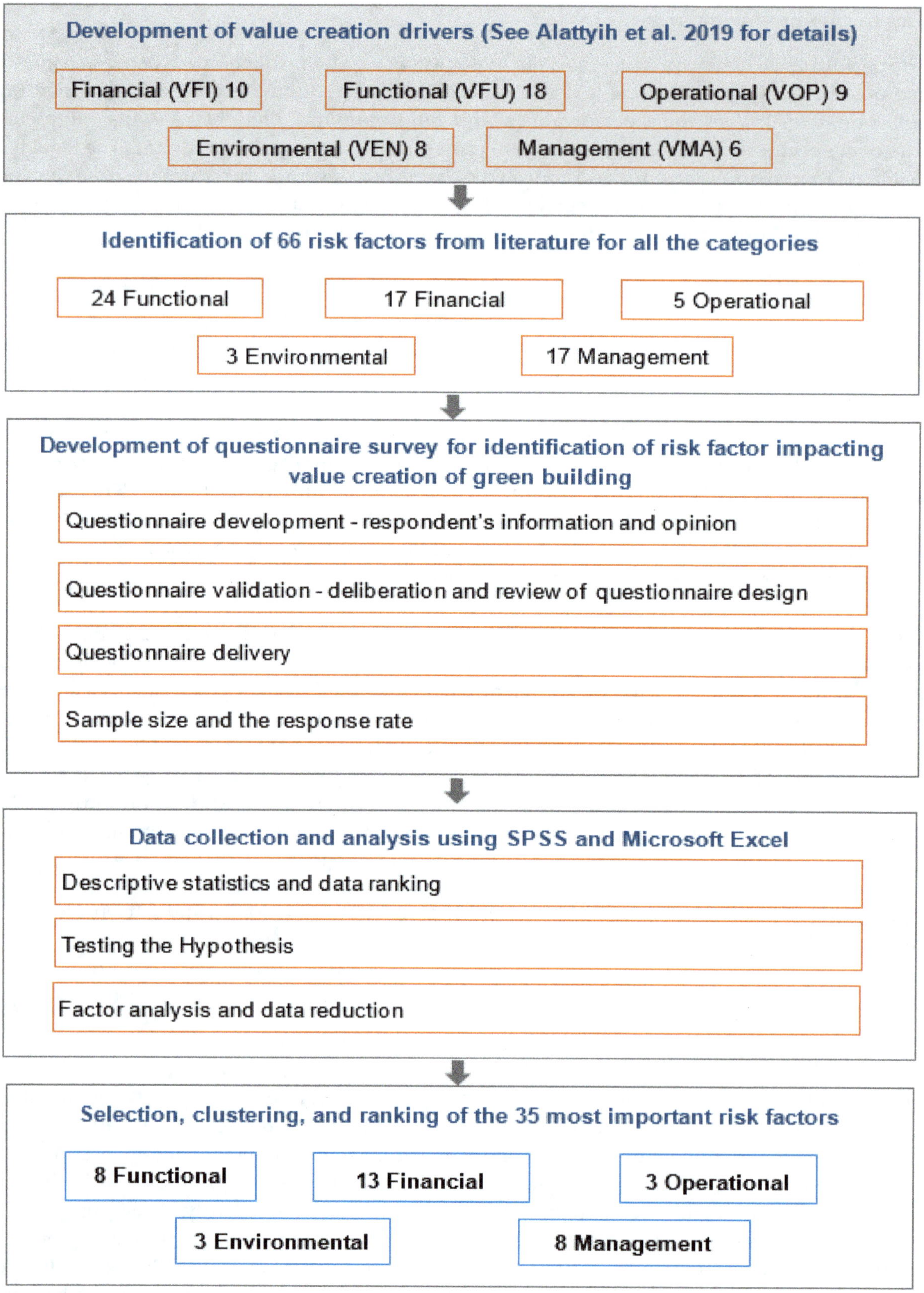

Figure 1. Framework for development of risk factors impacting value creation of green buildings.

3.2. Identification of Risk Factors

A detailed literature review developed a list of risks that may have a negative impact on value creation. The extracted risk factors were classified according to value driver groups. A set of 66 risk factors was classified into financial risks, functional risks, operational risks, environmental risks, and management risks. These risk factors, listed in Table 1, were used for developing the questionnaire to assess how each risk will impact value creation in green building development.

3.3. Questionnaire Design and Development

The questionnaire form began by giving an overview and objectives of the research. The first part of the questionnaire gathered the general information of the respondents, while the second part asked the respondents to evaluate and rate the list of identified risk factors. Part 1 obtained two types of information. In the first type, the respondents were asked to provide some general information (optional), such as their names, organization names, email addresses, phone numbers, and postal addresses. The second type of information was related to the job titles and the years of experience of the respondents. Based on the job titles, respondents were categorized into three groups, including manger, engineer, or architect. In order to facilitate subsequent statistical analysis, types of jobs were coded by ordinary numbers for discretion.

Table 1. Risk Factors impacting the project value created by the green buildings design.

Group	Code	Risk Factors	References
Functional Risks	R20	Failure to consider construction implications during design	Isa et al. [29] Al-Yousefi [28] Zou and Couani [25] Wiedemer et al. [27] Green Building Council [30] Boussabaine and Kirkham [21] Muldavin [46] Durmus-Pedini and Ashuri [36]
	R19	Failure to design to brief/specification	
	R17	Failure to examine specifications due to unnecessary expense	
	R33	Failure to integrate the various systems to achieve the lowest life cycle costs	
	R21	Design changes	
	R22	Redesign/rework	
	R35	Failure to identify low-value, long-lead-time items	
	R36	Failure to consider design risks	
	R15	Failure to consider constructability issues	
	R16	Failure to appreciate design uncertainty	
	R18	Failure to translate specification into the design	
	R23	Failure to consider changes to current design standards	
	R24	Failure to examine attributes which have no useful function	
	R25	Failure to identify and understand functions	
	R26	Unsuitable functional analysis	
	R27	Failure to identify appropriate alternative function	

Table 1. *Cont.*

Group	Code	Risk Factors	References
	R29	Unproven construction techniques	
	R30	Failure to consider legislative/regulation issues	
	R31	Failure to consider new/change in technology	
	R32	Failure to meet performance or design standards	
	R34	Failure to choose/specify reliable materials	
	R37	Failure to consider inter-dependence between components in design	
	R40	Insufficient space and capacity	
	R41	Failure to consider construction health and safety	
Financial Risks	R11	Failure to recognize cost–value mismatches	Isa et al. [29] Zurich [31] Zou and Couani [25] NAO [33] Haghnegahdar and Asgharizadeh [32] Boussabaine and Kirkham [21] Green Building Council [30] Cajias and Bienert [47] Eichholtz et al. [48] Parker [49] Davies [35] Muldavin [46] Durmus-Pedini and Ashuri [36]
	R10	Failure to identify cost–value relationships	
	R8	Failure to consider the cost of losing potential revenue	
	R12	Failure to appropriately locate cost-to-function allocation	
	R9	Uncertainty about prices	
	R5	Inappropriate cost evaluation criteria	
	R6	Failure to consider future operational costs	
	R3	Failure to consider implication of economic conditions	
	R1	Insufficient funding	
	R7	Failure to recognise cost as resource expenditure	
	R65	Incorrect estimated cost of maintenance	
	R63	Incorrect cost estimate	
	R66	Incorrect estimated cost of energy used	
	R2	Funding is unavailable	
	R4	Inadequate inflation forecasts	
	R13	Failure to identify cost-to-worth relationships	
	R14	Failure to link or identify performance-to-cost relationships	
Operational Risks	R44	Failure to consider increase in routine maintenance	Lutkendorft and Lorenz [37] Zurich [31] Muldavin [46] Boussabaine and Kirkham [21]
	R45	Failure to consider increase in life cycle replacement	
	R39	Failure to consider design impact on operating efficiency	
	R42	Failure to consider component repair and replacement	
	R46	Limited knowledge of maintenance issues	

Table 1. *Cont.*

Group	Code	Risk Factors	References
Environmental Risks	R38	Failure to consider obsolescence of equipment impact	Boussabaine and Kirkham [21]; Green Building Council [30]; Muldavin [46]; Zurich [31]; Lützkendorfa and Lorenza [37]
	R43	Failure to consider maintainability and reparability impact	
	R47	Failure to consider implication of environmental risks	
Management Risks	R49	Poor project management	Wi and Junge [45] Jha and Iyer [50] NAO [33] Gritzo et al. [44] Muldavin [46] Durmus-Pedini and Ashuri [36] Shen and Liu [51] Zou and Couani [25]
	R50	Poor definition of the scope and objectives of projects	
	R51	Project scope unscheduled items	
	R57	Improper project planning and budgeting	
	R53	Lack of coordination and decision-making	
	R64	Incorrect time estimate	
	R54	Poor team relationships	
	R28	Poor design that may lead to higher operation costs	
	R48	Lack of background information	
	R52	Failure to consider construction techniques	
	R55	Programming issues	
	R56	Failure to consider procurement method	
	R58	Failure to build to design	
	R59	Failure to identify complex items or processes with little or no value added to the facility	
	R60	Failure to consider rate of deterioration of components	
	R61	Lack of considering early failure of components and equipment	
	R62	Failure to consider delay and higher cost due to bad weather conditions during construction	

In part 2, a Likert scale ranging from 1 to 5 (very unlikely, unlikely, neutral, likely, and very likely) was used for rating the list of risk factors, because it is easy to construct, modify, and direct use for statistical inference of the numerical measurement for reliable results. The questionnaire acquired the rate the likelihood of risk factors impacting on the project value created by green building design. A sample of the questionnaire is given in Appendix A. Details on questionnaire development and validation can be seen in Alattyih et al. [19].

The population and the sample size were based on the number of professionals having knowledge about the application of value engineering approaches in the Saudi Arabian construction industry. As per SAVE International, more than 1356 people have obtained value engineering certificates in Saudi Arabia [52,53]. Approximately 30 of them are Certified Value Specialist (CVS), i.e., 16% of global certified population. Annually, 60–80 value engineering training workshops and more than 80 VE study programs are being offered in Saudi Arabia and the Arab Gulf countries. Using a confidence interval of 10% and confidence level of 95% from the population of 1356, the research needs at least 76 respondents. Based on the perceived response of less than 50%, the questionnaires were hand-delivered, in person, to a sample size of 300 professionals with value engineering knowledge and experience in the Saudi Arabian construction industry. The participants were randomly selected from various cities in Saudi Arabia, in order to develop research data covering all the country's major cities.

3.4. Statistical Analysis

SPSS software and Microsoft Excel software performed factor analysis and data reduction. The collected data were processed through two statistical procedures, (i) the data ranking process, that is based on mean weighted rating, standard deviation, severity index and coefficient of variation for the risk factors; and (ii) factor analysis and the data reduction process were utilized to decrease the number of variables in order to handle the task more easily and efficiently.

3.4.1. Descriptive Statistics and Data Ranking

The degree of significance on risk factors, in the context of value created by green building designs in construction projects, is described in the following sections. The comparison of the data ranking took place based on severity indices, average weighted mean, and standard deviation of each risk factor. Further analysis of the data were conducted for ranking based on respondents' answers and their experience (0–5 years, 6–10 years, and more than 10 years of experience) and their professional job (manager, engineer, or architect).

A mean weighted rating for each value attribute and risk factor was computed to indicate the importance of each indicator, using Equation (1). Meanwhile, the range varies from 1 to 5, therefore, the moderate point for value attributes and natural point for risk factors is 3.

$$Mean\ weighted\ rating = \frac{\sum R \times F}{n} \quad (1)$$

where R is the rating of each value attribute and risk factor (1, 2, 3, 4, 5), F is the frequency of responses, and n is the total number of responses ($n = 89$).

A severity index ($S.I.$) measure is employed in order to rank the indicators according to their significance, which is the higher the percentage (%), the more significant the attribute/factor. Equation (2) shows how the $S.I.$ is calculated

$$S.I. = \left(\frac{\sum W \times F}{n}\right) \times 100 \quad (2)$$

where W is the weight of each rating (1/5, 2/5, 3/5, 4/5, 5/5).

The coefficient of variation (COV) is an expression of standard deviation as a percentage (%) of the mean, which is for comparison of the relative variability of various responses. Better variability is shown by a lower variation coefficient. The COV was computed as a ratio between standard deviation and mean.

3.4.2. Testing the Hypotheses

The means and standard deviations and the coefficient of variations' values were quite close together for the three groups of respondents (mangers, engineers, and architects). Therefore, SPSS software was used for further analysis. ANOVA test analysis was conducted in order to justify the statistical differences between the groups' responses.

The SPSS software was used with a significance level of 0.05 to examine the differences between the groups regarding the likelihood of the risk impacting on the project value creation of green building development using the following hypothesis testing:

H_0: *$p > 0.05$—There is no significant difference among the respondents' ratings for the likelihood of risk factors impacting the value created by green building design.*

H1: $p < 0.05$—*There is significant difference among the respondents' ratings for the likelihood of risk factors impacting the value created by green building design (at least one of the groups is significantly different from other groups).*

After that, a follow-up test was conducted to make multiple comparisons if there is significant difference among the respondents. The follow-up test used in this research was the post hoc multiple comparison test. The Turkey test is one of the post hoc tests used as the sample size is uneven.

3.4.3. Factor Analysis and Data Reduction

A number of risk factors with the highest degree of likelihood of risk factors impacting on project value creation might be considered as representative of the whole set of data. Therefore, the most significant factors are extracted and treated as representative of the whole set of risk indicators. Based on the factors' relationship and correlations, the outcome of the data reduction develops a few clusters that consist of the most important risk factors of the original large group of 66 risk factors. A clear understanding of new risk clusters and their implications will be instrumental in assessing and evaluating the value creation and their performances in green building design.

The factor analysis technique investigated if there is an underlying relationship between the different factors within a questionnaire. Figure 2 illustrates the overall analysis process. The factor analysis process determines the strength of the relationship between the variables, extracts a matrix of correlation coefficients, and finally extracts the components that have an eigenvalue of 1 or more from this matrix, which is the most common method of extraction in principal component analysis. The data reduction process identifies the variables that correlate highly with a set of other variables for clustering them in a meaningful way. The next stage generates a rotated component matrix to find out the risk factors that have a more effective influence on each component. Through identifying redundant data, the existing 66 risk factors were reduced down to 35 most important risk factors for green building design. Figure 2 shows that, through the use of data reduction in SPSS, the risk factors have been categorized into five clusters.

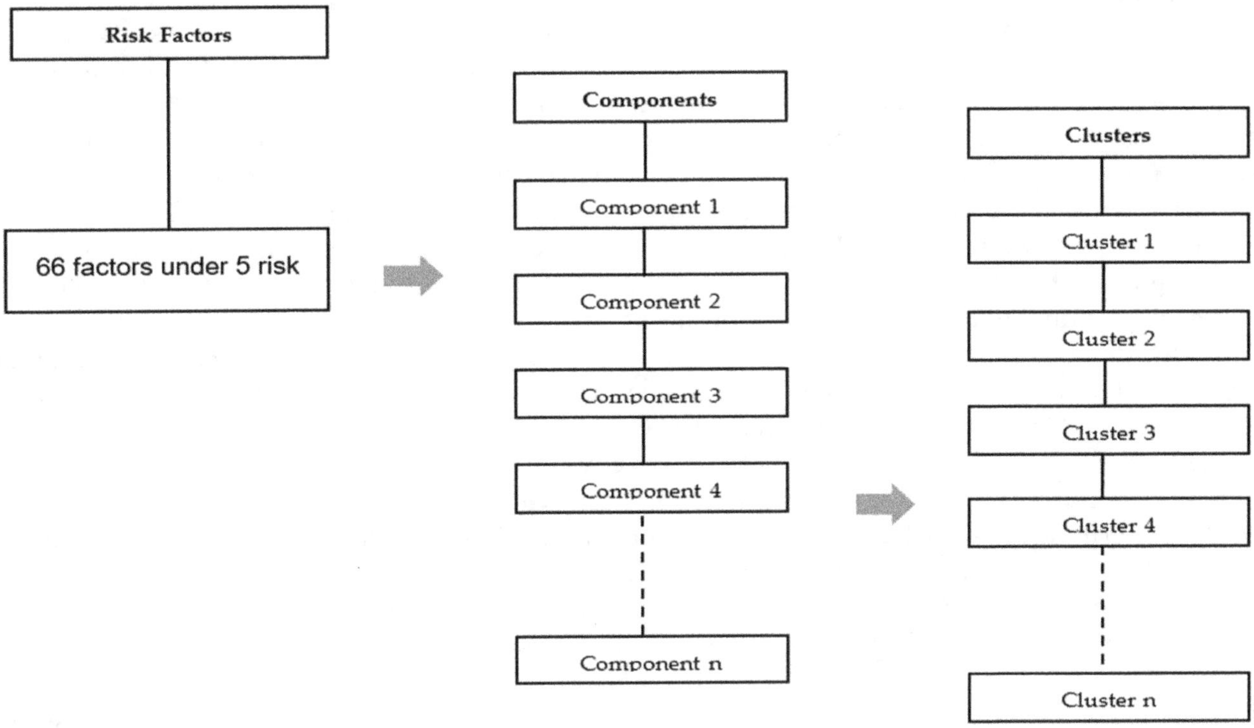

Figure 2. Factor analysis and data reduction process.

Usually, a few components account for most of the variation, and these components can be considered to replace the original variables [54–56]. The results presented in the following section extract the most important risk factors that are essentially the representatives of the whole set of risk factors. The degree of significance of each risk factor in green building design varies according to its impact on a construction project.

Reliability analysis practically validates the properties of a measurement scale and checks the reliability of the items. Low reliability shows that the items that make up the scale do not correlate strongly enough; thus, they might not be selecting the same construct domain. As a measure of reliability, Cronbach's Alpha was calculated to check the consistency of the research items and to identify problem items that need to be excluded from the scale [57]. Based on George and Mallory's [58] measures, Cronbach's Alpha is assessed in Table 2. The reliability of the data was checked for the risk factors by using Cronbach's test. The Cronbach's Alpha values of risk factors was 0.969 and shows good to excellent internal consistency of the components.

Table 2. Cronbach's Alpha measures (source: George and Mallory [58]).

Cronbach's Alpha	Internal Consistency
>0.90	Excellent
0.80–0.89	Good
0.70–0.79	Acceptable
0.60–0.69	Questionable
0.50–0.59	Poor
<0.50	Unacceptable

4. Results and Discussion

4.1. Descriptive Statistics and Data Ranking

Identification of risk factors, associated with value and evaluation of their level of influence, plays a pivotal role in the project value created by the green building design. The risk factors were ranked based on their likelihood of impacting the project value created. Out of 300 professionals, 89 returned their fully completed questionnaire. This is an acceptable response rate (29.7%) as the typical response rate for a postal questionnaire survey in the construction industry is 20–30% [59]. The survey response revealed than 45% of the respondents were managers, 38% were engineers, and 17% were architects. Around 16% of the respondents were young, having 0–5 years of experience, 28% had experience ranged between 6 and 10, while the largest contribution was from senior respondents possessing more than 10 years of experience in the profession. Among the top 30 factors, the following discussion is limited to the top 3 and the bottom 10 risk factors.

As the detailed statistical ranking results cannot be presented due to space limitations, a summary of the overall ranking and the ranking by each expert along with their years of experience is presented in Appendix B. The appendix shows that the average weighted mean for the risk factors varies from 3.33 to 4.24, with an overall mean of 3.78. The severity indices range within 66.52% to 84.72%. The highest-ranked factor is R35 (failure to identify low-value, long-lead-time items) with a mean of 4.24 and severity index of 84.72%. An overall examination of the first 30 ranked risk factors indicates that all of them have a minimum mean value of 3.81 (which is higher than the average overall mean of 3.78) and severity indices of 76.18%. This means that the first 30 ranked risk factors seem to be important as viewed by the respondents. They are namely: R1, R3, R5, R6, R7, R8, R9, R12, R17, R20, R21, R22, R28, R33, R35, R36, R39, R44, R45, R49, R50, R51, R53, R54, R57, R63, R64, R65, and R66.

Factor R64 (i.e., incorrect time estimate) has the second highest rank for the risk factors with a mean of 4.18 and a severity index of 83.60% and it is ranked as first out of 66 by the engineers and architects, and also by the experts with more than 10 years' experience. The managers ranked it fifth out of 66, and the other rankings based on years of experience are: 0 to 5 and 6 to 10 years ranked it ninth and second out of 66 respectively. The third overall ranking was for R28 (i.e., poor design that may lead to higher operation costs), and the six groups of respondents also ranked it as one of their top 20 highest-ranked risk factors. R28 has a mean of 4.10, severity index of 82.02% and low coefficient of variation of 22.97%.

The last 10 factors, amongst the top 30, have average weighted means for the risk factors that vary from 3.81 to 3.84 and severity indices that range between 76.18% and 76.85%. They have the following overall ranking, R54 (poor team relationships) is ranked 21st out of 66; r50 (poor definition of the scope and objectives of projects) is ranked 22nd; r20 (failure to consider construction implications during design) is ranked 23rd; r9 (uncertainty about prices) is ranked 24th; r33 (failure to integrate the various systems to achieve the lowest life-cycle costs) is ranked 25th; r66 (incorrect estimated cost of energy used) is ranked 26th; r12 (failure to appropriately locate cost-to-function allocation) is ranked 27th; r45 (failure to consider increase in life cycle replacement) is ranked 28th; R3 (failure to consider implication of economic conditions) is ranked 29th; and factor R11 (failure to recognize cost–value mismatches) is ranked 30th out of 66.

4.2. Testing the Hypothesis

The section examines the mind-sets of construction project professionals who were involved in value management/engineering in relation to how they value and perceive the likelihood of the risk factors having an impact on value creation in the green building design.

Overall, the ANOVA results showed no significant difference ($p > 0.05$) amongst different participants (i.e., managers, engineers, and architects) for all the factors and there is no need for the H_1 hypothesis. Figure 3 shows the overall perspectives of the average rating for the likelihood of risk factors having an impact on the project value created by green building design. The architects give a higher ranking for the overall average mean than the engineers and managers. All of the top 20 factors ranked by the architects have a mean of over 4.13 and the average overall mean for all the factors is 3.91. In contrast, the mean average for the managers' responses is 3.79, and for the engineers it is 3.72. Based on these results, it is clear that all three groups of respondents agree that most of the risk factors have a high impact on the project value created by green building design.

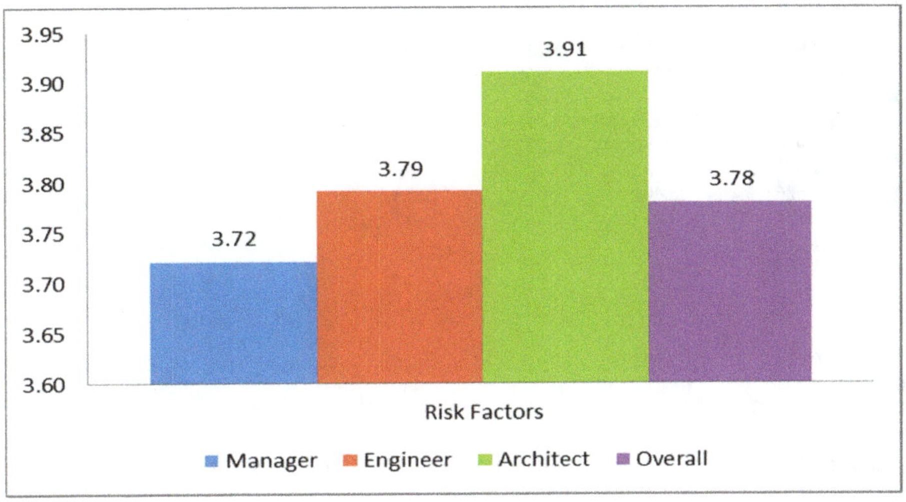

Figure 3. Average rating for the likelihood of risk factors having an impact on value creation.

4.3. Factor Analysis

Two statistical tests were carried out on data before conducting factor analysis to indicate the suitability of the data for structure detection. The first test is Kaiser–Meyer–Olkin (KMO) that measures sampling accuracy to indicate the proportion of variance of the variables that might cause by the underlying factors. High values close to 1.0 in the KMO test indicate that a factor analysis is useful for the data while the values less than 0.50 indicate that the results of the factor analysis are not useful. Also, KMO values between 0.5 and 0.7 are good, between 0.7 and 0.8 are very good, and above 0.8 are excellent. The Bartlett Test of Sphericity tests the hypothesis that the correlation matrix is an identity matrix, i.e., a small significance level of less than 0.05 indicates the need for factor analysis [55–57]. Field [55] mentions that a value close to 1 indicates that the patterns of correlations are relatively compact and the factor analysis will provide distinct and reliable factors. Kaiser [60] recommended accepting values greater than 0.5 as acceptable. In the present research, the corresponding value for KMO was 0.702 and Bartlett test found significance value of '0'. As the KMO value was found close to 1, the factor analysis is likely to be appropriate and acceptable.

Bartlett's test measures the null hypothesis (H_0: $p > 0.05$) and shows that the original correlation matrix is an identity matrix. Therefore, factor analysis needs some relationships between variables and the significance value to be ($p < 0.05$). By considering the significance level of 0.05, Bartlett's test showed the p-values for the likelihood of risk impact were highly significant. This test shows that the correlation matrix is not an identity matrix, so there are some relationships between the variables. Both KMO and Bartlett tests demonstrated that factor analysis is appropriate for these data.

Table 3 presents the components extracted by the principal component analysis (PCA). The likelihood of risk factors impacting on the value created by green building design shows that just 16 components carry an eigenvalue of more than 1 and account for nearly 79.939% of the whole variance. Consequently, the 16 components can be considered as representative of the 66 factors included in this study.

Table 3. Total variance explained for risk factors.

Component	Initial Eigenvalues			Extraction Sums of Squared Loadings			Rotation Sums of Squared Loadings		
	Total	% of Variance	Cumulative %	Total	% of Variance	Cumulative %	Total	% of Variance	Cumulative %
1	22.380	33.909	33.909	22.380	33.909	33.909	6.027	9.132	9.132
2	4.154	6.294	40.203	4.154	6.294	40.203	5.499	8.332	17.464
3	3.418	5.179	45.382	3.418	5.179	45.382	5.036	7.630	25.094
4	2.988	4.527	49.909	2.988	4.527	49.909	4.364	6.613	31.707
5	2.640	4.000	53.909	2.640	4.000	53.909	3.506	5.312	37.019
6	2.599	3.938	57.846	2.599	3.938	57.846	3.341	5.062	42.081
7	2.200	3.334	61.180	2.200	3.334	61.180	3.195	4.841	46.921
8	1.975	2.993	64.173	1.975	2.993	64.173	3.143	4.762	51.683
9	1.721	2.608	66.781	1.721	2.608	66.781	3.039	4.605	56.289
10	1.540	2.333	69.113	1.540	2.333	69.113	2.863	4.338	60.627
11	1.403	2.126	71.240	1.403	2.126	71.240	2.432	3.686	64.312
12	1.312	1.988	73.228	1.312	1.988	73.228	2.386	3.615	67.927
13	1.273	1.930	75.157	1.273	1.930	75.157	2.311	3.501	71.428
14	1.101	1.668	76.826	1.101	1.668	76.826	2.039	3.090	74.518
15	1.040	1.576	78.401	1.040	1.576	78.401	1.844	2.793	77.311
16	1.015	1.538	79.939	1.015	1.538	79.939	1.735	2.628	79.939
17	0.918	1.390	81.330						
18	0.832	1.260	82.590						
.	.	.	.						
.	.	.	.						
.	.	.	.						
65	0.009	0.014	99.991						
66	0.006	0.009	100.000						

In the subsequent phase of factor analysis, a rotated component matrix was extracted to find out the risk factors with the highest level of influence on project value creation. Table 4 presents the summary of factor analysis results. The matrix loading scores in Table 3 extracted the most effective factors of each component. The factors with the highest scores and correlation values were chosen for each component. For example, the risk factor (R1; 0.808) has greater influence on component 13 compared to other components, whereas the risk factor R46 (0.843) has more influence on component 4 in relation to other components, and R57 (0.762) has more influence on component 9 in relation to other components.

Table 4. Summary of factor analysis for risk factor components.

Risk Factor Components	Extracted Eigenvalue	Extraction Sum of Squared Loadings: Variance %	Rotation Sum of Squared Loadings: Variance %	Factor Loading Score	Code	Risk Factor
1	22.380	33.909	9.132	0.801	R13	Failure to identify cost-to-worth relationships
				0.796	R11	Failure to recognize cost–value mismatches
				0.781	R10	Failure to identify cost–value relationships
				0.780	R14	Failure to link or identify performance-to-cost relationships
				0.600	R8	Failure to consider the cost of losing potential revenue
				0.598	R12	Failure to appropriately locate cost-to-function allocation
				0.528	R9	Uncertainty about prices
				0.425	R5	Inappropriate cost evaluation criteria
2	4.154	6.294	8.332	0.858	R27	Failure to identify appropriate alternative function
				0.788	R26	Unsuitable functional analysis
				0.744	R25	Failure to identify and understand functions
				0.691	R24	Failure to examine attributes which have no useful function
				0.539	R23	Failure to consider changes to current design standards
				0.481	R59	Failure to identify complex items or processes with little or no value added to the facility
				0.529	R47	Failure to consider implication of environmental risks
3	3.418	5.179	7.630	0.822	R61	Lack of considering early failure of components and equipment
				0.726	R60	Failure to consider rate of deterioration of components
				0.709	R65	Incorrect estimated cost of maintenance
				0.641	R62	Failure to consider delay and higher cost due to bad weather conditions during construction
				0.626	R63	Incorrect cost estimate
				0.527	R66	Incorrect estimated cost of energy used
4	2.988	4.527	6.613	0.843	R46	Limited knowledge of maintenance issues
				0.734	R48	Lack of background information
				0.639	R44	Failure to consider increase in routine maintenance
				0.592	R45	Failure to consider increase in life cycle replacement
				0.540	R39	Failure to consider design impact on operating efficiency

Table 4. *Cont.*

Risk Factor Components	Extracted Eigenvalue	Extraction Sum of Squared Loadings: Variance %	Rotation Sum of Squared Loadings: Variance %	Factor Loading Score	Code	Risk Factor
5	2.640	4.000	5.312	0.723	R42	Failure to consider component repair and replacement
				0.717	R41	Failure to consider construction health and safety
				0.602	R43	Failure to consider maintainability and reparability impact
				0.423	R40	Insufficient space and capacity
6	2.599	3.938	5.062	0.724	R50	Poor definition of the scope and objectives of projects
				0.699	R51	Project scope unscheduled items
				0.580	R58	Failure to build to design
				0.494	R52	Failure to consider construction Techniques
				0.465	R49	Poor project management
7	2.200	3.334	4.841	0.785	R4	Inadequate inflation forecasts
				0.649	R6	Failure to consider future operational costs
				0.598	R3	Failure to consider implication of economic conditions
				0.454	R15	Failure to consider constructability issues
8	1.975	2.993	4.762	0.789	R20	Failure to consider construction implications during design
				0.732	R19	Failure to design to brief/specification
				0.502	R18	Failure to translate specification into the design
				0.497	R17	Failure to examine specifications due to unnecessary expense
9	1.721	2.608	4.605	0.762	R57	Improper project planning and budgeting
				0.728	R56	Failure to consider procurement method
				0.586	R64	Incorrect time estimate
				0.493	R53	Lack of coordination and decision making
10	1.540	2.333	4.338	0.633	R31	Failure to consider new/change in technology
				0.597	R33	Failure to integrate the various systems to achieve the lowest LCC
				0.577	R29	Unproven construction techniques
				0.498	R16	Failure to appreciate design uncertainty
				0.407	R32	Failure to meet performance or design standards

Table 4. *Cont.*

Risk Factor Components	Extracted Eigenvalue	Extraction Sum of Squared Loadings: Variance %	Rotation Sum of Squared Loadings: Variance %	Factor Loading Score	Code	Risk Factor
11	1.403	2.126	3.686	0.773	R21	Design changes
				0.702	R22	Redesign/rework
12	1.312	1.988	3.615	0.614	R38	Failure to consider obsolescence of equipment impact
				0.589	R37	Failure to consider inter-dependence between design components
13	1.273	1.930	3.501	0.808	R1	Insufficient funding
				0.499	R2	Funding is unavailable
14	1.101	1.668	3.090	0.762	R35	Failure to identify low-value, long-lead-time items
				0.550	R36	Failure to consider design risks
15	1.040	1.576	2.793	0.612	R7	Failure to recognize cost as resource expenditure
				0.456	R30	Failure to consider legislative/regulation issues
16	1.015	1.538	2.628	0.625	R54	Poor team relationships
				0.468	R28	Poor design that may lead to higher operation costs

Table 3 also shows the strength of the relationship between the variables and the extracted matrix of correlation coefficients, and then extracts the components that have an eigenvalue of more than 1 from the matrix of correlation coefficients. The results present the variables that highly correlate with a set of other variables. The eigenvalues for the components varied between 33.91 and 1.54, and the rotated variance load varied between 9.13% and 2.63%. Each component has more than two factors with a loading score of more than 0.4. The value attributes that have a loading score of more than 0.4 were reduced and redundant data were eliminated in the clustering stage in order to obtain a few variables that present the risk characteristics and their impacts on value created by green building design. Further reduction was carried out in the subsequent section based on their ranking by the professionals.

Table 4 presents the factor analysis and data reduction results for the five new clusters that were molded based on the 16 extracted components and their most important risk factors in Table 3. The new clusters were considered as risk indicator clusters that impact the value creation and can be used for managing the project risks at tolerance level. The variance percentage of each risk factor was extracted from Table 3 while the variance of each cluster was calculated by summation of each component's variance in the same generated cluster. For example, the functional risk cluster in Table 4 is one of five clusters for impact of risk factors on value creation. The cluster encompasses component 8 (variance of 4.76%) that presents R20, R19, and R17; component 11 (variance of 3.69%) that presents R21 and R22; component 10 (variance of 4.34%) that presents R33; and component 14 (variance of 3.501%) that presents R35 and R36 as the main indicators of its set. Consequently, the percentage of variance for this cluster (Functional risk) in Table 5 was calculated as 4.762 + 3.686 + 4.338 + 3.501 = 15.876%.

Table 5. Factor reduction for the risk factors: five new categories.

Cluster	Variance %	Component	Code	Risk Factors
Financial Risks	27.897	1	R11	Failure to recognize cost–value mismatches
			R10	Failure to identify cost–value relationships
			R8	Failure to consider the cost of losing potential revenue
			R12	Failure to appropriately locate cost-to-function allocation
			R9	Uncertainty about prices
			R5	Inappropriate cost evaluation criteria
		7	R6	Failure to consider future operational costs
			R3	Failure to consider implication of economic conditions
		13	R1	Insufficient funding
		15	R7	Failure to recognize cost as resource expenditure
		3	R65	Incorrect estimated cost of maintenance
			R63	Incorrect cost estimate
			R66	Incorrect estimated cost of energy used
Functional Risks	15.876	8	R20	Failure to consider construction implications during design
			R19	Failure to design to brief/specification
			R17	Failure to examine specifications due to unnecessary expense
		10	R33	Failure to integrate the various systems to achieve the lowest life-cycle costs
		11	R21	Design changes
			R22	Redesign/rework
		14	R35	Failure to identify low-value, long-lead-time items
			R36	Failure to consider design risks
Operational Risks	6.613	4	R44	Failure to consider increase in routine maintenance
			R45	Failure to consider increase in life cycle replacement
			R39	Failure to consider design impact on operating efficiency
Environmental Risks	17.259	12	R38	Failure to consider obsolescence of equipment impact
		5	R43	Failure to consider maintainability and reparability impact
		2	R47	Failure to consider implication of environmental risks

Table 5. *Cont.*

Cluster	Variance %	Component	Code	Risk Factors
Management Risks	12.295	6	R49	Poor project management
			R50	Poor definition of the scope and objectives of projects
			R51	Project scope unscheduled items
		9	R57	Improper project planning and budgeting
			R53	Lack of coordination and decision making
			R64	Incorrect time estimate
		16	R54	Poor team relationships
			R28	Poor design that may lead to higher operation costs

In Table 5, the risk factors grouped in five clusters are highly manageable without losing a large amount of data, and therefore just 100% − 79.9% = 20.1% of the existing information was compromised. Using the method of factor analysis and data reduction, the questionnaire's 66 factors were reduced to 16 components and then grouped/categorized into five fundamental clusters. The five fundamental clusters include just 35 original factors from the questionnaire that represent the most relevant data on risk indicators that impact value creation.

4.3.1. Cluster 1: Financial Risk

The cluster of financial risk comprises components 1, 3, 7, 13, and 15 and represents 28.44% of the total explained variance. The cluster represents 13 risk indicators. The likelihood of the impact of these risks would have a large influence on financial investment, especially the capital expenditure cost (CAPEX) and operating expenditure cost (OPEX). The risk indicators in this cluster should be considered during the early stage of a design because they reflect the impact on economic and monetary aspects for the project's life. The selected indicators are, failure to recognize cost–value mismatches, failure to identify cost–value relationships, failure to consider the cost of losing potential revenue, failure to appropriately locate cost to function allocation, and failure to consider future operational costs and economic conditions as well as incorrect estimated cost and/or insufficient funding.

4.3.2. Cluster 2: Functional Risk

The functional risk cluster identifies the risk indicators that impact the functional performance and affect the asset's functional reliability. The cluster consists of eight risk indicators distributed into four components—i.e., 8, 10, 11, and 14—with a total variance of 15.876%. The risks relate to design considerations such as construction implication, specification, systems, and/or changes in design.

4.3.3. Cluster 3: Operational Risk

The operational risk cluster variance is 6.613% and comprises three risk indicators, including failure to increase routine maintenance, failure to consider increase in life cycle replacement and failure to consider design impact on operating efficiency. The impacts of these indicators might have a large effect on project life efficiency.

4.3.4. Cluster 4: Environmental Risk

The environmental risk cluster consists of components 2, 5, and 12 with a variance of 17.259% and three risk indicators. These risks could have a negative effect on the building's efficiency and might also have an unsafe environmental impact. The selected risk indicators in the environmental cluster include failure to consider implication of environmental risks, failure to consider the impact of maintainability and reparability, and failure to consider the impact of obsolete equipment.

4.3.5. Cluster 5: Management Risk

The management risks influence the project management performance, which reduces the project's ability to deliver within the required objectives. In this cluster, eight risk indicators were distributed into components 6, 9, and 16 with a total variance of 12.295%. The management risk indicators should be considered at an early stage of a project in order to avoid any obsolescence and to manage the risk at tolerance level. The risk indicators in this cluster concern poor project management, project definition, planning, team relationships, and design; incorrect time estimates; as well as lack of coordination and decision-making.

4.4. Impact of Risk Factors to Value Creation

4.4.1. Financial Risks

The financial value or economic value from green assets can be undermined if the risk factors are not factored into the design and operation of these assets [30]. Furthermore, financial risks in green buildings are attributed to the additional capital cost for the inclusion of green strategies in the design. From a value engineering analysis point of view, financial risks relate to the fact that stakeholders are unable to take into consideration the risks and the opportunities associated with the risks shown in Figure 4. The figure indicates that engineers perceived R11 "Failure to recognize cost–value mismatches", R10 "Failure to identify cost–value relationships", and R12 "Failure to appropriately locate cost-to-function allocation" risks are less important in identifying value. This result is not in keeping with the literature because these three risks are the foundation on which value engineering analysis is based. One plausible explanation of this anomaly is that, compared to Western practices, the Kingdom of Saudi Arabia (KSA)'s practices do not consider this as important.

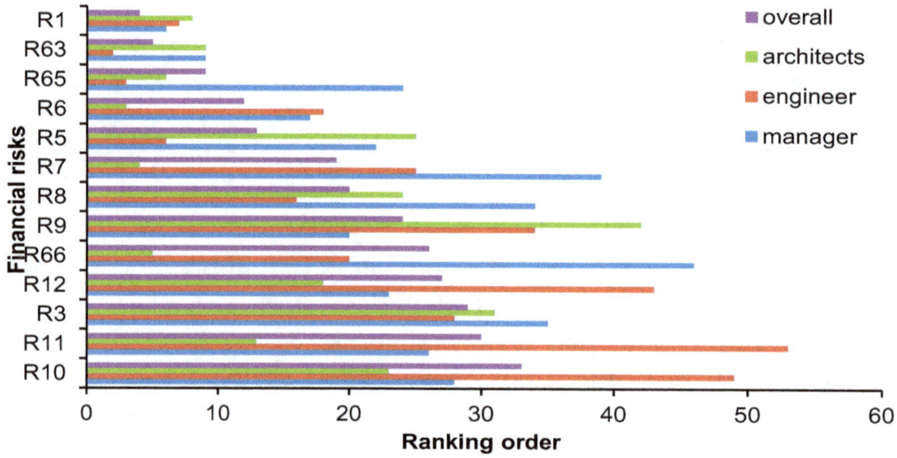

Figure 4. Ranking the financial risk factors impacting on the project value.

The results in Figure 4 show that managers did not highly rank R7 "Failure to recognize cost as resource expenditure", R8 "Failure to consider the cost of losing potential revenue", R66 "Incorrect estimated cost of energy used" risks. Although these risks are important in value entering analysis, the KSA managers perceived these risks differently. This could be attributed to the fact that most buildings in KSA are publicly owned, hence issues like revenue and energy costs are not considered important. Although future energy cost, performance of new green technologies and functional performance are uncertain, architects thought R9 "Uncertainty about prices" risk may not have much impact on value. All respondents agreed that R1 "Insufficient funding", R65 "Incorrect estimated cost of maintenance", and R63 "Incorrect cost estimate" risks will influence value analysis. Other risk factors such as R5 "Inappropriate cost evaluation criteria, R6 "Failure to consider future operational costs", and R3 "Failure to consider implication of economic conditions" were perceived as important by the respondents. This reiterate the view expressed in the literature that the investment appraisal must address various options for creating the value required from green buildings. Failure to consider implications of economic conditions and to recognize cost as a resource expenditure will certainly lead to the risk of affordability and to the risk that whole-life cost estimates are not realistic and are based on unreliable evidence or assumptions. This in turn will lead to income and the future value of real estate assets being affected.

4.4.2. Functional Risks

Functional risks are associated to the building asset condition over its life span. The building design function and its components' specifications are generally based on assumptions. Changes in the assumptions of these parameters may lead to different levels of performance and reliability. Thus, the design of green buildings must be subject to risk analysis to assess the uncertainties associated with the function and design parameters. Figure 5 portrays the respondents' perceptions about the functionality risks. The figure below clearly indicates that architects' views on R19 "Failure to design to brief/specification", R20 "Failure to consider construction implications during design", R21 "Design changes", R22 "Redesign/rework", and R36 "Failure to consider design risks" are in keeping with the other respondents' views. This result may suggest that because these risk factors are design-related and architects generally tended to approach their design with the bias.

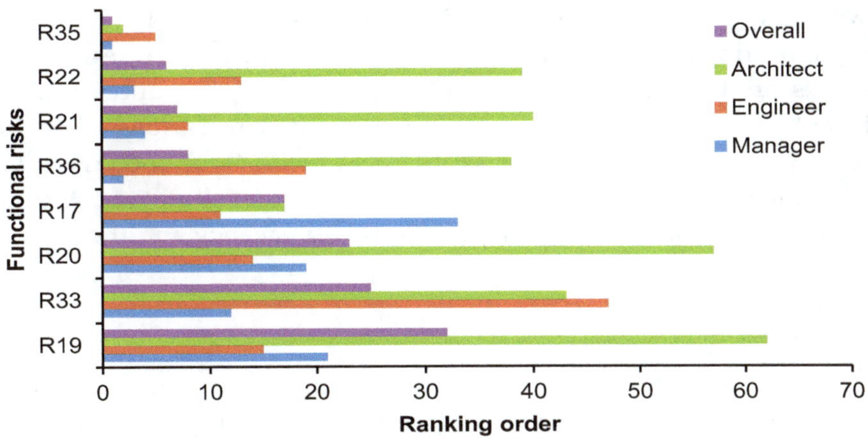

Figure 5. Ranking the functional risk factors impacting on the project value.

The results also showed that there is total agreement on the importance of R35 "Failure to identify low-value, long-lead-time items". However, the engineers' view was in accordance with that of the architects for R33 "Failure to integrate the various systems to achieve the lowest life-cycle costs". Again, this result is not in keeping with literature regarding Western economies, where lowest life cycle costs are considered an important value generator [21]. This reaffirms that the risk of obsolescence of the green building functions or those of its components will result in larger OPEX and may undermine the value of the asset, leading to lower income, smaller capital receipts, higher costs, and the possibility of legal action [30].

4.4.3. Operational Risks

The operational performance of green buildings has a significant impact on their rental and market value. The investors are protecting and increasing the value of their investment in green real estates by incorporating initiatives to improve the energy efficiency and sustainability of their portfolios [30]. Nevertheless, many in the construction industry view the performance of green products, systems, and buildings as a risk [31]. Thus, considering operational features during the early stages of value planning will probably go a long way to protect and increase the investment value in green real estate by taking into consideration risks and initiatives to improve the energy efficiency and sustainability of assets throughout their entire life cycle. Figure 6 below portrays the respondents' perceptions about the operational risks that might have an impact on value creation in green building development if not taken into consideration during the value engineering analysis process.

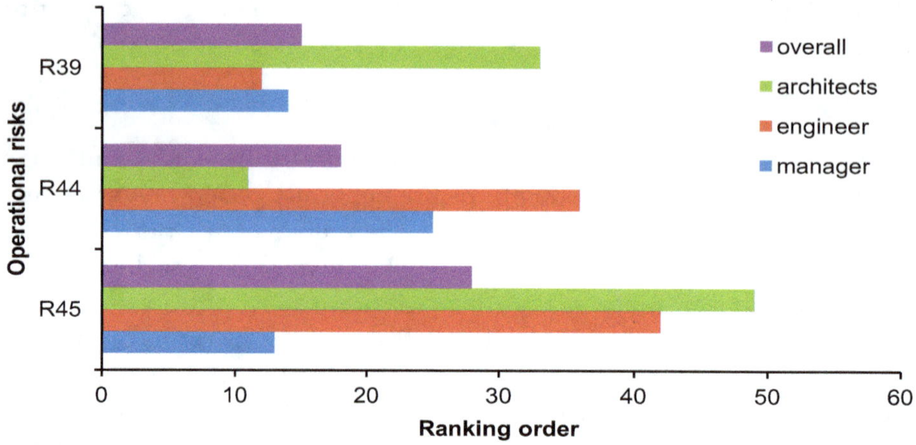

Figure 6. Ranking the operational risk factors impacting on the project value.

Figure 6 shows that only R45 "failure to consider increase in life cycle replacement", R44 "failure to consider increase in routine maintenance", and R39 "failure to consider design impact on operating efficiency" risks were scored highly by the respondents, whereas R42 "failure to consider component repair and replacement" and R46 "limited knowledge of maintenance issues" risks were viewed as not important. In fact, it is surprising to see that risk R42 is ranked 64 out of 66 risks. Component repair and replacement is an integral part of whole life cycle management strategies. Both have an impact on OPEX budget and asset availability (directly associated with vacancy and the rental value risks).

4.4.4. Environmental Risks

The environmental risks are associated with minimizing the impacts on the environment throughout the development and operation of green building facilities. The ethos behind green building development is to create facilities that must minimize waste and energy use throughout their life cycle stages. To cope with the rapid development in the green building sector, green materials were developed without robust testing of their performance and environmental credentials that may lead to liability litigation over unfit for purpose specification, material or product failure, obsolescence, durability, etc.

Figure 7 illustrates that architects thought R47 "failure to consider implication of environmental risks" risk ought to be taken into consideration in value engineering analysis, whereas engineers and managers considered R38 "failure to consider obsolescence of equipment impact" as an important risk to consider. The results might have been influenced by the professional bias. Architects ranked R43 "failure to consider maintainability and reparability impact" slightly higher than the other respondents. The way in which green buildings are conceived, constructed, operated, and disposed of will influence (or increase) the environmental impacts during their lifecycle, including global warming, acidification, eutrophication, winter smog, heavy metals, and energy [42]. The ANOVA results in Table 6 show that there were significant differences between the respondents regarding R38: failure to consider obsolescence of equipment impact risk.

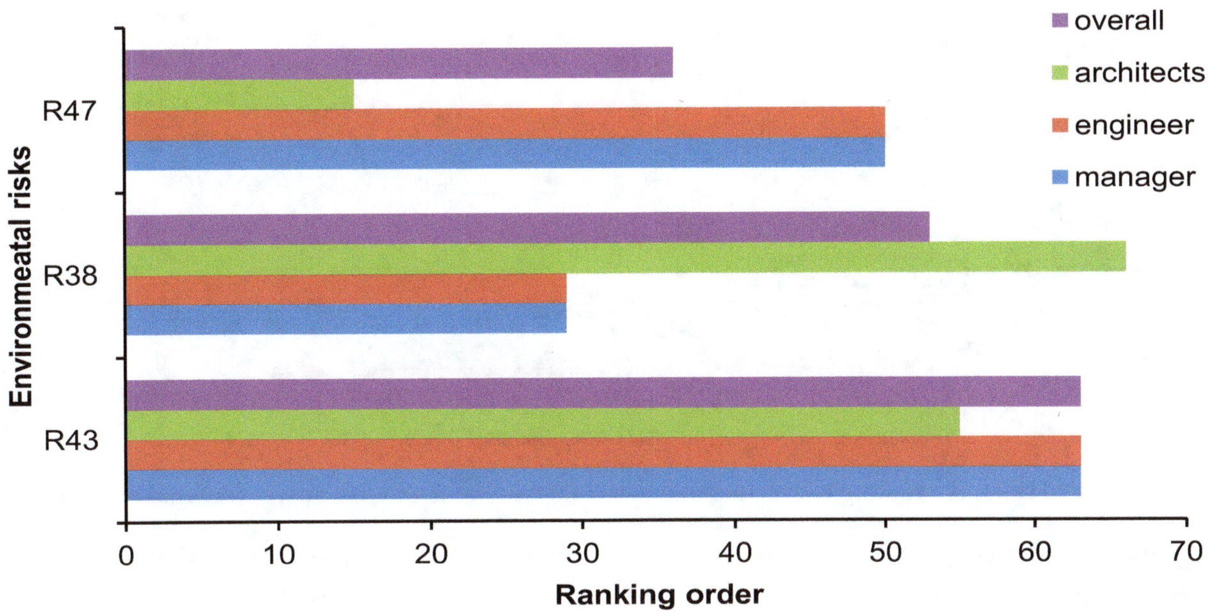

Figure 7. Ranking the environmental risk factors impacting on the project value.

Table 6. Environmental risks and hypothesis test.

Research Question	**Environmental Risks:**
Hypothesis	H_0: $p > 0.05$; H_1: $p < 0.05$
Results	The (ANOVA) results indicated that: There were significant differences between the survey participants regarding Risk factors: R38: Failure to consider obsolescence of equipment impact
Researcher's observation	It appears that architects in KSA are not acquainted with the fact that whole-life performance of green buildings is highly linked to the obsolescence of equipment and components.
Conclusion	The null hypothesis was rejected for R38 risk. The null hypothesis (H_0: $p > 0.05$) was retained for other risks.

4.4.5. Management Risks

There is ample evidence in the literature to suggest that it is essential to have a skills and integration project team coupled with effective management processes to unlock value during the early stage of value planning as well as during the development operation of green facilities. The endeavor of the project should be geared towards identifying risks and opportunities to maximize value and minimize waste at every stage of the construction and procurement process, from the minute that the need for a building is identified to when it is ready for operation [33,61]. Figure 8 illustrates the respondents' perceptions about the management risks that might have an impact on value created by green building development if not taken into consideration during the value engineering analysis process. Out of the 17 managerial risk factors identified in the literature, only half of them were found to be having a negative impact on value creation attributes of green buildings, i.e., R28, R49, R50, R51, R53, R54, R57, and R64.

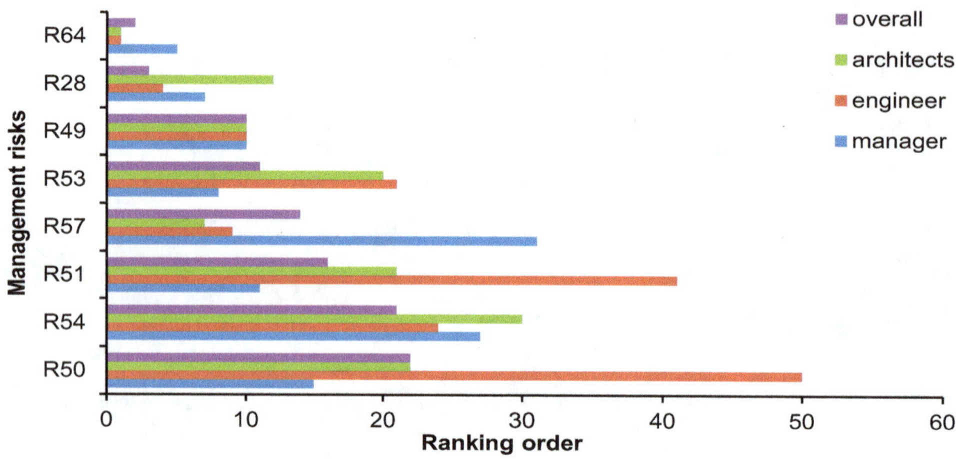

Figure 8. Ranking the management risk factors impacting on the project value.

All respondents agree on considering R28 and R49 risks during the appraisal of green buildings. However, engineers viewed R51 and R50 risks different to other participants. This might suggest that engineers in KSA are not often involved in the very early stages of the design process. However, the findings here are consistent with Shen and Liu [51] who listed the factors that might influence the success of using value analysis in construction projects.

5. Conclusions

Present research analyzed various aspects of risk to optimize the value creation in the development of green buildings. Sixty-six (66) risk factors were classified in 5 risk categories, including functional, financial, operational, environmental, and management, with an objective to evaluate the impacts of these risks on the value creation of green buildings design. A questionnaire compiling a list of identified risk factors was hand-delivered to 300 practitioners working for value engineering of the construction industry in Saudi Arabia. The overall response rate was 29.7%. Among the respondents, 45% were managers, 38% were engineers, and 17% were architects who participated in the survey. The participation from young professionals with 0–5 years of experience was 16%, and from the middle-career professionals with 6–10 years of experience was 28%. Interestingly, senior professionals with more than 10 years of experience held the largest contribution (56%) among all the respondents.

Based on the participants responses, descriptive statistics identified important risk factors with a minimum mean value of 3.81 (i.e., higher than the average overall mean of 3.78) and severity indices of 76.18%. Furthermore, the principal component analysis (PCA) extracted 16 components, based on the likelihood of risk factors impacting the value created by green building design, that carry an eigenvalue of more than 1 and account for nearly 79.939% of the whole variance. Finally, the factor analysis grouped 35 most significant risk factors in 5 clusters, i.e., 8 in functional risk, 13 in financial risk, 3 in operational risk, 3 in environmental risk, and 8 in management risk cluster.

Due to the differences in perception regarding the risk factors, there is a need for improved communication between the decision-makers' groups for developing a shared understanding of project value creation and associated risks. Absence of such understanding may raise the possibility of conflicts amongst different groups that ultimately effects the expected outcomes of the project. Future work can establish the interaction between various value drivers (identified in the authors' previous work) and the risk factors (selected in the present study) using an effective framework to enhance value creation in green buildings. Furthermore, the impact of the risk factors on project constrains—i.e., quality, cost, and time—can also be investigated.

A clear understanding of new risk clusters and their implications will be instrumental in assessing the design indicators and evaluating the impact of risk factors on value creation of green building in Saudi Arabia and elsewhere in the world.

Author Contributions: W.A.—conceptualization, methodology, survey, performed detailed data collection and analysis, software, and original manuscript. H.H.—conceptualization, methodology, and paper writing. H.B.—conceptualization, methodology, administration, supervision, visualization, and paper review. All authors have read and agreed to the published version of the manuscript.

Acknowledgments: The author gratefully acknowledges respondents from the construction industry in Saudi Arabia. The first author acknowledges the financial support secured at Qassim University that was provided by the Saudi Cultural Bureau in United Kingdom under the Higher Education Scholarship Program of Saudi Arabia. Authors also like to thank the School of Engineering, Faculty of Science and Engineering, University of Liverpool, UK for providing the technical support for this research.

Appendix A

Sample of the second part of the design questionnaire for the risk factor

What is the likelihood of the following risk factors having an impact on the project value created by green building design?						
No.	**Risk Factors**	**Rate the Likelihood Impact of the Risk Factors on Project Value Creation (Please Tick One Box)**				
		Very Likely	**Likely**	**Neutral**	**Unlikely**	**Very Unlikely**
1	Funding is unavailable					
2	Failure to consider implication of economic conditions					
3	Inadequate inflation forecasts					
4	Inappropriate cost evaluation criteria					
...	...					

Appendix B

Rating the likely impact of the risk factors

Code	Mean	Std. Deviation	Severity index	Coefficient of variation	Ranking						
					Manager	Engineer	Architect	0 to 5 years	6 to 10 years	More than 10 years	Overall Ranking
R1	4.09	0.861	81.80	21.06	6	7	8	18	4	6	4
R2	3.63	1.274	72.58	35.10	59	35	44	24	36	62	56
R3	3.81	0.940	76.18	24.68	35	28	31	3	51	33	29
R4	3.76	0.966	75.28	25.65	54	17	26	33	41	32	35
R5	3.94	0.871	78.88	22.08	22	6	25	45	34	5	13
R6	3.96	1.065	79.10	26.93	17	18	3	21	18	16	12
R7	3.85	0.995	77.08	25.81	39	25	4	44	35	15	19
R8	3.85	0.886	77.08	22.99	34	16	24	29	50	12	20
R9	3.83	1.069	76.63	27.89	20	34	42	43	46	14	24
R10	3.78	1.020	75.51	27.01	28	49	23	42	54	20	33
R11	3.81	0.952	76.18	24.99	26	53	13	20	59	18	30
R12	3.82	0.860	76.40	22.52	23	43	18	22	45	21	27
R13	3.72	0.965	74.38	25.95	30	60	35	41	63	23	41
R14	3.63	0.981	72.58	27.04	45	63	41	59	62	35	55
R15	3.54	1.139	70.79	32.17	56	66	34	51	65	52	61
R16	3.70	1.172	73.93	31.70	38	48	48	40	40	45	43
R17	3.88	0.939	77.53	24.23	33	11	17	2	26	30	17
R18	3.69	0.984	73.71	26.70	18	59	61	50	25	56	48
R19	3.79	0.959	75.73	25.33	21	15	62	39	43	24	32
R20	3.83	0.991	76.63	25.87	19	14	57	25	31	27	23
R21	4.04	0.928	80.90	22.95	4	8	40	23	3	11	7
R22	4.04	0.952	80.90	23.55	3	13	39	12	10	9	6
R23	3.69	0.937	73.71	25.41	37	40	60	38	37	47	46

Code	Mean	Std. Deviation	Severity index	Coefficient of variation	Ranking						
					Manager	Engineer	Architect	0 to 5 years	6 to 10 years	More than 10 years	Overall Ranking
R24	3.76	0.966	75.28	25.65	32	30	50	55	8	51	34
R25	3.53	1.098	70.56	31.13	53	52	65	65	52	60	62
R26	3.43	1.096	68.54	31.99	65	57	63	66	29	65	65
R27	3.56	1.022	71.24	28.69	62	38	59	57	22	63	59
R28	4.10	0.942	82.02	22.97	7	4	12	17	7	4	3
R29	3.64	0.944	72.81	25.94	58	27	53	32	39	58	54
R30	3.33	1.106	66.52	33.24	66	65	54	62	48	66	66
R31	3.69	0.984	73.71	26.70	52	46	16	31	21	59	47
R32	3.58	1.009	71.69	28.16	49	62	52	61	61	40	58
R33	3.82	0.960	76.40	25.13	12	47	43	11	38	28	25
R34	3.74	1.028	74.83	27.48	40	23	51	37	44	34	38
R35	4.24	0.739	84.72	17.44	1	5	2	1	1	3	1
R36	4.03	0.885	80.67	21.94	2	19	38	5	6	13	8
R37	3.80	0.894	75.96	23.54	16	33	56	36	30	29	31
R38	3.65	1.078	73.03	29.51	29	32	66	58	47	44	53
R39	3.92	0.944	78.43	24.08	14	12	33	6	20	22	15
R40	3.54	1.159	70.79	32.73	61	37	64	56	53	61	60
R41	3.60	1.063	71.91	29.56	57	51	47	16	64	53	57
R42	3.44	0.953	68.76	27.72	64	61	58	60	58	64	64
R43	3.51	0.978	70.11	27.91	63	56	55	47	66	57	63
R44	3.87	0.855	77.30	22.13	25	36	11	8	9	41	18
R45	3.82	0.886	76.40	23.20	13	42	49	10	32	31	28
R46	3.70	1.005	73.93	27.18	55	29	36	4	56	46	44
R47	3.76	0.930	75.28	24.70	50	22	15	35	28	39	36

Code	Mean	Std. Deviation	Severity index	Coefficient of variation	Ranking						
					Manager	Engineer	Architect	0 to 5 years	6 to 10 years	More than 10 years	Overall Ranking
R48	3.73	1.042	74.61	27.93	36	58	14	19	24	54	39
R49	4.00	0.953	80.00	23.84	10	10	10	28	11	8	10
R50	3.83	0.980	76.63	25.57	15	50	22	15	14	42	22
R51	3.89	0.845	77.75	21.74	11	41	21	30	15	25	16
R52	3.72	1.087	74.38	29.23	44	45	32	53	19	50	42
R53	3.96	0.838	79.10	21.19	8	21	20	14	13	17	11
R54	3.84	1.010	76.85	26.29	27	24	30	48	55	7	21
R55	3.65	1.035	73.03	28.33	41	64	37	13	60	48	52
R56	3.75	0.945	75.06	25.18	47	26	29	46	16	49	37
R57	3.92	1.068	78.43	27.25	31	9	7	26	17	19	14
R58	3.72	0.953	74.38	25.63	43	54	19	52	23	43	40
R59	3.66	0.965	73.26	26.34	42	55	46	49	33	55	51
R60	3.66	0.916	73.26	25.02	60	31	28	64	42	37	50
R61	3.66	0.988	73.26	26.97	48	44	45	63	57	26	49
R62	3.69	1.083	73.71	29.38	51	39	27	54	49	36	45
R63	4.08	0.815	81.57	19.98	9	2	9	34	12	2	5
R64	4.18	0.899	83.60	21.51	5	1	1	9	2	1	2
R65	4.01	0.872	80.22	21.75	24	3	6	27	5	10	9
R66	3.82	1.061	76.40	27.78	46	20	5	7	27	38	26

References

1. Green Building. What is Green Building. Available online: https://greenbuildingcanada.ca/green-building-guide/what-is-green-building (accessed on 15 October 2020).
2. He, B.J. Towards the next generation of green building for urban heat island mitigation: Zero UHI impact building. *Sustain. Cities Soc.* **2019**, *50*, 101647. [CrossRef]
3. Zinzi, M.; Agnoli, S. Cool and green roofs. An energy and comfort comparison between passive cooling and mitigation urban heat island techniques for residential buildings in the Mediterranean region. *Energy Build.* **2012**, *55*, 66–76. [CrossRef]
4. Alotaibi, K.; Ghumman, A.R.; Haider, H.; Ghazaw, Y.M.; Shafiquzzaman, M. Future predictions of rainfall and temperature using GCM and ANN for arid regions: A case study for the Qassim Region, Saudi Arabia. *Water* **2018**, *10*, 1260. [CrossRef]
5. Alkolibi, F.M. Possible effects of global warming on agriculture and water resources in Saudi Arabia: Impacts and responses. *Clim. Chang.* **2020**, *54*, 225–245. [CrossRef]
6. Shin, M.H.; Kim, H.Y.; Gu, D.; Kim, H. LEED, its efficacy and fallacy in a regional context—An urban heat island case in California. *Sustainability* **2017**, *9*, 1674. [CrossRef]
7. Dwaikat, L.N.; Ali, K.N. The economic benefits of a green building–Evidence from Malaysia. *J. Build. Eng.* **2018**, *18*, 448–453. [CrossRef]
8. MacNaughton, P.; Cao, X.; Buonocore, J.; Cedeno-Laurent, J.; Spengler, J.; Bernstein, A.; Allen, J. Energy savings, emission reductions, and health co-benefits of the green building movement. *J. Expo. Sci. Environ. Epidemiol.* **2018**, *28*, 307–318. [CrossRef]
9. Connaughton, J.N.; Green, S.D. *Value Management in Construction: A Client's Guide*; CIRIA: London, UK, 1996.
10. Banaitiene, N.; Banaitis, A. Risk Management in Construction Projects. Risk Management–Current Issues and Challenges. In *Risk Management–Current Issues and Challenges*; Banaitiene, N., Ed.; IntechOpen: London, UK, 2012; pp. 429–448.
11. Godfrey, P.S. *Control of Risk a Guide to the Systematic Management of Risk from Construction*; CIRIA: London, UK, 1996.
12. Langford, D. Revaluing Construction-Hard and Soft Values. In *CIB Priority Theme-Revaluing Construction: A W065 'Organisation and Management of Construction' Perspective*; Sexton, M., Kaehkoenen, K., Lu, S., Eds.; Ciblibrary: Ottawa, ON, Canada, 2007; pp. 66–75.
13. Gao, Y.; Yang, G.; Xie, Q. Spatial-Temporal Evolution and Driving Factors of Green Building Development in China. *Sustainability* **2020**, *12*, 2773. [CrossRef]
14. ICE. *Design and Practice Guide, Creating Value in Engineering*; LoThomas Telford: London, UK, 1996.
15. Leek, S.; Christodoulides, G. A framework of brand value in B2B markets: The contributing role of functional and emotional components. *Ind. Mark. Manag.* **2012**, *41*, 106–114. [CrossRef]
16. Boussabaine, A. *Cost Planning of PFI and PPP Building Projects*; Taylor & Francis: Abingdon, UK, 2007.
17. Alalshikh, M. The Development of a Value Management Approach for the Saudi Public Sector. Unpublished. Ph.D. Thesis, University of Leeds, Leeds, UK, 2010. in press.
18. Snoj, B.; Pisnik Korda, A.; Mumel, D. The relationships among perceived quality, perceived risk and perceived product value. *J. Prod. Brand Manag.* **2004**, *13*, 156–167. [CrossRef]
19. Alattyih, W.; Haider, H.; Boussabaine, H. Development of Value Creation Drivers for Sustainable Design of Green Buildings in Saudi Arabia. *Sustainability* **2019**, *11*, 5867. [CrossRef]
20. Tiong, R.L. BOT projects: Risks and securities. *Constr. Manag. Econ.* **1990**, *8*, 315–328. [CrossRef]
21. Boussabaine, H.A.; Kirkham, R.J. *Whole Life-Cycle Costing: Risk and Risk Responses*; Blackwell Publishing Ltd.: Oxford, UK, 2008.
22. Medda, F. A game theory approach for the allocation of risks in transport public private partnerships. *Int. J. Proj. Manag.* **2007**, *25*, 213–218. [CrossRef]
23. Zou, P.X.; Zhang, G.; Wang, J. Understanding the key risks in construction projects in China. *Int. J. Proj. Manag.* **2007**, *25*, 601–614. [CrossRef]
24. Yang, J.; Zou, P.; Keating, B. Analysing Stakeholder-Associated Risks in Green Buildings: A Social Network Analysis Method. In *WBC13*; Queensland University of Technology: Brisbane, Australia, 2013; pp. 1–12.

25. Zou, P.; Couani, P. Risks in green building development: A supply chain perspective. In Proceedings of the 36th Australasian University Building Educators Association (AUBEA) Conference, Gold Coast, Queensland, Australia, 27–29 April 2011.
26. National Research Council. *The Owner's Role in Project Risk Management*; The National Academies Press: Washington, DC, USA, 2005.
27. Wiedemer, J.P.; Goeters, J.E.; Graham, J.E. *Real Estate Investment: South.-Western Cengage Learning*; South-Western Educational Pub: Mason, OH, USA, 2011.
28. Al-Yousefi, A. *Value Management: Concept & Techniques*, 4th ed.; Twaig Publishing: Riyadh, UAE, 2004.
29. Isa, M.; Rahman, M.M.G.M.A.; Sipan, I.; Hwa, T.K. Factors Affecting Green Office Building Investment in Malaysia. *Procedia Soc. Behav. Sci.* **2013**, *105*, 138–148. [CrossRef]
30. Green Building Council. The Bussine Case Green Building: A Review of the Costs and Benefits for Developers Investors and Occupants. Available online: http://www.worldgbc.org/files/1513/6608/0674/Business_Case_For_Green_Building_Report_WEB_2013-04-11.pdf (accessed on 20 August 2015).
31. *Zurich Green Building: What are the Risks?* Zurich Services Corporation: Schaumburg, IL, USA, 2011.
32. Haghnegahdar, L.; Asgharizadeh, E. The Risk and Value Engineering Structures and their Integration with Industrial Projects Management (A Case Study on IK Corporation). *World Acad. Sci. Eng. Technol.* **2008**, *40*, 375–383.
33. NAO. *Getting Value for Money from Construction Projects through Design: How Auditors Can. Help, Guidelines to Help Auditors Take Account of Good Design in Public Sector Built Environment Projects*; Davis Langdon & Everest. Stairway Communications Ltd.: London, UK, 2004.
34. Kim, S.; Lim, B.T.H.; Kim, J. Green Features, Symbolic Values and Rental Premium: Systematic Review and Meta-analysis. *Procedia Eng.* **2017**, *180*, 41–48. [CrossRef]
35. Davies, R. *Green Value Report: Green Building, Growing Assets*; Royal Institution of Chartered Surveyors: London, UK, 2005.
36. Ashuri, B.; Durmus-Pedini, A. An overview of the benefits and risk factors of going green in existing buildings. *Int. J. Facil. Manag.* **2010**, *1*, 1.
37. Lützkendorf, T.; Lorenz, D. Integrating Sustainability Issues into Property Risk Assessment–An Approach to Communicate the Benefits of Sustainable Design. In Proceedings of the Rethinking Sustainable Construction Conference, Sarasota, FL, USA, 19–22 September 2006.
38. Zhao, R.J.; Moh, W.H. Value management practices on major construction projects and green building. *Front Eng. Manag.* **2016**, *3*, 147–157. [CrossRef]
39. Novais, R.M.; Senff, L.; Carvalheiras, J.; Labrincha, J.A. Bi-Layered Porous/Cork-Containing Waste-Based Inorganic Polymer Composites: Innovative Material towards Green Buildings. *Appl. Sci.* **2020**, *10*, 2995. [CrossRef]
40. Mutani, G.; Azzolino, C.; Macrì, M.; Mancuso, S. Straw Buildings: A Good Compromise between Environmental Sustainability and Energy-Economic Savings. *Appl. Sci.* **2020**, *10*, 2858. [CrossRef]
41. Shao, B.; Du, X.; Ren, Q. Numerical investigation of energy saving characteristic in building roof coupled with PCM using lattice Boltzmann method with economic analysis. *Appl. Sci.* **2018**, *8*, 1739. [CrossRef]
42. Fawaz, A.N.; Ruparathna, R.; Chhipi-Shrestha, G.; Haider, H.; Hewage, K.; Sadiq, R. Sustainability assessment framework for low rise commercial buildings: Life cycle impact index-based approach. *Clean Technol. Environ. Policy* **2016**, *1*, 2579–2590.
43. Orsi, A.; Guillén-Guillamón, I.; Pellicer, E. Optimization of Green Building Design Processes: Case Studies within the European Union. *Sustainability* **2020**, *12*, 2276. [CrossRef]
44. Gritzo, L.; Doerr, W.; Bill, R.; Nong, S.; Krasner, L. *The Influence of Risk Factors on Sustainable Development*; FM Global: Boston, MA, USA, 2009.
45. Wi, H.; Jung, M. Modeling and analysis of project performance factors in an extended project-oriented virtual organization (EProVO). *Expert Syst. Appl.* **2010**, *37*, 1143–1151. [CrossRef]
46. Muldavin, S.R. *Value beyond Cost Savings: How to Underwrite Sustainable Properties*; Green Building FC: Fort Collins, CO, USA, 2010.
47. Cajias, M.; Bienert, S. Does sustainability pay off for European listed real estate companies? The dynamics between risk and provision of responsible information. *J. Sustain. Real Estate* **2011**, *3*, 211–231. [CrossRef]
48. Eichholtz, P.; Kok, N.; Quigley, J.M. *Sustainability and the Dynamics of Green Building: New Evidence of the Financial Performance of Green Office Buildings in the USA*; Research Report by RICS: London, UK, 2010.

49. Parker, D. Valuation of Green Buildings: Is Greed Becoming Fear? *Aust. N. Z. Prop. J.* **2008**, *1*, 552–560.
50. Jha, K.; Iyer, K. Critical determinants of project coordination. *Int. J. Proj. Manag.* **2006**, *24*, 314–322. [CrossRef]
51. Shen, Q.; Liu, G. Critical success factors for value management studies in construction. *J. Constr. Eng. Manag.* **2003**, *129*, 485–491. [CrossRef]
52. Al-Yousefi, A. *Application of Value Engineering in Saudi Arabia and Arabian Gulf Countries*; Powerpoint Slides; Ministry of Water and Electricity: Riyadh, Saudi Arabia, 2010.
53. Al-Yousefi, A. *Total Value Management*; Powerpoint Slides; Alyousefi Value Engineering: Riyadh, Saudi Arabia, 2011.
54. Punch, K.F. *Introduction to Social Research: Quantitative & Gualitative Approaches*, 2nd ed.; SAGE Publications: London, UK, 2005.
55. Morgan, G.A.; Leech, N.L.; Gloeckner, G.W.; Barret, K.C. *SPSS for Introductory Statistics: Use and Interpretation*, 2nd ed.; Lawrence Erlbaum Associate Publishers: Mahwah, NJ, USA, 2004.
56. Field, A. *Discovering statistics using SPSS*, 2nd ed.; SAGE Publications: London, UK, 2005.
57. SPSS 22. *0.0.1. SPSS 22.0.0.1 for Windows*; SPSS Inc.: Chicago, IL, USA, 2014.
58. George, D.; Mallery, P. *SPSS for Windows Step by Step: A Simple Guide and Reference*, 4th ed.; Allyn & Bacon: Boston, MA, USA, 2003.
59. Akintoye, A. Analysis of factors influencing project cost estimating practice. *Constr. Manag. Econ.* **2000**, *18*, 77–89. [CrossRef]
60. Kaiser, H. An index of factorial simplicity. *Psychometrika* **1974**, *39*, 31–36. [CrossRef]
61. Nykamp, H. Policy Mix for a Transition to Sustainability: Green Buildings in Norway. *Sustainability* **2020**, *12*, 446. [CrossRef]

Impact of Heat Pump Flexibility in a French Residential Eco-District

Camille Pajot [1,*], Benoit Delinchant [1], Yves Maréchal [1] and Damien Frésier [2]

1 University Grenoble Alpes, CNRS, Grenoble INP, G2Elab, 38000 Grenoble, France; benoit.delinchant@g2elab.grenoble-inp.fr (B.D.); yves.marechal@g2elab.grenoble-inp.fr (Y.M.)
2 Gaz et Electricité de Grenoble, 57 Rue Pierre Semard, 38000 Grenoble, France; d.fresier@geg.fr
* Correspondence: camille.pajot1@g2elab.grenoble-inp.fr

Abstract: This paper investigates the impact of load shedding strategies on a block of multiple buildings. It particularly deals with the quantification of the factors i.e., peak shaving, occupants' thermal comfort or CO_2 emission reduction and how to quickly quantify them. To achieve this goal, the paper focuses on a new residential district, thermally fed by heat pumps. Four modeling approaches were implemented in order to estimate buildings' response towards load shedding. Two schemes were combined in order to study an overall load shedding. This strategy for the neighborhood has proved itself efficient for both peak shaving and thermal comfort. Most of the clipped heating load during the peak period is shifted to low-consumption periods, providing an effective peak shaving. The thermal comfort is guaranteed for at least 96% of the time. For CO_2 emissions reduction, the link between consumption reduction and CO_2 emissions savings should be realized carefully, since shifting the consumption could increase these emissions.

Keywords: peak shaving; demand response; block of buildings; thermal model; TEASER

1. Introduction

1.1. Background

Within three years of the 21st United Nations Climate Change Conference (COP21), a number of energy transition policies have been carried out in order to respect the Paris Agreement in keeping the global average temperature below 2 °C of pre-industrial levels [1]. The massive integration of renewable energy sources, together with the electrical peak consumption augmentation put load flexibility in a central position in regards to energy transition strategies, as it could help to guarantee grid stability [2].

Many new stakeholders, as well as new markets, appear in order to modulate electrical consumption [3]. However, aggregators mostly apply these demand response strategies on electricity-intensive industries, excluding lower power level sites such as buildings. Nevertheless, the latter represents a large share of global energy consumption. Owing to their thermal inertia, different load shedding schemes (recurring or non-recurring) can be implemented on the buildings; yet the flexibility of these schemes is hard to evaluate quickly.

Therefore, it became one of the key points studied in the European project City-zen [4]. Our collaborative effort with the local Distribution System Operator (DSO), Grenoble Electricity and Gas (GEG), takes place in this context and focuses on a new residential eco-district. This district consists of 23 residential buildings having 264 apartments and is thermally fed by ground source heat pumps (GSHP); thus emphasizing the primary use of electrical energy for heating purposes. Indeed, GSHP represent a significant research field [5,6], so that they are at the cutting edges of research with the demand side management (DSM) in order to manage electric grid constraints [7].

1.2. Literature Review

On the one hand, peak-shaving strategies are widely studied in order to deal with electric grid constraints [8]. On the other hand, DSM becomes more studied at a local scale [9], focusing on local energy integration [10], demand curve smoothing [11] or economic purposes [12]. These two combined lead to an increase of research papers on the field of peak-shaving for better management of local electric grid constraints through DSM [13].

The idea of using buildings' thermal inertia in order to modulate the heating load is also vastly studied. While buildings' flexibility is studied with several aims such as increasing district heating efficiency [14,15] or for a better integration of local production sources [16], it is also investigated with the objective to evaluate what could be their future impact in smartgrids [17] and to compare them to storage solutions [18]. Several methodologies have been developed in order to quantify this flexibility but only three are commonly applied using building structural mass [19]. Moreover, only a few of these papers evaluate the impact on thermal comfort as they mostly consider it as a constraint [20]. In our case, the possibility to be out of the comfort zones will be considered, usually defined by set-point temperature ranges [18,21], while estimating this impact by standing on comfort ranges as defined in [22].

Most of the time, estimating the building temperature is possible as the building thermal flexibility quantification is based on thermal models. These models can be from low-order RC models [23] to higher-detailed models often based on widespread tools such as EnergyPlus [18,24] or based on the Modelica language such as the library IDEAS [25]. Even the district scale becomes more and more widespread in the energetic dynamic simulation software, like DIstrict MOdeller and SIMulator (DIMOSIM) developed by CSTB [26], City Energy Analyst (CEA) developed by ETH Zurich [27] or Tool for Energy Analysis and Simulation for Efficient Retrofit (TEASER) developed by RWTH Aachen [28]. It was observed that this scale change could be impractical by being relevant to annual heating needs, but not anymore when focusing on power analysis [29]. For this reason, models with different levels of detail have been studied.

1.3. Context and Aims of the Study

In the local context of Grenoble, an electrical consumption peak appears between 5 a.m. and 10 a.m. GEG is interested in peak shaving during this morning period by implementing effective load shedding scheme. However, it could be hard to quickly quantify the impact of load shedding strategies through peak shaving while avoiding thermal discomfort for occupants and an increase in carbon footprint simultaneously.

Indeed, the problem is complex to model. In order to maintain occupants' thermal comfort, load shedding can be realized after an over-heating so that the building could store heat before disconnecting the heating system. As it induces an over-consumption, this over-heating should be performed before the peak period (i.e., between 5 a.m. and 10 a.m. in our case), with the purpose of reducing the consumption peak. In order to study a strategy of district peak reduction by deliberate load shedding building by building, two load shedding strategies (with or without over-heating) will be compared.

This paper aims to quickly quantify the influence of this district heat load shedding strategy on the heat load curve, thermal comfort and greenhouse gases emissions reduction. Moreover, the study will try to quantify the impact of heat load profiles modeling on the results.

1.4. Paper Structure

The first section of this paper describes the methods used for load shedding impact quantification. At first, the impact indicators in terms of peak shaving, thermal comfort and CO_2 reduction will be defined. Then, the heat load profiles modeling will be presented. Finally, the load shedding scenarios studied in the paper will be introduced. In the 'Results' section, the two load shedding strategies will be analyzed on a building with respect to the indicators introduced earlier. The results will show the impact on three aspects: peak-shaving, thermal comfort, and CO_2 emissions while analyzing the effect of the heat load profiles modeling. A conclusion will be drawn at the end of the paper based on the observed results.

2. Methods for Impact Quantification of Load Shedding

2.1. Indicators for Impact Quantification of Load Shedding

2.1.1. Peak Shaving

In order to quantify the amount of load reduction in a district, the key indicators will be defined first. Many studies show some rebound effects after a load shedding, related to the restart of the consumption [30]. This behavior could not only affect the energy conservation by providing any (or few) consumption reductions at a daily scale, but it could also lead to failure of peak shaving strategy. Indeed, if concentrated during a short period, a load shifting could cause bigger grid constraints during this time period. For this reason, in order to quantify the impact of load shedding strategies for peak shaving purpose, the consumption behaviour after the load shedding period has to be analyzed. To do so, the study is based on one indicator used by the French TSO RTE [31] (cf. Equations (1)–(4) and (8)). The daily Load Shifting rate (LS^d_{rate})) is defined in Equation (1) from the ratio of the addition of anticipated energy and delayed energy, by the cut-off energy. These three energies are defined in Equations (2)–(4) by the integral in a given period of the consumption power (P_t) and its reference value without operation (P_t^{ref}). These energies are also visible in Figure 1.

$$LS^d_{rate} = \frac{E_{anticipated} + E_{delayed}}{E_{cut\_off}} \tag{1}$$

where:

$$E_{cut\_off} = \int_{\tau^b_{ls}}^{\tau^e_{ls}} P_t^{ref}\, dt \tag{2}$$

$$E_{anticipated} = \int_{\tau^b_{ls}-1}^{\tau^b_{ls}} (P_t - P_t^{ref})\, dt \tag{3}$$

$$E_{delayed} = \int_{\tau^e_{ls}}^{\tau^e_{ls}+23} (P_t - P_t^{ref})\, dt \tag{4}$$

By taking into account the energy reported during the 23 h following the load shedding, this indicator gives information at a daily scale. However, in order to get a better understanding of the dynamic behavior of this energy report, the study will rather focus on an adapted form of this load shifting rate, that we proposed, defined in Equation (5). This indicator will be calculated on an hourly basis in order to quantify the distribution of energy report, hour by hour, as shown in Figure 1.

$$LS^h_{rate} = \frac{\int_h^{h+1} (P_t - P_t^{ref})\, dt}{E_{cut\ off}} \tag{5}$$

Defined as such, the load shifting rate can be used to study dynamically the load variations and gives information on the efficiency of the load shedding strategy to reduce a long peak period (more than one hour).

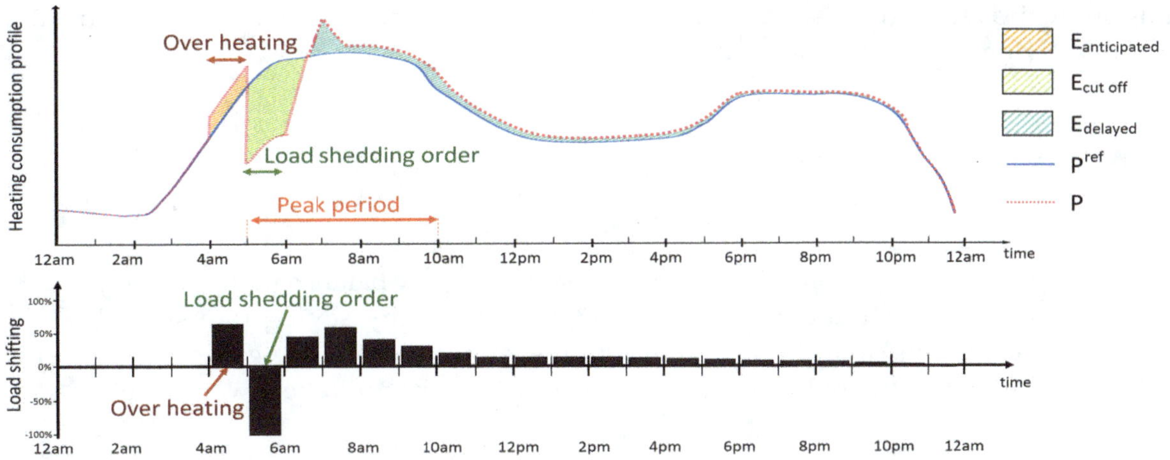

Figure 1. Representation of a daily heating load curve modification with a load shedding order and associated load shifting rates.

2.1.2. Thermal Comfort

It is important to keep in mind that turning off the heat supply can affect the thermal comfort so that this aspect has to be estimated too. When the thermal supply of a building is turned off, the internal temperature does not decrease instantaneously to the level of external temperature. This building dynamic can be explained by the possibility for buildings to store heat in their heavy components, such as walls. Indeed, due to their significant inertia, walls will cool later than the air, in cases of heating load shedding. The phenomena are important in terms of comfort, as walls and air temperatures respectively reflect radiation and convection effects perceived by the occupants of the building. Since the feeling of thermal comfort is related to this perception of both air and wall temperature, studies analyzed the relationship between thermal comfort and the operative temperature (T_{op}, defined Equation (6)) [32].

$$T_{op} = \frac{T_{walls} + T_{air}}{2} \tag{6}$$

In this study, this operative temperature will be taken as an indicator in order to estimate the comfort level, according to levels defined in [22]:

- Comfortable: A range of $+/-$ 1 °C about the temperature set-point (T_{set})
- Slightly uncomfortable: A range of $+/-$ 1 °C and $+/-$ 2 °C about T_{set}
- Uncomfortable: A difference of more than 2 °C with T_{set}

2.1.3. CO_2 Emissions Reduction

Finally, the impact of the different load shedding methods on CO_2 emissions will be studied. To do so, the work will rely on the actual CO_2 emissions from the French power generation in January 2016 [33]. This will allow us to estimate the gross CO_2 emissions variation when the load shedding strategies will be applied while taking into account hourly and daily variation (see Equation (7)).

$$CO_2S^m = \frac{\int_{month}(CO_{2_t}^{ref} - CO_{2_t})\,\mathrm{d}t}{\int_{month}(CO_{2_t}^{ref})\,\mathrm{d}t} \tag{7}$$

This variation not only takes into account the load shifting, but also the consumption reduction. Moreover, it is very common to conclude that CO_2 emissions will obviously decrease with load shedding strategies when there is energy saving. A widespread indicator for energy saving [31] is the daily Energy Saving rate (ES^d_{rate}) defined as follows:

$$ES^d_{rate} = \frac{E_{cut\_off} - E_{anticipated} - E_{delayed}}{E_{cut\_off}} \tag{8}$$

This energy saving rate is commonly used to quantify energy balance in the long-run by showing the amount of non-reported energy 23 h after [31]. The energy saving rate should be put into perspective, as it represents saving in regard to cut-off energy. Since the energy saving of an entire day is much lower than ES^d_{rate}, the reduction of daily CO_2 emissions would be lowered too.

$$ES^m = \frac{\int_{month}(P^{ref}_t - P_t)\,dt}{\int_{month}(P^{ref}_t)\,dt} \tag{9}$$

Nevertheless, this does not prevent us from expecting that CO_2 emissions would decrease as much as daily consumption. To deeply examine the impact on district carbon footprint, it is crucial to consider the daily and intra-day CO_2 emission variability for the electrical energy generation. Shifting the electrical consumption from one time period to another could increase the CO_2 emissions if local peaks do not match the total electricity generation.

For all these reasons, the paper will compare the total consumption reduction (ES^m) to the total CO_2 emissions reduction during the month (CO_2S^m). Doing this will help us in combining the two reduction factors (energy consumption and CO_2 emissions) into a single indicator, the Expected Gain reduction (EG_{red}), defined as follows:

$$EG_{red} = \frac{ES^m - CO_2S^m}{ES^m} \tag{10}$$

2.2. Heat Load Profiles Models

According to the previously defined indicators, the thermal load variation between no-load shedding and the applied load shedding strategy need to be assessed in order to quantify the impact on peak shaving and CO_2 emissions. In the present study, two modeling approaches will be compared. At first, a standard load shifting profile will be established with experimental data. Then, thermal models with several levels of details will be introduced in order to assess thermal comfort.

2.2.1. Experimental Load Shifting Profile

A first estimation can rely on experimental results from similar buildings and load shedding strategies. The main advantage of this method is to assess very quickly the peak shaving indicator. To do so, a standard load shifting rate profile was defined, based on experimental results from the French GreenLys project [34] and from a study led by the French TSO RTE [31]. As the experimental building stock contains two new eco-districts [34], the use of the resulting standard profile is considered suitable for our new residential district. The experimental results show an energy saving rate around 90%. The load shifting rate profiles are plotted below.

Figure 2a is representing the hourly load shifting rates for a one-hour residential heat load shedding without pre-heating, while Figure 2b shows it after an over-heating of one hour, consuming 50% of the cut-off energy.

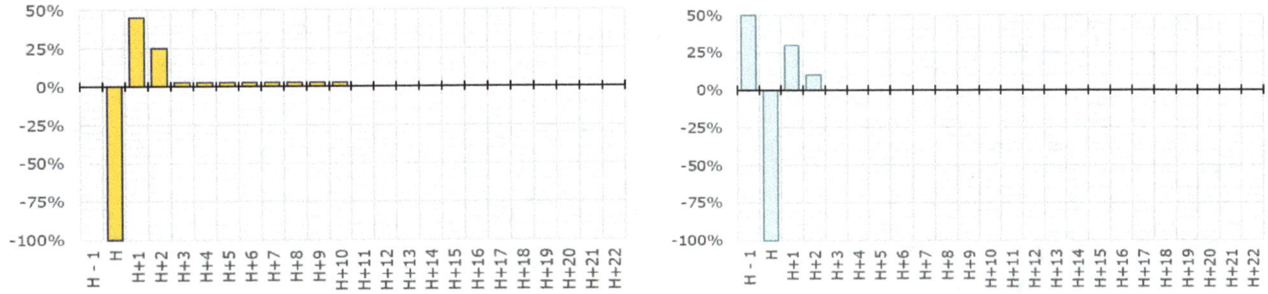

(a) **(b)**

Figure 2. Load shifting rate profiles. (**a**) One-hour residential heat load shedding without pre-heating (**b**) One-hour residential heat load shedding after one-hour pre-heating.

2.2.2. Thermal Models

In order to quantify the impact on thermal comfort, it is also necessary to estimate T_{op} (cf. Equation (6)). For this reason, two thermal models with identified parameters were used in order to simulate the building dynamic during and after the load shedding.

In order to manage both modelings (thermal model of the building and electric model of the grid), an open-source tool, TEASER [35] was chosen for the first thermal model creation. It allows to automatically generate RC thermal models in the Modelica language for the AixLib [36] and the Annex60 [37] libraries.

The thermal model can be created thanks to building envelope information (wall areas, orientations, thickness, materials, etc.). As it can be difficult to access the specific data about each building envelope at district level, it can be realized with at least 5 input data: year of construction, net area, type of use, number of floors and height of each floor [35], making the tool very useful for saving time. If only this data is given, the tool enriches the dataset based on pre-defined statistical data, whose use in a French context will be analyzed in this study case.

For the same thermal model structure (see Figure 3), two precision levels can be achieved depending on the input data. In order to compare the impact of dataset enrichment, both the model generated with the 5 minimal parameters and the model enriched with the building envelope data were analyzed. The first one will further be referred to as the 'Simple' model, while the second one enriched with data used for the regulatory Building Energy Simulation (BES) will further be referred to as the 'Enriched' model.

The second one is the fully detailed thermal model used for the mandatory study. Indeed, in a French context, since each building construction requires an energy requirement study based on a fully detailed thermal model, existing thermal models from this mandatory study can also be re-used. In this study, a Pleiades tool [38] has been used to build a detailed model (each room is considered as a thermal zone), that will be called 'Complex' model afterward.

For all heat load profiles from simulation models, the result from a thermal dynamic simulation of a building in our district was considered as reference heat load profile (P_t^{ref}). The building behavior in the case of a temperature set-point of 20 °C was simulated during the month of January. All other data: weather, occupancy schedules, internal gains for lightning etc. have been set to the same values in order to obtain a better comparison between simulation results. However, the set-point temperatures for the 'Simple' and 'Enriched' models are ambient temperatures, while the one in 'Complex' model is on the operative temperature, and cannot be changed. Therefore, small differences between the results could still be expected.

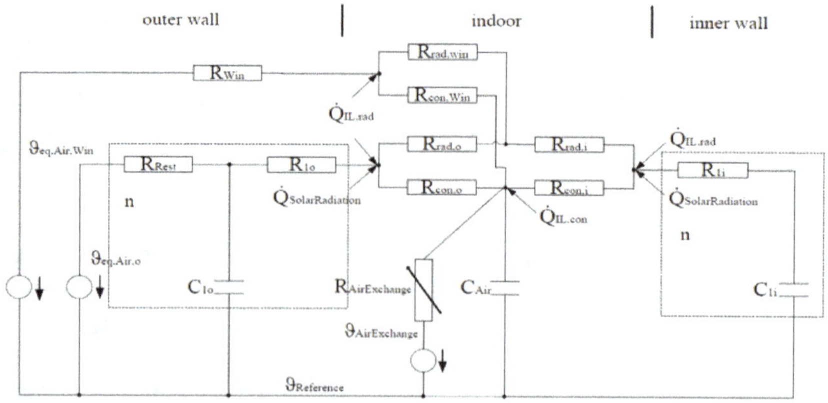

Figure 3. Scheme of the RC (Resistance Capacity) equivalent model generated by Tool for Energy Analysis and Simulation for Efficient Retrofit (TEASER).

2.2.3. Summary of Modeling Approaches

To summarize, four modeling approaches are considered in order to estimate the load shifting rates is this paper:

- The 'Standard' model: the statistical model from experimental data
- The 'Simple' model: thermal model generated by TEASER (Figure 3) with database
- The 'Enriched' model: thermal model generated by TEASER (Figure 3) with building envelope data
- The 'Complex' model: multi-zone thermal model created with a Pleiades tool.

2.3. Modeling of Load Shedding Scenarios

For the aforementioned models, two scenarios will be studied:

- One-hour thermal load shedding after one-hour over-heating: In order to over-heat the building outside of the peak period (5 a.m. to 10 a.m.), the load shedding order will be applied from 5 a.m. to 6 a.m.
- Simple one-hour thermal load shedding: In this scenario, the load shedding will be applied in the middle of the peak period, from 7 a.m. to 8 a.m. without any over-heating.

Each load shedding will be obtained by a very low set-point temperature (around 0 °C) in order to cut the heat supply. The aim of these two scenarios is to represent the two load shedding approaches that could be made building by building in the district. For instance, by applying the first strategy (with over-heating) to a first building and then applying the second strategy to each of the four other buildings hour by hour, the resulting load shedding on the entire district would be obtained by adding the individual effects (see Figure 4).

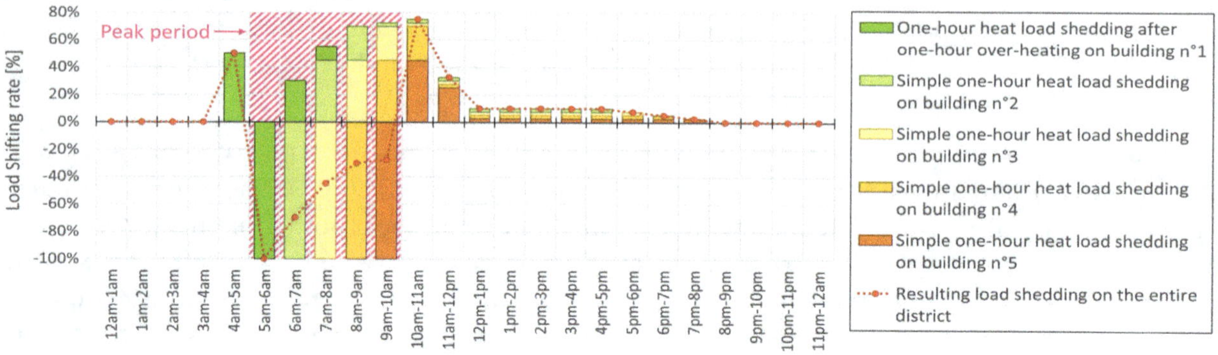

Figure 4. Load shifting rate profiles during a day for the multiple one-hour thermal load shedding from 5 a.m. to 10 a.m. Example of the 'Standard' model for load shifting rate profiles for 5 buildings.

3. Results and Discussion

The load shedding impact in terms of peak shaving, thermal comfort and CO_2 emissions will be analyzed in this section. As explained previously, the district load-shedding strategy consists in shedding thermal load for the buildings one by one for one hour. The first load shedding would integrate a one-hour over-heating during the previous hour, while all the following hours will face simple one-hour thermal load shedding. Two different building behaviors are thus expected for these two strategies.

3.1. *Results for Peak Shaving*

In order to reduce the mean power during the peak period, the load shifting rate would have to be the most diffused as possible to shift most of the consumption out of the peak period. Thus, the lower load shifting rates are, the higher the efficiency of peak shaving. As two dynamic responses are expected from the building depending on whether it has been previously over-heated or not, results will be demonstrated for both load shedding strategies. At first, results for a simple thermal load shedding happening from 7 a.m. to 8 a.m. each day of the month of January will be drawn. Then, the second load shedding strategy with a one-hour over-heating before cutting the heat supply will be shown. This load shedding will be applied from 5 a.m. to 6 a.m., in order to shift the anticipated consumption before the peak period.

3.1.1. Thermal Load Shedding from 7 a.m. to 8 a.m.

The results of the simple one-hour load shedding applied all days of January are displayed in Figure 5. The mean load shifting can be found for all of the four modeling approaches ('Standard', 'Simple', 'Enriched' and 'Complex'). The mean operative temperatures are also drawn with their variation range during the month.

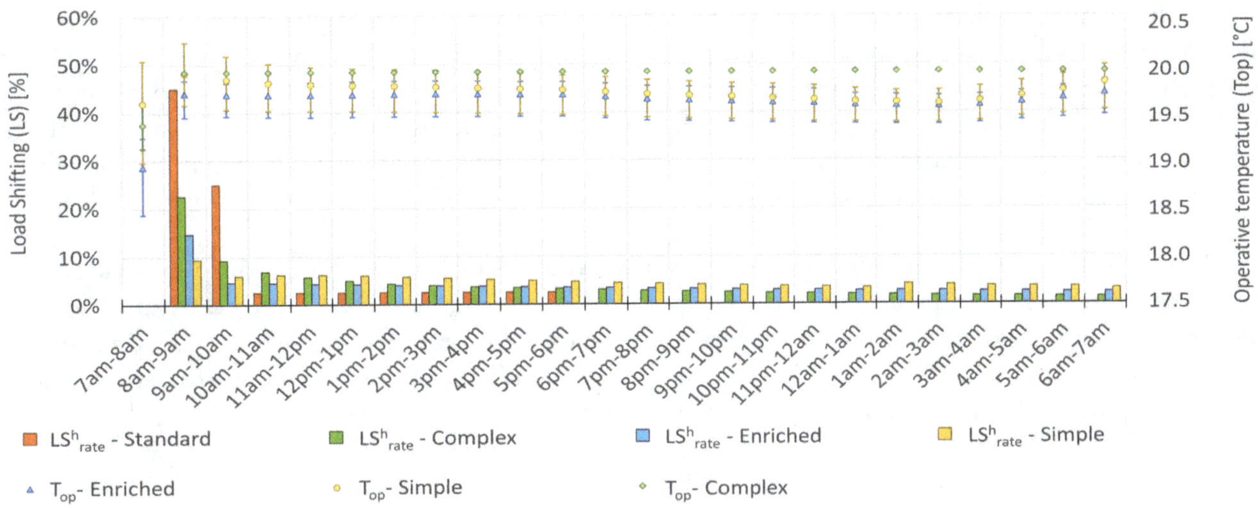

Figure 5. One-hour thermal load shedding.

Two different building behaviors can be observed for this strategy:

- Most of the rebound effect appears within the two hours following the load shedding (experimental load shifting profile)
- The shed consumption is shifted during the entire day (load shifting profiles from thermal models)

Here, it can be noticed that all the simulation results from thermal models show a slower dynamic than the experimental load shifting profile.

3.1.2. Heat Load Shedding from 5 a.m. to 6 a.m. after an Over-Heating in Preceding Hour

The results of the complex one-hour thermal load shedding strategy, taking place from 5 a.m. to 6 a.m. after a one-hour over-heating are presented in Figure 6. The strategy was applied each day of January and results are the average value for each time slot and each modeling, like those presented in Figure 5. As seen before, there is a big difference between experimental and simulated buildings behaviors for simple heat load shedding strategy. This gap between models is therefore confirmed for this complex strategy too.

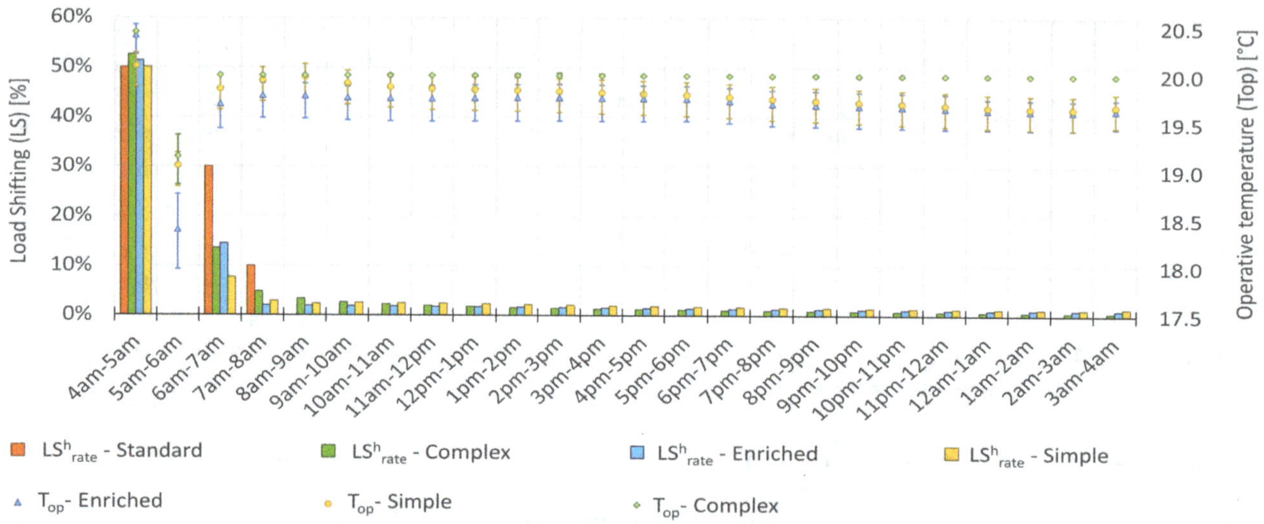

Figure 6. One-hour thermal load shedding after one-hour over-heating.

Although this standard profile was established from real measurement, it remains difficult to consider it as more reliable than simulated profiles. Indeed, weather and occupancy data differ and several types of buildings are aggregated. Even with fewer old buildings than new ones in the experiment, consumption values for heating tend to be higher for the first category and mean values could be strongly affected. In spite of this conclusion, the standard model reminds that occupancy behavior could play a great role in the rebound effect. In our study case, the thermal load is entirely controlled by the building manager, so this effect can be neglected.

Thus, conclusions on the effect of the overall strategy (starting with the complex strategy and continuing, building by building, with the simple load shedding) could be realized by analyzing the building behaviors predicted by thermal models. These three modeling approaches all lead to diffused energy reports so that it can be concluded that this load shedding strategy could be efficient in regards to a peak shaving objective, even if thermal comfort still needs to be looked at.

3.2. Results for Thermal Comfort

As explained above, the study focuses on the building operative temperature as thermal comfort indicator. With a set-point temperature fixed to 20 °C, the comfort zone is reached over 19 °C and below 21 °C. Results obtained for the month of January are presented in Figures 5 and 6. Mean operative temperatures are represented, with their variation intervals, so that it can be noticed that:

- On average, load shedding hours are slightly uncomfortable
- On average, other hours of the days are comfortable

Distribution of comfort level for occupants for the load shedding strategy after an over-heating is presented in Table 1:

Table 1. Thermal comfort levels distribution for the load shedding after over-heating strategy.

	Load Shedding after over-Heating		
Comfort level/Models	Reduced	Enriched	Complex
Comfortable	98.5%	96%	100%
Slightly uncomfortable	1.5%	3.6%	0%
Uncomfortable	0%	0.4%	0%

Similar results are obtained for the load shedding strategy without over-heating. Most of the hours are "comfortable" (at least 96% of the time for all models) and "uncomfortable" level is rarely reached (0.4% of the time and only with the 'Enriched' model). However, further studies on the impact of heating systems (as shown in [16]) or on the perception of this comfort could be needed to consolidate these results (as studied in [39]). As internal gains and external temperatures differ each hour, each time slot should be considered individually for further studies in the entire district impact. Finally, requiring less energy for heating purposes, new low-consumption buildings could get a large part of their heating needs from internal gains (lightning, devices, occupation, etc.). This would also have to be considered carefully in order to avoid an under-evaluation of thermal discomfort in buildings.

3.3. Results for CO_2 Emissions Reduction

In this last section, the results for the impact on the carbon footprint are presented. The chosen indicator for the effect is the expected gain reduction (EG_{red}), representing the difference between CO_2 emission reduction expected by looking at the energy consumption savings (ES^m) and the effective CO_2 emission diminution (CO_2S^m). In order to put into perspective the feeling of savings obtained by looking at the mean ES^d_{rate}, the following Table 2 will also integrate it in comparison to the effective consumption reduction during the month (ES^m).

Table 2. Consumption and CO_2 emission reduction in January for the simple load shedding strategy (a) and for the load shedding after over-heating strategy (b).

	(a) Load Shedding		
Models	Simple	Enriched	Complex
ES^d_{rate}	−10.0%	13.1%	5.5%
ES^m	0.40%	0.50%	0.22%
CO_2S^m	0.38%	0.50%	0.16%
EG_{red}	3.2%	0.70%	28%
	(b) Load Shedding after over-Heating		
Models	Simple	Enriched	Complex
ES^d_{rate}	3.1%	2.1%	2.4%
ES^m	0.21%	0.22%	0.14%
CO_2S^m	0.01%	0.05%	−0.06%
EG_{red}	94%	76%	144%

For the load shedding between 7 a.m. and 8 a.m. (simple load shedding), the intuition is confirmed since the overall consumption reduction leads to the reduction of the CO_2 emission, though a little less than expected. However, for the second strategy with overheating, reduction of CO_2 emission is not so obvious anymore. Depending on the models, the expected CO_2 emission reduction is lower than the calculated one. It goes from 76% less than expected to an increase of 44% of CO_2 emissions. In terms of avoided CO_2 emissions, the savings go from 245 g (CO_2S^m = 0.01%) to 7.82 kg (CO_2S^m = 0.38%) The reason for these differences is that the load is transferred from a low-CO_2 emissions time slot to higher CO_2 emissions times of the day. In order to be consistent with energy transition strategies, it is important to consider this aspect into load shedding impact evaluation to avoid a local amelioration to the detriment of general interest.

4. Conclusions

With often few data available at the district level, thermal models with parameters from the statistical database could provide coherent load shedding impacts results, in respect to those available with detailed thermal simulation models. However, a comparison between simulation results and measurements would be necessary to validate the accuracy of the results, although it was unfortunately impossible due to lack of data on the considered buildings and external factor such as occupation schedules and weather data. For our study case, thermal models are considered as reliable for trend estimation on the effects on peak shaving, thermal comfort and CO_2 emissions reduction for a first approach at a district scale. The overall strategy for the district studied in this paper relies on two load shedding approaches:

- An overheating from 4 a.m. to 5 a.m. before load shedding from 5 a.m. to 6 a.m.
- One-hour load shedding building by building beginning from 6 a.m. to 9 a.m.

The two approaches were analyzed separately for these three aspects:

- Peak shaving
  Turning off the heating supply for one hour successively for building by building in an entire district seems to be effective for peak-shaving. Indeed, the transferred load is very diffused ($LS^h_{rate} < 25\%$ the first hour and $LS^h_{rate} < 10\%$ the following hours) so that the rebound effects of the previous buildings do not cancel the peak reduction obtained by the current load shedding. These results are crucial in the case of a long peak (more than an hour), offering the possibility to shift the load outside consumption peak period.
- Thermal comfort
  Thermal comfort is reduced during the load-shedding hours. Measurements would have to be realized in order to determine if operative temperature evaluation is more reliable when based on the 'Simple' model, the 'Enriched' model or the 'Complex' model. Indeed, the 'Complex' model assessed only 0.8% of the time as not comfortable, while this discomfort could cover up to 4% of the time with the 'Enriched' model. Moreover, the 'Enriched' model gives a minimal operative temperature of 18 °C, while the operative temperatures estimated by the 'Simple' and the 'Complex' models never reach values below 18.8 °C. The different modeling approaches used do not allow to estimate precisely how much thermal comfort can be reduced and how it will be perceived by occupants but they help the stakeholder understand what could be the issue. In all cases, one solution to investigate the reduction of thermal discomfort could be to reduce heat loads instead of shedding them, or to turn off the thermal load during shorter duration.
- CO_2 emission reduction
  In the case of CO_2 emission reduction, estimation cannot be based only on consumption reduction as CO_2 emission for electrical systems have dynamic variations that have to be taken into account. Only by considering dynamic CO_2 variations and by calculating the difference between emissions with or without load shedding strategy could lead to a reliable estimation of CO_2 emissions variations. Indeed, even with effective consumption diminution, a load shedding strategy could shift consumption from low-CO_2 periods to higher-CO_2 time slots, increasing the overall CO_2 emissions. For instance, in the case of the load shedding after over-heating, the 'Complex' model assessed 0.14% of energy saving during the month, while the CO_2 emissions increased from 0.06%. Therefore, the link between energy saving and CO_2 emission reduction has to be realized carefully.

Finally, the modeling approach will depend on the accuracy required, the data available and the time for the study design, so that mixing modeling approaches for a study at the district scale may be required. A further work will consist in coupling the reduced thermal models together with generation parameters tools into an optimization library. This optimization point of view could allow stakeholders such as DSOs to define the best load shedding sequences in a district in order to maximize peak-shaving while minimizing both the occupants' thermal discomfort and CO_2 emission.

Author Contributions: Resources, D.F.; Supervision, B.D. and Y.M.; Writing—original draft, C.P.; Writing—review & editing, B.D., Y.M. and D.F.

Abbreviations

The following abbreviations are used in this manuscript:

COP21	21th Conference of the Parties
CSTB	French Scientific and Technical Center for Building
DSM	Demand Side Management
DSO	Distribution System Operator
GEG	Grenoble Gas and Electricity
GSHP	Ground Source Heat Pumps
RTE	French transmission system operator
TEASER	Tool for Energy Analysis and Simulation for Efficient Retrofit
TSO	Transmission System Operator
UNFCCC	United Nations Framework Convention on Climate Change

Nomenclature

CO_{2_t}	[kg]	CO_2 emissions at time t (with load shedding)
$CO_{2_t}^{ref}$	[kg]	Reference CO_2 emissions at time t (without load shedding)
CO_2S^m	[%]	CO_2 Saving in a month (Reduction of CO_2 emission on the month)
$E_{anticipated}$	[kWh]	Anticipated energy consumption during the hour before the load shedding
E_{cut_off}	[kWh]	Cut-off energy consumption during the load shedding
$E_{delayed}$	[kWh]	Delayed energy consumption during the 23 h after the load shedding
EG_{red}	[%]	Expected Gains Reduction (CO_2 emissions diminution expected by looking at the energy consumption reduction)
ES_{rate}^d	[%]	Energy Saving rate defined 23 h after the load shedding
ES^m	[%]	Energy Saving in a month (Reduction of energy consumption on the month)
LS_{rate}^d	[%]	Load Shifting rate defined during a day
LS_{rate}^h	[%]	Load Shifting rate defined during an hour
P_t	[kW]	Power consumed at time t (with load shedding)
P_t^{ref}	[kW]	Reference power consumed at time t (without load shedding)
T_{air}	[°C]	Ambient temperature
T_{set}	[°C]	Set-point temperature
T_{op}	[°C]	Operative temperature
T_{walls}	[°C]	Walls temperature
τ_{ls}^b	[h]	Beginning of the load shedding
τ_{ls}^e	[h]	End of the load shedding

References

1. The Paris Agreement | UNFCCC. Available online: https://unfccc.int/process-and-meetings/the-paris-agreement/the-paris-agreement (accessed on 10 August 2018).
2. Lund, P.D.; Lindgren, J.; Mikkola, J.; Salpakari, J. Review of energy system flexibility measures to enable high levels of variable renewable electricity. *Renew. Sustain. Energy Rev.* **2015**, *45*, 785–807. [CrossRef]
3. Ponds, K.; Arefi, A.; Sayigh, A.; Ledwich, G.; Ponds, K.T.; Arefi, A.; Sayigh, A.; Ledwich, G. Aggregator of Demand Response for Renewable Integration and Customer Engagement: Strengths, Weaknesses, Opportunities, and Threats. *Energies* **2018**, *11*, 2391. [CrossRef]

4. City-zen | New Urban Energy. Available online: www.cityzen-smartcity.eu (accessed on 19 June 2018).
5. Congedo, P.M.; Colangelo, G.; Starace, G. CFD simulations of horizontal ground heat exchangers: A comparison among different configurations. *Appl. Therm. Eng.* **2012**, *33–34*, 24–32. [CrossRef]
6. Congedo, P.; Lorusso, C.; De Giorgi, M.; Laforgia, D.; Congedo, P.M.; Lorusso, C.; De Giorgi, M.G.; Laforgia, D. Computational Fluid Dynamic Modeling of Horizontal Air-Ground Heat Exchangers (HAGHE) for HVAC Systems. *Energies* **2014**, *7*, 8465–8482. [CrossRef]
7. Razmara, M.; Bharati, G.R.; Hanover, D.; Shahbakhti, M.; Paudyal, S.; Robinett, R.D. Building-to-grid predictive power flow control for demand response and demand flexibility programs. *Appl. Energy* **2017**, *203*, 128–141. [CrossRef]
8. Uddin, M.; Romlie, M.F.; Abdullah, M.F.; Halim, S.A.; Bakar, A.H.A.; Kwang, T.C. A review on peak load shaving strategies. *Renew. Sustain. Energy Rev.* **2018**, *82*, 3323–3332. [CrossRef]
9. Müller, D.; Monti, A.; Stinner, S.; Schlösser, T.; Schütz, T.; Matthes, P.; Wolisz, H.; Molitor, C.; Harb, H.; Streblow, R. Demand side management for city districts. *Build. Environ.* **2015**, *91*, 283–293. [CrossRef]
10. Salpakari, J.; Mikkola, J.; Lund, P.D. Improved flexibility with large-scale variable renewable power in cities through optimal demand side management and power-to-heat conversion. *Energy Convers. Manag.* **2016**, *126*, 649–661. [CrossRef]
11. Zhang, Y.; He, Y.; Yan, M.; Guo, C.; Ding, Y.; Zhang, Y.; He, Y.; Yan, M.; Guo, C.; Ding, Y. Linearized Stochastic Scheduling of Interconnected Energy Hubs Considering Integrated Demand Response and Wind Uncertainty. *Energies* **2018**, *11*, 2448. [CrossRef]
12. Behrangrad, M. A review of demand side management business models in the electricity market. *Renew. Sustain. Energy Rev.* **2015**, *47*, 270–283. [CrossRef]
13. Baeten, B.; Rogiers, F.; Helsen, L. Reduction of heat pump induced peak electricity use and required generation capacity through thermal energy storage and demand response. *Appl. Energy* **2017**, *195*, 184–195. [CrossRef]
14. Kensby, J.; Trüschel, A.; Dalenbäck, J.O. Potential of residential buildings as thermal energy storage in district heating systems—Results from a pilot test. *Appl. Energy* **2015**, *137*, 773–781. [CrossRef]
15. Romanchenko, D.; Kensby, J.; Odenberger, M.; Johnsson, F. Thermal energy storage in district heating: Centralised storage vs. storage in thermal inertia of buildings. *Energy Convers. Manag.* **2018**, *162*, 26–38. [CrossRef]
16. Reynders, G.; Nuytten, T.; Saelens, D. Potential of structural thermal mass for demand-side management in dwellings. *Build. Environ.* **2013**, *64*, 187–199. [CrossRef]
17. De Coninck, R.; Helsen, L. Bottom-Up Quantification of the Flexibility Potential of Buildings. In Proceedings of the 13th Conference of International Building Performance Simulation Association, Chambéry, France, 26–28 August 2013.
18. Le Dréau, J.; Heiselberg, P. Energy flexibility of residential buildings using short term heat storage in the thermal mass. *Energy* **2016**, *111*, 991–1002. [CrossRef]
19. Reynders, G.; Amaral Lopes, R.; Marszal-Pomianowska, A.; Aelenei, D.; Martins, J.; Saelens, D. Energy flexible buildings: An evaluation of definitions and quantification methodologies applied to thermal storage. *Energy Build.* **2018**, *166*, 372–390. [CrossRef]
20. Reynders, G.; Diriken, J.; Saelens, D. Generic characterization method for energy flexibility: Applied to structural thermal storage in residential buildings. *Appl. Energy* **2017**, *198*, 192–202. [CrossRef]
21. Aduda, K.O.; Labeodan, T.; Zeiler, W.; Boxem, G.; Zhao, Y. Demand side flexibility: Potentials and building performance implications. *Sustain. Cities Soc.* **2016**, *22*, 146–163. [CrossRef]
22. Faria Neto, A.; Bianchi, I.; Wurtz, F.; Delinchant, B. *Thermal Comfort Assessment*; Final ELECON Workshop Federal Institute of Santa Catarina—IFSC: Florianópolis, Brazil, 2016. [CrossRef]
23. Goy, S.; Ashouri, A.; Maréchal, F.; Finn, D. Estimating the Potential for Thermal Load Management in Buildings at a Large Scale: Overcoming Challenges Towards a Replicable Methodology. *Energy Procedia* **2017**, *111*, 740–749. [CrossRef]
24. Hurtado, L.A.; Rhodes, J.D.; Nguyen, P.H.; Kamphuis, I.G.; Webber, M.E. Quantifying demand flexibility based on structural thermal storage and comfort management of non-residential buildings: A comparison between hot and cold climate zones. *Appl. Energy* **2017**, *195*, 1047–1054. [CrossRef]
25. Reynders, G.; Diriken, J.; Saelens, D. Quality of grey-box models and identified parameters as function of the accuracy of input and observation signals. *Energy Build.* **2014**, *82*, 263–274. [CrossRef]

26. Perez, N. Contribution à la Conception éNergéTique De Quartiers: Simulation, Optimisation et Aide à la Décision. Ph.D. Thesis, Université de La Rochelle, Rochelle, France, 2017.
27. Fonseca, J.A.; Nguyen, T.A.; Schlueter, A.; Marechal, F. City Energy Analyst (CEA): Integrated framework for analysis and optimization of building energy systems in neighborhoods and city districts. *Energy Build.* **2016**, *113*, 202–226. [CrossRef]
28. Remmen, P.; Lauster, M.; Mans, M.; Fuchs, M.; Osterhage, T.; Müller, D. TEASER: An open tool for urban energy modelling of building stocks. *J. Build. Perform. Simul.* **2018**, *11*, 84–98. [CrossRef]
29. Frayssinet, L.; Kuznik, F.; Hubert, J.L.; Milliez, M.; Roux, J.J. Adaptation of building envelope models for energy simulation at district scale. *Energy Procedia* **2017**, *122*, 307–312. [CrossRef]
30. Morales, D.X.; Besanger, Y.; Sami, S.; Alvarez Bel, C. Assessment of the impact of intelligent DSM methods in the Galapagos Islands toward a Smart Grid. *Electr. Power Syst. Res.* **2017**, *146*, 308–320. [CrossRef]
31. RTE. Évaluation des Economies D'éNergie Et Des Effets De Bord AssociéS aux Effacements de Consommation. 2016. Available online: https://clients.rte-france.com/htm/fr/mediatheque/telecharge/20160401_Rapport_report_complet.pdf (accessed on 19 October 2018).
32. De Dear, R.; Brager, G.S. *Developing an Adaptive Model of Thermal Comfort and Preference*; UC Berkeley: Berkeley, CA, USA, 1998.
33. Eco2mix CO2. 2014. Available online: https://www.rte-france.com/fr/eco2mix/eco2mix-co2 (accessed on 20 May 2018).
34. Annuaire des Projets en France / Greenlys. Available online: http://www.smartgrids-cre.fr/index.php?p=greenlys (accessed on 20 June 2018).
35. Fuchs, M.; Teichmann, J.; Lauster, M.; Remmen, P.; Streblow, R.; Müller, D. Workflow automation for combined modeling of buildings and district energy systems. *Energy* **2016**, *117*, 478–484. [CrossRef]
36. AixLib: A Modelica Model Library for Building Performance Simulations. 2018. Available online: https://github.com/RWTH-EBC/AixLib (accessed on 10 July 2014).
37. Modelica-Ibpsa: Modelica Library for Building and District Energy Systems Developed within IBPSA Project 1. 2018. Available online: https://github.com/ibpsa/modelica-ibpsa (accessed on 21 September 2013).
38. Logiciel Pleiades—Izuba énergies. Available online: http://www.izuba.fr/logiciels/outils-logiciels/ (accessed on 20 September 2018).
39. Amasuomo, T.; Amasuomo, J.; Amasuomo, T.T.; Amasuomo, J.O. Perceived Thermal Discomfort and Stress Behaviours Affecting Students' Learning in Lecture Theatres in the Humid Tropics. *Buildings* **2016**, *6*, 18. [CrossRef]

5

Impact of Service Life on the Environmental Performance of Buildings

Shahana Y. Janjua [1,*], Prabir K. Sarker [1] and Wahidul K. Biswas [2]

1 School of Civil and Mechanical Engineering, Curtin University, Perth 6102, Australia; P.Sarker@curtin.edu.au
2 Sustainable Engineering Group, Curtin University, Perth 6102, Australia; W.Biswas@curtin.edu.au
* Correspondence: s.janjua@postgrad.curtin.edu.au

Abstract: The environmental performance assessment of the building and construction sector has been in discussion due to the increasing demand of facilities and its impact on the environment. The life cycle studies carried out over the last decade have mostly used an approximate life span of a building without considering the building component replacement requirements and their service life. This limitation results in unreliable outcomes and a huge volume of materials going to landfill. This study was performed to develop a relationship between the service life of a building and building components, and their impact on environmental performance. Twelve building combinations were modelled by considering two types of roof frames, two types of wall and three types of footings. A reference building of a 50-year service life was used in comparisons. Firstly, the service life of the building and building components and the replacement intervals of building components during active service life were estimated. The environmental life cycle assessment (ELCA) was carried out for all the buildings and results are presented on a yearly basis in order to study the impact of service life. The region-specific impact categories of cumulative energy demand, greenhouse gas emissions, water consumption and land use are used to assess the environmental performance of buildings. The analysis shows that the environmental performance of buildings is affected by the service life of a building and the replacement intervals of building components.

Keywords: building; environmental life cycle assessment; service life; environmental performance

---

1. Introduction

A building is a complex product of different components of variable materials, structural importance, functional life, exposure constraints, and damage mechanisms. Each component of a building has a typical functional requirement and it should perform as per the prescribed function in its service life. Life cycle assessment (LCA) studies that have been conducted, to date, consider the service life of building and building components between 30 and 70 years with a most commonly used value of 50 years (Table 1). However, the real picture is quite contradictory to these assumptions as the service life of buildings varies with materials, operation and maintenance and the surrounding environment [1,2]. This discrepancy may lead to inaccuracy of LCA analyses, and material and energy balance. Any building needs regular maintenance and replacement of its non-structural components to keep the building in performing conditions. In second half of the 20th century, a considerable number of buildings were constructed that need annual inspections and maintenance, influencing the national economy and competitive position of the construction industry [3]. The maintenance and replacement intervals of existing buildings need to be optimized to achieve environmental, social and economic benefits. For new constructions, the estimated intervals of maintenance and replacements should be planned as concisely and wisely as possible. The integration of knowledge of building component durability and its structural and functional performance into building LCA could help conduct a realistic assessment of the environmental performance of building components [4]. Due to the uncertainty associated with the use of assumed service life of a building, as well as the unavailability of service life data of building components, LCA studies have not frequently addressed the real energy consumed during maintenance and replacement activities. However, this energy (hereafter, named replacement energy) may be as much as 7% to 110% of the initial embodied energy, if the service life of building materials is not properly implemented in the design phase of a building [4–7]. The building life span, whether short or long, has discretionary effects on a building's environmental performance. Short service life of buildings results in excessive solid waste, embodied energy and subsequent greenhouse gas (GHG) emissions during pre-use stage (extraction of material to construction). Long service life of buildings increases replacement of building components, resulting in an increase of replacement energy and prolonged use stage, increasing operational energy and GHG emissions [1]. These two constraints need to be taken into account during material selection by considering the service life of the whole building, as well as its components, and is essential to achieve environmental performance while fulfilling social and economic objectives.

Table 1. Existing case studies.

Author	Life Span (Years)	Impact Indicators
Ramesh et al. [8]	75	Life cycle energy demand
Allacker K. [9]	60	External costs
Audenaert A. [10]	50	Waste generation
Carre A. [11]	60	Global warming potential (GWP), Cumulative energy demand (CED), water use, solid waste, photochemical oxidation, eutrophication, land use, and resource depletion
Iyer-Raniga U. [12]	100	Carbon emission, energy, photochemical oxidation, eutrophication, land use and water use
Rouwette R. [13]	50	GHG, CED
Cuellar-Franca R.M. [14]	50	GWP, acidification, eutrophication, abiotic depletion, ozone depletion, photochemical ozone creation, human toxicity
Nemry F. [15]	40	GWP, primary energy, acidification, eutrophication, ozone depletion, photochemical, ozone creation
Ortiz O. [16]	50	GWP, acidification, human toxicity, abiotic depletion, ozone depletion
Crawford et al. [6]	50	Embodied energy, cost, operational energy
Cabeza et al. [17]	30 to 100 mostly 50	Life cycle energy
Biswas W.K. [18]	50	GHG emissions, Embodied energy (EE)
Islam H. [19]	50	Life cycle energy, life cycle cost (LCC)
Atmaca A. [20]	30 to 100 mostly 50	GHG emissions
Lawania K.K. [21]	50	GHG emissions, life cycle energy
Grant A. [1]	Estimated	GWP, atmospheric ecotoxicity, atmospheric acidification
Dixit M.K. [22]	50	Embodied energy
Vitale P. [23]	50	GWP, respiratory inorganics potential, non-renewable energy potential, waste framework directive
Vitale P. [24]	50	Respiratory inorganics, GWP, non-renewable energy
Balasbaneh A.T. [25]	50	GWP, human toxicity, acidification, eutrophication, LCC, labor wage rate, job creation

According to ISO 15686-1, "Service life is the period of time after construction, in which a building and its parts meet or exceed the acceptable minimum requirements of performance established" [26]. The service life of building components largely depends on the materials' properties, damage mechanisms, environment and quality of design, and work execution. This study aims to estimate the service life of buildings and building components and expected replacement intervals of non-structural components, and to assess the impact of this service life on life cycle environmental performance of buildings.

1.1. Service Life Estimation

Service life (SL) estimation of buildings is quite a complicated process that involves intensive data analysis as there is no proto-type in buildings. Each building is unique in its composition, material specification and architectural and structural design. Therefore, the SL estimation cannot be generalized and needs to be carried out on a component to component basis. Construction materials have different properties and damage mechanisms and behave differently in different climates. User requirements, degradation agents, and building performance against these agents are important factors to consider for service life planning [27]. The state-of-the-art report on performance-based methods on service life prediction states that "Prediction of durability is subject to many variables and cannot be an exact science" [28]. Therefore, efforts should be made to achieve the most likely estimate by considering the most reliable data sources.

SL prediction methods should be generalized, easy to apply to a variety of materials, user friendly and give clear boundary limitations [29]. SL was first studied by Legget and Hutcheon in 1958. However, SL estimation has been under the limelight since the 1990s by different standard institutes. The Guidelines for Service Life Planning were first published by the Architectural Institute of Japan (AIJ) in 1989 followed by British Standard Institute (BSI) in 1992 and Canadian Standards Association (CSA) in 1995. International standard organizations (ISO) published ISO 15686-1, Building and constructed assets—Service Life Planning—Part 1 in 2000 [30]. A series of publications on ISO 15686 were published afterwards, covering different aspects and procedures of service life predictions.

Service life can be estimated by deterministic, engineering and probabilistic methods. The probabilistic method is the research approach considering degradation probability of a building during a prescribed time. The deterministic method is a simple approach utilizing factors influencing the degradation of a building under certain conditions. The factor method, described in standard, ISO 15686-2 [31], is the well-known deterministic approach. Engineering methods lie somewhere in between deterministic and probabilistic methods. Engineering methods are easy, and use the time-based degradation mechanism for interpretation [32]. SL estimation needs a wide range of data from different sources and under different conditions. These information resources may be existing building data, information collected by surveys, manufacturer data, service life modelling, insurance companies and real estate data, and expert opinion [33]. The engineering approach depends on structural properties of materials, loading conditions, chemical composition, and damage mechanisms in a buildings' life time. However, there is a huge variety of chemical compositions in materials, degradation in different environments, and variable human influences, to treat all materials just the same. Accelerated life tests carried out on building components to predict SL give reasonably accurate results. It is still a big challenge to depict the realistic conditions for life tests. In addition, the accelerated tests are quite expensive. There are also some other approaches to predict SL by considering service life models and obsolescence factors [1,34]. This method can be used for existing buildings or to be built buildings with the same material. This method requires empirical data that cannot be collected for innovative materials. Acquiring data for service life models and time constraint can pose a challenge for the SL prediction approach.

The factor method is the deterministic method that uses seven factors to predict the service life behavior of the building in different climatic conditions and geographic locations. The factor method uses reference service life (RSL) of a building component as a baseline and seven factors to modify the RSL to estimated service life (ESL). The service life estimation is different from service life prediction in the sense that the first is meant for particular conditions, and the second is recorded performance over a prescribed time or referenced SL [30,35]. The factor method helps to estimate service life of building and building components using Equation (1) [30].

$$\mathrm{ESL} = \mathrm{RSL} \times \mathrm{A} \times \mathrm{B} \times \mathrm{C} \times \mathrm{D} \times \mathrm{E} \times \mathrm{F} \times \mathrm{G}, \quad (1)$$

where,

ESL = Estimated service life of building components

RSL = Reference service life of building components

Factor A = Quality of components including manufacturing, storage, transport and protective coating etc.

Factor B = Design level including incorporation, sheltering by rest of structure and surrounding buildings

Factor C = Work execution level, site management, workmanship level, weather condition during work

Factor D = Indoor environment conditions, humidity, ventilation, and condensation etc.
Factor E = Outdoor environment, microenvironmental conditions, weathering factors, building elevation etc.
Factor F = In-use conditions, mechanical impact, wear and tear, category user etc.
Factor G = Maintenance level, quality and frequency.

The method incorporates the material behavior, human involvement and degradation mechanism to interpret the ESL. The factor method is flexible, and it considers the combined effect of different deteriorating factors. The method needs judgement of factors as protective or deteriorating and requires fair and definite limitations on factors to avoid complexity [36]. Reliable data is required for the RSL and factors for each building component. The availability of data and reliability of data sources play an important role in SL estimation. The data sources may be manufacturers of building products, test laboratories, government agencies reports, existing studies etc., [37]. The most challenging issue in SL estimation is how to use effectively the available data to predict the SL of a structure that is to be built. In this study, the service life was estimated for most likely values (± 5 years) using the factor method.

1.2. Environmental Life Cycle Assessment

The environmental life cycle assessment (ELCA), frequently known as life cycle assessment is a comprehensive tool to assess the environmental impacts of a product or system or service, in pre-use, use, and post-use stages [38]. The ELCA was studied for the first time in the 1960s and up until the 1970s, it was used only to compare the packaging options of consumer goods. In 1969, the Midwest Research Institute conducted a study on LCA for a Coca Cola Company for different types of beverage containers [39]. The studies in this period revolved around policy making and enterprises with a focus on solid wastes, energy consumption, and air pollutant impacts. In the 1990s, SETAC, conducted various workshops and published the first code of practice for life cycle assessment in 1993 [40]. Afterwards, the international standards organization (ISO), was involved actively and published generalized procedures and methods for LCA in ISO 14040-44 in 1997–2000 [38].

In the construction sector, ELCA was first applied in 1980s by Bekker to study the environmental implications of the use of renewable resources in buildings [41]. ELCA was used in buildings to assess the environmental impacts of construction materials and is a credible solution to compare material sustainability [42–45]. Now, the ELCA covers a wide range of areas from building materials (i.e., bricks, cement etc.) to urban planning [46]. The life cycle stages that are usually considered from life cycle assessment of buildings and building components include pre-construction, construction, use and end of life stages. Environmental product declarations (EPDs) involved the use of LCA to estimate environmental impacts for environmental declaration purposes for certification purposes [47]. ELCA helps to improve the performance of building in its entire life span by first identifying hotspots and then by applying mitigation strategies [47,48]. However, the system boundaries, functional units and scope definition are unique for each building LCA study, resulting in variation in results among studies [49–51].

Environmental performance of buildings is also defined as a quantified relationship between occupant's comfort level and environmental impacts [52–54]. Embodied and operational impacts are usually two main categories of environmental impacts. Embodied impacts are static and further divided into pre-use embodied impacts and replacement embodied impacts [6]. Pre-use embodied impacts are the impacts due to extraction, manufacturing and construction of buildings and replacement embodied impacts are a result of renovations, replacements and maintenance in the active service life of buildings. The operational or use stage impacts are dynamic in nature and occur in the service life of building [55,56]. Better building performance can be achieved by considering factors including material selection, construction techniques, cost factors, and cleaner production strategies (CPS).

Whilst Australia accounts for only 0.32% of the world's population, its per capita GHG emission is extremely high compared to countries with similar economies (UK, Mexico, South Korea) i.e., 26 tonnes GHG emissions per capita per year as opposed to 13 tonnes per capita GHG emissions for South Korea, 10 tonnes per capita GHG emissions for UK and 20.3 tonnes per capita GHG emissions for Canada [57]. Australia is the second driest continent after Antarctica [58]. The annual rainfall is highly variable and central Australia is mostly arid with only 6% arable land in coastal areas [59]. Water is the most precious commodity and its scarcity is covered by desalination of sea water [60–62]. Water mapping in the construction industry helped to identify need for reducing the life cycle water demand/footprint of buildings by using renewable resources. In addition, Australia's per capita waste generation is 2.6 tonnes per year as compared to 0.706 tonne per capita per year for US, out of which 0.8 tonnes per capita per year is construction and demolition waste [63]. Therefore, these two issues are inevitable for assessment of the environmental impacts of building and construction industry at the planning stage of buildings using an ELCA to discern strategies to avoid these environmental consequences. This study thus considered these impact categories, including cumulative energy demand, GHG emissions, water consumption and land use to assess the environmental performance of buildings.

2. Method

This study focuses on the impact of service life on environmental performance of buildings. The methodology consists of four main steps (Figure 1). Step 1: Twelve residential buildings were selected. All specifications of the buildings including the architectural design, covered area, orientation, and utility were the same except for the difference in building materials. The residential buildings were modelled using three main systems of roof, wall and footing.

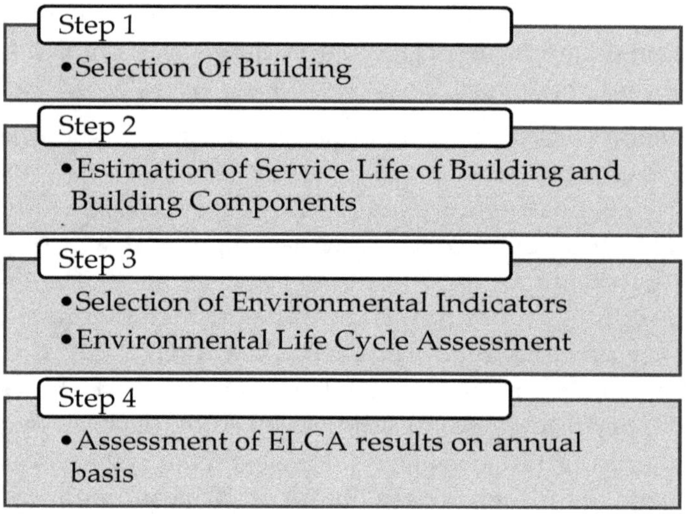

Figure 1. Building environmental performance assessment procedure.

The roof system comprised of roof cladding, roof frame, and suspended ceiling. The wall system comprised of exterior render, wall frame and interior plaster and the footing system comprised of footing slab and flooring. The variation in buildings was created only in materials of structural components. The materials for non-structural components were unchanged as replacements for these components are considered easy and does not affect the service life of the whole building. Two types of roof frames, two wall frames and three types of slab footings, resulted in 12 combinations of buildings (Table 2) and a conventional building named Building-50 composed of a timber roof frame, double brick walls and conventional concrete was considered as a reference case for comparison with the aforementioned 12 buildings.

Table 2. Building components of building combinations and the reference building.

Building	Building Specifications	Building Components		
		Roof Frames	Wall Frames	Slab Footing
Building-50	TF-DB-CC	Timber frame	Double Brick	Conventional concrete
1	TF-CB-CC	Timber frame	Concrete Block	Conventional concrete
2	TF-CB-FAGC	Timber frame	Concrete Block	30% FA Green concrete
3	TF-CB-GGBFS	Timber frame	Concrete Block	30% GGBFS Green concrete
4	TF-DB-CC	Timber frame	Double Brick	Conventional concrete
5	TF-DB-FAGC	Timber frame	Double Brick	30% FA Green concrete
6	TF-DB-GGBFS	Timber frame	Double Brick	30% GGBFS Green concrete
7	SF-CB-CC	Steel Frame	Concrete Block	Conventional concrete
8	SF-CB-FAGC	Steel Frame	Concrete Block	30% FA Green concrete
9	SF-CB-GGBFS	Steel Frame	Concrete Block	30% GGBFS Green concrete
10	SF-DB-CC	Steel Frame	Double Brick	Conventional concrete
11	SF-DB-FAGC	Steel Frame	Double Brick	30% FA Green concrete
12	SF-DB-GGBFS	Steel Frame	Double Brick	30% GGBFS Green concrete

Step 2: The service life of each building component was estimated using the factor method [37]. The service life of a system was taken as the service life of structural components i.e., ESL of the roof system was the value of service life for the roof frame. The least value of service life among building systems i.e., roof system, wall system and footing system, was taken as the estimated service life of the building [3]. The service life estimation of components was required, not only to determine the service life of the whole building, but also, to find the replacement intervals of non-structural components during the service life of a building. The service life of a reference building, Building-50 was assumed 50 years based on a literature review (Table 1).

Step 3: The indicators for environmental objective were selected by consulting existing studies. ELCA of the building was carried out as per ISO 14040-44 [38]. A quantitative life cycle inventory for building materials and transportation was compiled for construction, the subsequent replacements and demolishing activities. The ELCA considered a cradle to grave approach including pre-use (mining to material, transport of material to site and construction), use, post-use (demolition and disposal) and replacement (replacement of building components throughout the active service life of building) stages. ELCA software, SimaPro 8.4 [64], was used to determine the environmental indicators for impact categories of energy, GHG emissions, water consumption and land use.

Step 4: The impact values were presented on an annual basis for a service life of a building as estimated in the second step in order to investigate the impact of SL on environmental performance of buildings.

3. Case Studies

A typical house of four bedrooms and two bathrooms, with a covered area of 245.5 m^2 located in Perth WA, was selected for the case study. Twelve building combinations were created based on the roof, wall and footing systems, keeping architectural design, orientation, location and covered area constant (Table 2).

Table 3. Criteria and Coefficients of service life estimation.

Factor Description		Roof System				Wall System				Footing System			
Factor	Criteria	Timber Truss	Steel Truss	Terracotta Tiles	Gypsum board	Concrete Block	Double Brick	Plaster	Render	CC	30% FAGC	30% GGBFS	Ceramic Tiles
A	A = 1.1, Best Available Material; A = 1.05, Good Material; A = 1.0, N/A-No effects; A = 0.95, Slightly low Standard material; A = 0.90, Low Standard Material	1.10	1.10	1.10	1.10	1.10	1.10	1.05	1.05	1.05	1.05	1.10	1.10
B	B = 1.1, Best Design with special considerations to strengthen the structure; B = 1.05, Good Design, (as per standards approach); B = 1.0, N/A-No effects; B = 0.95, low Standard design; B = 0.90, poor design	1.05	1.05	1.00	1.00	1.10	1.10	1.05	1.05	1.05	1.05	1.05	1.00
C	C = 1.1, Satisfaction Level ≥ 90%; C = 1.05, 80% < Satisfaction Level > 90%; C = 1.0, N/A-No effect; C = 0.95, 70% < Satisfaction Level > 80%; C = 0.9, Satisfaction Level ≤ 70%	0.90	0.90	0.90	1.05	0.90	0.90	0.90	0.95	0.95	0.95	0.95	0.90
D	NOT CONSIDERED	1.00	1.00	1.00	1.00	1.00	1.00	1.00	1.00	1.00	1.00	1.00	1.00
E	E = 1.1, Supportive; E = 1.05, mild; E = 1.0, N/A-No effect; E = 0.95, Harsh; E = 0.9, Reactive	1.05	1.05	0.95	1.00	1.00	1.00	1.00	0.90	0.90	1.05	1.05	1.00
F	NOT CONSIDERED	1.00	1.00	1.00	1.00	1.00	1.00	1.00	1.00	1.00	1.00	1.00	1.00
G	G = 1.1, Best quality and interval as specified by manufacturer/designer; G = 1.05, good quality and interval as per requirement; G = 1.0, N/A-NO EFFECT; G = 0.95, low quality and as per required; G = 0.9, poor quality	1.05	1.05	1.05	1.05	1.00	1.00	1.05	1.05	1.00	1.00	1.00	1.05
RSL	Primary source [80% reliability] = manufacturers technical sheets, EPDs	50	75	50	25	60	75	25	15	60	60	60	50
ESL	Secondary source [60% reliability] = Databases like NAHB, BOMA; literature; experimental studies; codes and practices;	57	86	49	30	65	82	26	15	57	66	69	52

Each of the roof, wall and footing systems was modelled using structural and non-structural building components. Only the non-structural components were selected, that resulted in costly replacements and provided a thermal envelope to the building. However, this aspect of the building enveloping components will be assessed in future study. The roof system included two types of roof assemblies: TF—timber roof frame, terracotta tiles and gypsum board ceiling; and SF—steel roof frame, terracotta tiles and gypsum board ceiling. The wall system consisted of two types of wall assemblies: DB—double brick wall and interior plaster; and CB—concrete block wall, exterior render and interior plaster.

The footing system comprised of on-grade slab footing and ceramic tile flooring with three types of concrete mixes: CC—conventional concrete; 30% FA—Green concrete with 30% replacement of Ordinary Portland Cement (OPC) by class F fly ash; 30% GGBFS—Green concrete with 30% replacement of OPC by ground granulated blast furnace slag (GGBFS). The building systems were developed based on the most commonly used materials, in Western Australia with a design life of 50 years as proposed by National Building Codes. A conventional residential building with a timber roof, double brick walls and conventional concrete slab footing and 50-year service life was used as the reference building, Building-50.

A thorough study was conducted to collect the service life data of the building components used in the case study. Based on the gathered information, ranking criteria were set for each factor, to get most likely values (ESL $\pm$ 5 years) of ESL (Table 3). Factor A, B, C, E, G were assigned ranking values from 1.1 to 0.9 [37], and Factor D, F were not considered in the study as these are dependent on occupant behavior and vary greatly. These factors were assigned a value of 1.0 in service life estimation equation. The factor values were reduced to increase the confidence level as compared to previously used values to test the sustainability framework for Building 1 and 2 [65].

The manufacturer data, life expectancy databases of building components and existing case studies were used as data sources for RSL. The manufacturer's technical data sheets were consulted to set the component quality. The factor B values were assigned by considering commonly used practices in building design in Western Australia. Building commission WA annual reports were consulted to estimate the construction works execution level. The climatic conditions, reports of Bureau of Meteorology Western Australia, and inspection reports of residential buildings were considered for weighting outdoor climatic conditions and subsequent effect on the building components.

The life cycle assessment software SimaPro 8.4 was used to assess the environmental impacts of buildings with a grave to cradle approach. Materials required for each building were estimated for building construction and successive replacements. The transportation distances were calculated for nearest available materials retailers and manufacturers. The energy consumption during the use stage was estimated for thermal comfort, hot water, lighting and home appliances. AccuRate sustainability software [66] was used to estimate the annual cooling, heating, and hot water demand. The life cycle inventories for materials, energy consumption and transportation distances were compiled to incorporate in the SimaPro (Tables A1–A5). Table 4 shows the environmental impact categories, impact indicators and methods used to assess the environmental impacts. These impact indicators were selected based on literature review and relevance to the scope of the study.

Table 4. Environmental Impact indicators.

Impact Category	Impact Indicators	Impact Assessment Methods
Energy	Cumulative energy demand	Cumulative Energy Demand V1.09
GHG emissions	Life cycle GHG emissions	IPCC 2013 GWP 100a V1.02
Land use	Land use	Ecological footprints Australian V1.00
Water consumption	Resource depletion	Pfister et al. 2009 (Eco-indicator 99) V1.02

4. Results and Discussion

4.1. Estimated Service Life

The ESL of buildings and building components are presented in Table 5. The service life estimation shows that due to the large variation in service life of building components, enough life of building components is compromised. Approximately, 20% to 35% of ESL of structural components of 12 buildings, studied in this paper, is wasted. In buildings 4–7, the wall system has 82 years ESL that is 30.5% more than the ESL of the building. In buildings 10–12, the wall and roof systems both have higher ESL values than the footing system. In buildings 7–9, the roof system has a high ESL value compared to the wall and footing systems.

Table 5. Estimated service life of building systems and buildings.

Service Life (Years)	TF-CB-CC	TF-CB-FAGC	TF-CB-GGBFS	TF-DB-CC	TF-DB-FAGC	TF-DB-GGBFS	SF-CB-CC	SF-CB-FAGC	SF-CB -GGBFS	SF-DB-CC	SF-DB-FAGC	SF-DB-GGBFS
	1.	2.	3.	4.	5.	6.	7.	8.	9.	10.	11.	12.
Roof System	57	57	57	57	57	57	86	86	86	86	86	86
Wall System	65	65	65	82	82	82	65	65	65	82	82	82
Footing System	57	66	69	57	66	69	57	66	69	57	66	69
Building	57	57	57	57	57	57	57	65	65	57	66	69

In building combinations 1–3, all the three systems have comparable ESLs that makes these the combinations with less material wastage at the post-use stage.

Based on the ESL of building components, the replacements of each building component in the ESL of buildings are specified. These ESLs are calculated conservatively considering that the components maintenance is carried out, strictly on schedule as described by the manufacturer or designer. No replacement is considered in the study for the building components within a range of ±5 years of the ESL of buildings [37]. This difference is assumed to be covered by maintenance. The ceramic floor tiles have an ESL of 52 years. As the ESL of ceramic tiles is within a range of ±5 years of the ESL of buildings 1–7 and 10, therefore, no replacement for ceramic tiles is suggested in the study. However, in buildings 8–9 and 11–12, one replacement of the ceramic floor tiles is considered. One replacement for terracotta tiles is considered for all buildings. The ESL of gypsum board ceiling needs to be replaced once in the ESL of buildings 1–7, 10 and twice in buildings 8–9, and 11–12. However, exterior rendering and interior plaster of walls need regular replacements after 15 and 26 years respectively, to maintain the aesthetic looks of buildings and to strengthen the concrete block wall against its inherent porous structure. Figure 2 shows the total ESL of building components at post-use stage including replacements. The red line shows the ESL of building and above this line is the remaining ESL of building components at the post-use stage. The remaining ESL is the duration for which a component is still in serviceable condition at the time of demolition of the building. The remaining life for structural components is the ESL of the component, however, the remaining life for non-structural components is calculated by multiplying the ESL of building components with number of replacements and subtracting from the ESL of building. The estimated number of replacements and remaining service life of building components is presented in Tables A6 and A7 in Appendix A section. The study has shown that the lowest remaining life of building components at post-use stage (end of life of building), results in better environmental performance of building, due to reduced material wastage to landfill.

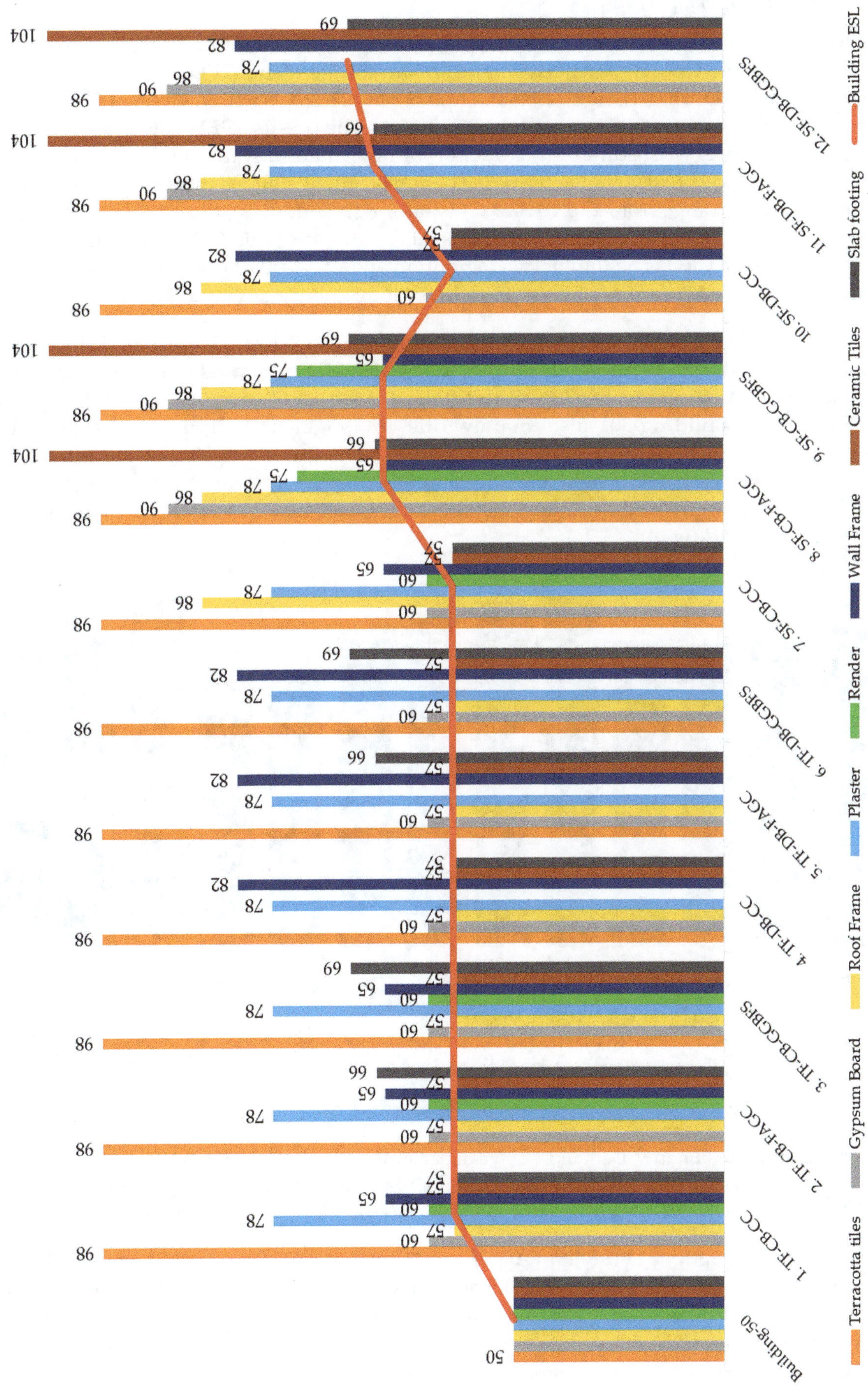

Figure 2. Remaining service life of building components at Post-Use Stage.

4.2. Cumulative Energy Demand

The cumulative energy demand (CED) is calculated on an annual basis to study the impact of service life on the environmental performance of buildings. The CED ranges from 97.799 to 101.813 GJ/year for 12 buildings. The CED is the highest in use stage with a value of 82.634 GJ/year and uniform in each case as no CPS is applied to buildings. The pre-use CED values are highest after the use stage due to energy consumed in extraction of raw materials, manufacturing of materials, transportation to site and construction activities. The replacement stage is the third main contributor to the CED of buildings (range from 2.02 to 3.73 GJ/year for 12 buildings), due to constant addition of embodied energy of replaced building components, at regular intervals (Table A6) during building service life. The post-use CED is negligible, as demolition and disposal of demolition waste to landfill ranges from 0.98 to 1.27 GJ/year for 12 buildings (1% of total energy demand) [4,67].

Figure 3 shows that CED is the lowest for building 8 (SF-CB-FAGC) with an ESL of 65 years i.e., 4.23% lower than the conventional Building-50 (TF-DB-CC) due to its longer ESL and the use of green concrete (OPC replaced by 30% FA). Similarly, the CED of building 9 (SF-CB-GGBFS) is 4.08% lower than Building-50, also mainly due to longer ESL. Buildings 4 (TF-DB-CC), 5 (TF-DB-FAGC), 6 (TF-DB-GGBFS) and building 10 (SF-DB-CC) have almost the same CED as Building-50 (102.118 GJ/year) with negligible differences between 0.37% and 0.63% owing to relatively shorter ESL than buildings 8–9, 11–12, and also because of the use of energy intensive structural material (e.g., double brick wall).

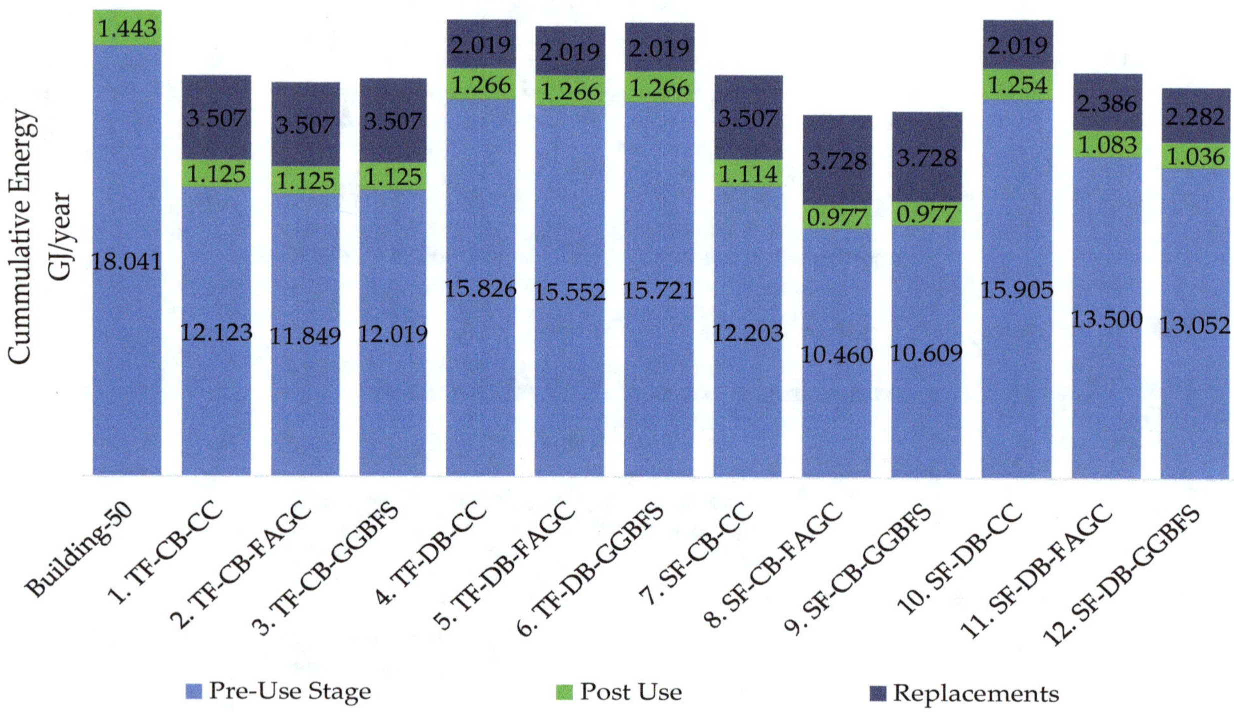

Figure 3. Cumulative Energy Demand per year per building, for 12 buildings and Building-50. (Use stage is omitted in the graph as the use-stage CED value 82.634 GJ/year is uniform in all cases and if plotted on same scale, other stages due to low values cannot be presented properly).

The longer ESL of building 12 (SF-DB-FAGC) and 11 (SF-DB-GGBFS) with ESL of 69 and 66 years, has reduced the share of pre-use energy consumption in comparison to buildings 4–6 and 10 that are composed of energy intensive brick walls and have an ESL of 57 years. Building 9 has the lowest energy demand in pre-use and post-use stage due to having low energy intensive concrete block wall and ESL of 65 years reducing per year share of CED of the building.

In the replacement stage, buildings with concrete block wall (1–3, 7–9) have high replacement embodied energy due to frequent replacements of energy intensive rendering and plastering. In buildings 8–9, the replacement embodied energies are highest due to ceramic tiles replacement in addition to rendering and plastering (Figure 2). Similarly, buildings 4–6, 10 with double brick walls have low replacement embodied energy as only interior plastering is replaced at a regular interval of 26 years (Table A6). Buildings 11 and 12, with longer ESL, have slightly higher replacement stage embodied energy due to replacement of ceramic tiles.

Although replacement embodied energy is higher in some buildings, the longer ESL of buildings reduces the impact of replacement embodied energy, as in buildings 8–9. In some cases, the use of energy intensive structural building components in fact reduced the overall CED of buildings. The steel frame roof is an energy intensive material, but its use had indirectly reduced the annual CED by increasing ESL of building combinations 8–9 and 11–12.

The results of this study were compared with similar studies in WA. Lawania and Biswas [68], estimated the annual CED for residential buildings across Western Australia showed slightly higher CED (138 GJ/year), which this value varies between 97.8 and 101.81 GJ/year under this current study. This variation happened due to the fact that Lawania and Biswas had used 18 different climatic locations and also one service life of 50 years was considered. In other studies of residential buildings that considered the embodied energy of building components replaced during ESL, embodied energy was found to increase by 20% to 40% due to increase of service life from 50 to 100 years [4,69,70]. Similar results were found for some buildings (buildings 8–9, 11–12) in the current study, where longer ESL had in fact increased the CED by 17% to 33% due to replacement of building components during ESL.

4.3. GHG Emissions

The GHG emissions in 12 buildings vary from 11.383 to 11.49 t CO_2 eq. The GHG emissions are highest in use stage with a value of 9.918 t CO_2 eq/year and uniform for all buildings like CED assessment due to use of electricity that is predominantly generated from fossil fuels (49% black coal and 36% gas) in WA [71]. The pre-use GHG emissions are highest after the use stage due to fossil fuel consumptions in extraction and manufacturing of materials, transportation of materials to site and construction equipment. The replacement stage is the third main contributor to the GHG emissions of buildings, due to the addition of the materials to building during ESL. The post-use CED is negligible as the demolition and disposal of demolition waste to landfill consumes only 1% of the total energy [4,67].

Building-50 (reference building) with a 50-year ESL, has annual GHG emissions of 11.455 t CO_2 eq/year, despite the absence of the replacement stage. Annual GHG emissions for building 2 (TF-CB-FAGC) with an ESL of 57-years, are the lowest among all the building combinations (i.e., 0.626% less than Building-50) due to use of low carbon intensive materials (timber, concrete blocks), less replacements of non-structural components (rendering, plastering) and most importantly due to having a similar ESL of building components as the whole building. These design specifications result in lower wastage of material or embodied energy at the post-use stage. Building 10 (SF-DB-CC) with an ESL of 57 years has the highest GHG emissions per year of 11.49 t CO_2 eq/year (i.e., 0.308% higher than Building 50) due to shorter ESL and use of energy intensive structural components (double brick wall, steel frame roof) with ESL longer than building ESL. It is worth mentioning that the building components with longer ESL than the whole building remains unused after the end of life demolition and disposal stage and are being considered as wastes.

Figure 4 shows that the double brick buildings 4 (TF-DB-CC), 5 (TF-DB-FAGC), 6 (TF-DB-GGBFS), and 10 (SF-DB-CC) have high pre-use stage annual GHG emissions of 1.265, 1.218, 1.230, and 1.279 t CO_2 eq/year, than Building-50, with 1.443 t CO_2 eq/year annual GHG emissions due to relatively longer ESL of 57 years and also due to use of energy intensive structural components i.e., brick walls (buildings 4–6, 10) and steel frame roof (building 10) with longer ESL (Figure 2) that is wasted to landfill at the post-use stage.

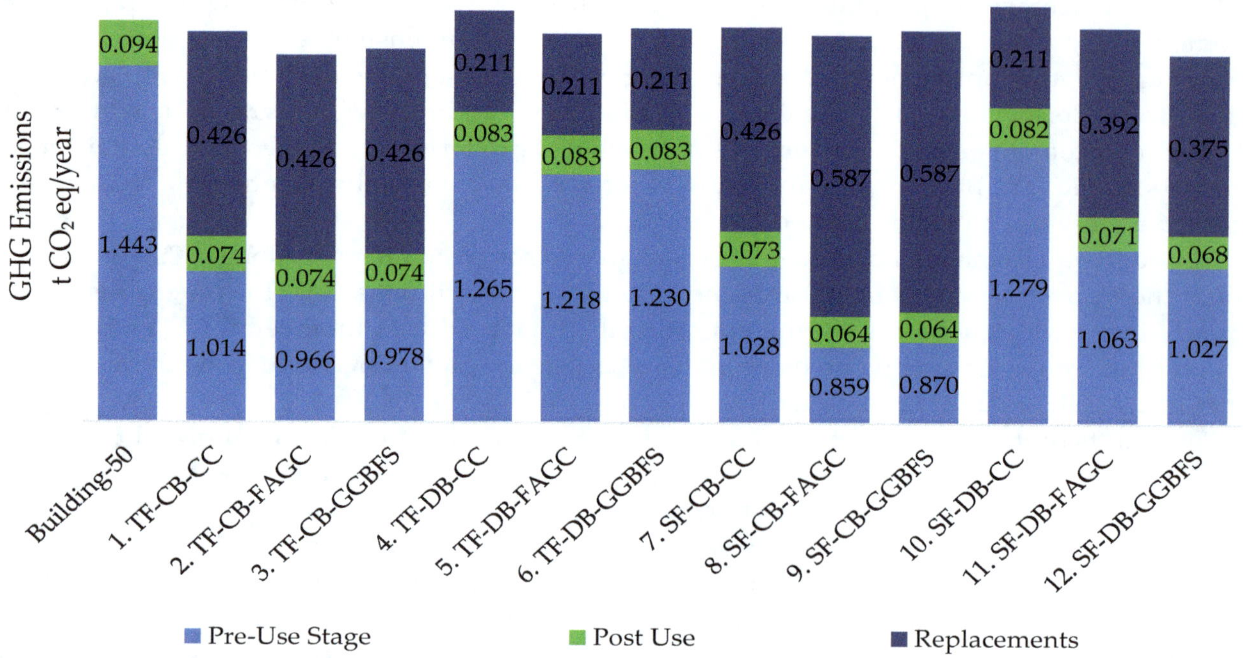

Figure 4. Life cycle greenhouse gas (GHG) emissions per year per building for 12 buildings and Building-50. (Use stage is omitted in the graph as the use-stage GHG value 9.918 t CO_2 eq/year is uniform in all cases and if plotted on same scale, other stages due to low values cannot be presented properly).

The GHG emissions associated with the replacement of components during the active service life are highest for building 8 (SF-CB-FAGC) and 9 (SF-CB-GGBFS) with a value of 0.587 t CO_2 eq/year. The reason for high replacement GHG emissions in these buildings is the use of carbon intensive ceramic tiles (13.07 t CO_2 eq) and rendering processes (10.56 t CO_2 eq). Additionally, the replaced ceramic tiles were not fully utilized as the ESL of buildings 8 and 9 expired at the 33% ESL of ceramic tile and therefore this valuable material was disposed along with other building materials into the landfill. The recovery of this carbon intensive material thus needs to be considered for use in similar applications during its remaining life (i.e., 67% of ESL). In addition, the rendering used large amounts of carbon intensive OPC (Ordinary Portland Cement). Nonetheless, due to the porous nature of concrete blocks, rendering or an alternative process is required to provide coverage to concrete blocks, which in fact increased the overall energy consumption as well as the GHG emissions.

Annual GHG emissions for case study buildings in the current study vary between 10.957 and 11.49 t CO_2 eq/year and is slightly higher than Lawania and Biswas [68,72] (9.4 t CO_2 eq/year). This is mainly due to differences in parameters like service life and climatic conditions affecting use stage GHG emissions. In a study by Carre A. [11] for Australian houses with a 50-year service life, pre-use GHG emissions (0.908 t CO_2 eq/year) are similar to current study (0.859 to 1.279 t CO_2 eq/year).

4.4. Land Use

Land use impact is the highest in the use stage (1.353 Ha_a/year) as standard energy input is used for all buildings without considering any greener choices such as wind mills, solar panels etc. The land use in the pre-use stage is the highest after the use stage as it is the summation of all land utilized during the extraction of raw materials, manufacturing of materials, transportation to site and construction equipment, followed by the replacement stage. The post-use stage land utilization is negligible as only energy consumption for the demolition and disposal of demolition waste to landfill is assessed for the study.

Figure 5 shows that the building 4 (TF-DB-CC) has the highest land use impact of 1.583 Ha_a (actual hectare) per year due to the timber frame roof (material acquired by plants) and shorter ESL.

Building 12 (SF-DB-GGBFS) has the lowest impact with a value of 1.558 Ha_a/year. The longer ESL of building 12 and use of industrial by-products like GGBFS [11] has contributed to the lower land use impact.

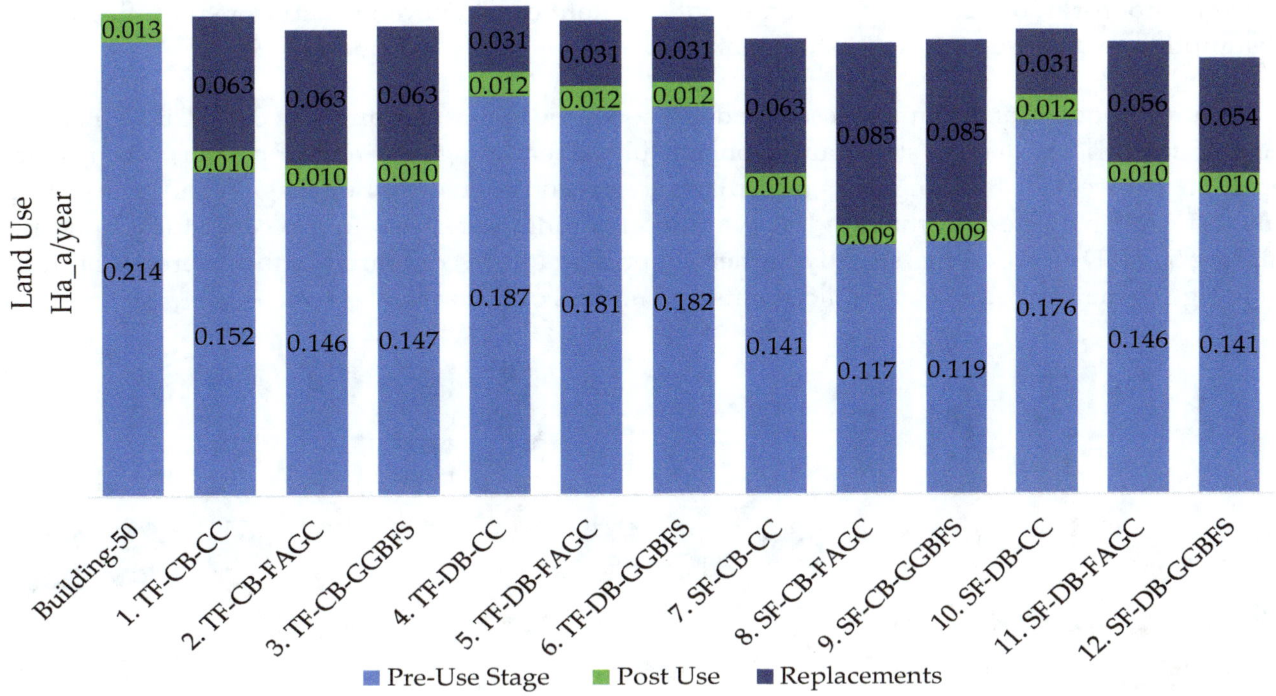

Figure 5. Life cycle land use per year per building for 12 buildings and Building-50. (Use stage is omitted in the graph as the use-stage Land use value 1.353 Ha_a/year is uniform in all cases and if plotted on same scale, other stages due to low values cannot be presented properly).

Building 4 (TF-DB-CC), 5 (TF-DB-FAGC), and 6 (TF-DB-GGBFS) have the highest land use in the pre-use and post-use stages among 12 buildings, due to the increased amount of land requirement associated with the production of a timber frame roof, and also because these materials have short ESL meaning that more land is required to make these materials to meet the demand for replacement. The land use impact for replacement stage is higher for buildings 8 and 9 with 65-year ESL, as more energy and carbon intensive materials (e.g., ceramic tiles, rendering, plastering) requiring more space for mining, processing and manufacturing are replaced during this long ESL (Figure 2). Replaced building components in buildings 8 and 9 have about 33% to 60% of the remaining life at the end of the building service life (Figure 2).

The building with components with similar ESLs to the whole building ESL generate less waste which means the diversion of waste from landfill or residue area, thus conserving land or reducing the land footprint. From the ecological footprint point of view, the buildings with a timber roof and double brick wall frame have higher impacts than building with building components of industrial material on an annual basis, which is consistent with the existing study of Allacker et al. [73]. In the post-use stage, ceramic tiles and rendering have the highest impacts as these materials are disposed of before their ESL was finished.

4.5. Water Consumption

Life time water consumption by case study buildings is calculated in terms of damage to resources. The resource depletion is the minimum time step to assess the water resource depletion in areas like Western Australia with fixed annual precipitation cycle [60].

Resource depletion is the highest in the use stage due to high water consumption in electricity generation [74]. The resource depletion in the pre-use stage is the second highest in buildings due to water consumption mainly in the extraction of materials and manufacturing processes. In the pre-use stage, onsite water consumption (construction works) is negligible (0.1%) as compared to upstream processing of materials. The replacement stage is contributing as the third major stage due to building component replacements. Like CED, GHG emissions and land use, the post-use stage has the least water footprint due to consideration of only demolition of buildings and transportation energy consumption to dispose of these wastes to landfill.

Figure 6 shows that the annual resource depletion is the highest for building 8 (SF-CB-GGBFS) due to use of industrial materials (steel frame, concrete block, rendering and interior plaster). The resource depletion is lowest for Building-50 as no replacement is considered for Building-50 and it has a brick wall and timber frame roof that are less water consuming materials. In the case study buildings, building 5 (TF-DB-FAGC) has slightly higher water demand as compared to the reference building (Building-50) mainly because of replacements and an ESL of 57 years.

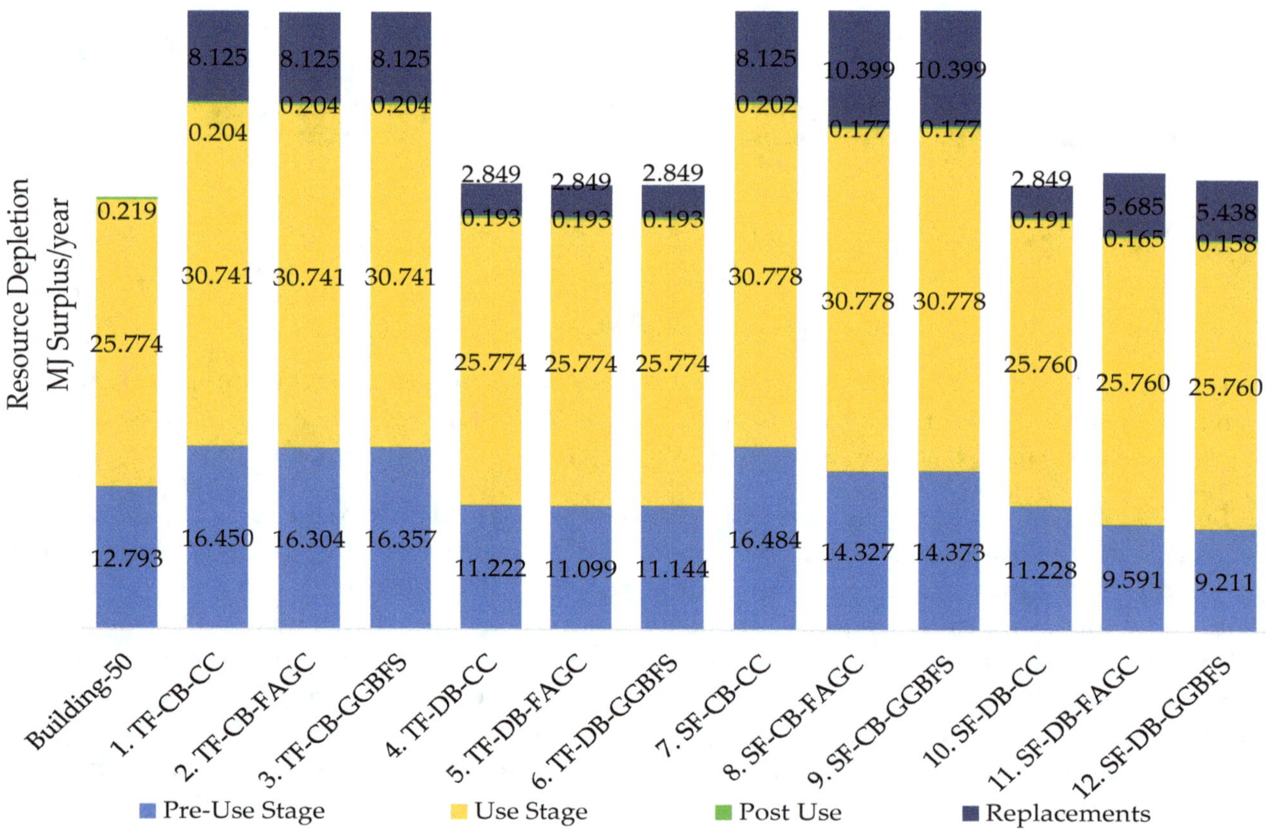

Figure 6. Life cycle resource depletion per year per building for 12 buildings and Building-50.

For pre-use stages, building 7 has the highest water footprint of 16.484 MJ surplus/year due to a shorter ESL and use of building components of industrial materials. The lowest water footprint in pre-use stage is for building 12 (SF-DB-GGBFS) with longer ESL and structural components like brick, green concrete, that have lower water demand and decreased overall water consumption by building [75,76]. In concrete block wall buildings (building 1 (TF-CB-CC), 2 (TF-CB-FAGC), 3 (TF-CB-GGBFS)), water consumption in concrete block and plaster production, as well as the rendering process and their shorter ESL, are the main contributing factors for water footprint [75]. In the replacement stage, the concrete block wall and steel frame roof in buildings 8 (SF-CB-FAGC) and 9 (SF-CB-GGBFS) with 65-year ESL have the highest mineral resource depletion due to frequent replacements of rendering, interior plaster and water intensive ceramic tiles.

5. Conclusions

This study used the life cycle assessment procedure (ISO 14040-44) [38] and factor method (ISO 15686-2) [31,37], to determine the impact of service life on the environmental performance of buildings. The service life of a building and building components have a direct relationship with environmental performance of building. Estimation of replacements intervals is important in the environmental life cycle assessment of buildings, as these are the third main contributing stage to environmental impacts, after use and pre-use stage. The cumulative energy demand for building 8 (SF-CB-FAGC) is 97.779 GJ/year, 4.23% lower than the reference building, due to the ESL of 65 years. The GHG emissions for building 2 (TF-CB-FAGC) is 11.383 t CO_2 eq/year, the lowest among the case study buildings, due to use of structural components with a comparatively similar service life as ESL of buildings. The land use for building 12 (SF-DB-GGBFS) is 1.558 Ha_a, the lowest among case study buildings (0.96% lower than reference building), due to longer ESL and the use of industrial by-products which reduces land use for residue storage. Building 5 (TF-DB-FAGC) has the lowest water footprint (39.915 MJ Surplus/year) amongst case study buildings (2.827% higher than the reference building), due to use of timber and brick.

This study showed that buildings 1–12 have better performance than Building-50 for CED, and buildings 1–3, 5–9 and 11–12 have lower GHG emissions than Building-50. In the land use impact category, buildings 1–3 and 5–12 have lower land use. However, the water footprint results are slightly different than CED, GHG and land use. Building-50 has the lowest water demand as compared to the 12 case study buildings.

Current research shows that building environmental performance is dependent on building component's materials and ESL, and the way these components are modeled into building. The use of industrial byproducts (concrete blocks, steel) could enhance performance for land use, while building materials like timber and brick have a better water footprint. Industrial byproducts (FA) have lower environmental impact for all indicators considered in this research. The service life of buildings and building components affect the environmental performance of buildings. The use of alternative, eco-friendly strategies in buildings like grey water, green concrete, renewable resources are effective only if these are aligned with building service life. Longevity of the service life of buildings can produce sustainable outcomes, if all of the building component's service life is nearest to the building's service life, which causes fewer replacements, as well as the GHG emissions from the transportation of waste during the post-use stage.

Author Contributions: Conceptualization and methodology, S.Y.J., P.K.S. and W.K.B.; Analysis, S.Y.J.; investigation, S.Y.J.; data curation, S.Y.J.; writing—original draft, S.Y.J.; visualization, S.Y.J.; writing—review and editing, S.Y.J., P.K.S. and W.K.B.; Supervision, S.Y.J., P.K.S. and W.K.B.

Acknowledgments: The Authors would like to thank four reviewers for their insightful comments to improve the manuscript.

Appendix A

Table A1. Life cycle inventory of materials for the construction stage of case study buildings.

Material (tonnes)	Building-50	1. TF-CB-CC	2. TF-CB-FAGC	3. TF-CB-GGBFS	4. TF-DB-CC	5. TF-DB-FAGC	6. TF-DB-GGBFS	7. SF-CB-CC	8. SF-CB-FAGC	9. SF-CB-GGBFS	10. SF-DB-CC	11. SF-DB-FAGC	12. SF-DB-GGBFS
1. Excavation of foundation	26.83	26.83	26.83	26.83	26.83	26.83	26.83	26.83	26.83	26.83	26.83	26.83	26.83
2. Sand: Sub-base	26.01	26.01	26.01	26.01	26.01	26.01	26.01	26.01	26.01	26.01	26.01	26.01	26.01
3. On grade slab (including water proofing membrane, mesh reinforcement, ready mix concrete-N20)	79.56	79.56	79.56	79.56	79.56	79.56	79.56	79.56	79.56	79.56	79.56	79.56	79.56
4. Floor Tiles (Ceramic)	5.47	5.47	5.47	5.47	5.47	5.47	5.47	5.47	5.47	5.47	5.47	5.47	5.47
5. Concrete Blocks wall (including mortar, steel rebar, metal lintels)		75.78	75.78	75.78				75.78	75.78	75.78			
6. Bricks (including mortar, metal lintels)	107.4				107.4	107.4	107.4				107.4	107.4	107.4
7. Plaster	10.54	10.54	10.54	10.54	10.54	10.54	10.54	10.54	10.54	10.54	10.54	10.54	10.54
8. Rendering		13.86	13.86	13.86				13.86	13.86	13.86			
9. Gypsum board ceiling	2.19	2.19	2.19	2.19	2.19	2.19	2.19	2.19	2.19	2.19	2.19	2.19	2.19
10. Roof Timber (including bat insulation)	4.61	4.61	4.61	4.61	4.61	4.61	4.61						
11. Roof Steel (including bat insulation)	x	x	x	x	x	x	x	2.33	2.33	2.33	2.33	2.33	2.33
12. Terracotta roof tiles	16.55	16.55	16.55	16.55	16.55	16.55	16.55	16.55	16.55	16.55	16.55	16.55	16.55
13. Metal door frames: 12 no.	0.18	0.18	0.18	0.18	0.18	0.18	0.18	0.18	0.18	0.18	0.18	0.18	0.18
Door shutters: 12 no.	0.37	0.37	0.37	0.37	0.37	0.37	0.37	0.37	0.37	0.37	0.37	0.37	0.37
14. Aluminum windows: Single glazed	1.43	1.43	1.43	1.43	1.43	1.43	1.43	1.43	1.43	1.43	1.43	1.43	1.43
Total material transported to construction site	254.3	236.5	236.5	236.5	254.3	254.3	254.3	234.3	234.3	234.3	252	252	252
Total material transported to landfill	26.83	26.83	26.83	26.83	26.83	26.83	26.83	26.83	26.83	26.83	26.83	26.83	26.83

Table A2. Life cycle inventory of materials for the replacement stage of case study buildings.

Material (tonnes)	Building-50	1. TF-CB-CC	2. TF-CB-FAGC	3. TF-CB-GGBFS	4. TF-DB-CC	5. TF-DB-FAGC	6. TF-DB-GGBFS	7. SF-CB-CC	8. SF-CB-FAGC	9. SF-CB-GGBFS	10. SF-DB-CC	11. SF-DB-FAGC	12. SF-DB-GGBFS
Terracotta tiles		16.55	16.55	16.55	16.55	16.55	16.55	16.55	16.55	16.55	16.55	16.55	16.55
Gypsum board		2.19	2.19	2.19	2.19	2.19	2.19	2.19	4.38	4.38	2.19	4.38	4.38
Plaster		21.08	21.08	21.08	21.08	21.08	21.08	21.08	21.08	21.08	21.08	21.08	21.08
Render		41.58	41.58	41.58	n/a	n/a	n/a	41.58	55.44	55.44	n/a	n/a	n/a
Ceramic tiles		0	0	0	0	0	0	0	5.47	5.47	0	5.47	5.47
Total material transported to site for replacement works		81.4	81.4	81.4	39.8	39.8	39.8	81.4	102.9	102.9	39.8	47.5	47.5
Total material transported to landfill		81.4	81.4	81.4	39.8	39.8	39.8	81.4	102.9	102.9	39.8	47.5	47.5

Table A3. Life cycle inventory of material transportation for the construction stage of case study buildings.

Carriage (tkm)	Building-50	1. TF-CB-CC	2. TF-CB-FAGC	3. TF-CB-GGBFS	4. TF-DB-CC	5. TF-DB-FAGC	6. TF-DB-GGBFS	7. SF-CB-CC	8. SF-CB-FAGC	9. SF-CB-GGBFS	10. SF-DB-CC	11. SF-DB-FAGC	12. SF-DB-GGBFS
Sand for levelling	1300.5	1300.5	1300.5	1300.5	1300.5	1300.5	1300.5	1300.5	1300.5	1300.5	1300.5	1300.5	1300.5
Material to site	6848.4	6316.2	6316.2	6316.2	6848.4	6848.4	6848.4	6247.8	6247.8	6247.8	6780	6780	6780
Construction waste	342.42	315.81	315.81	315.81	342.42	342.42	342.42	312.39	312.39	312.39	339	339	339
Dirt to landfill	1341.5	1341.5	1341.5	1341.5	1341.5	1341.5	1341.5	1341.5	1341.5	1341.5	1341.5	1341.5	1341.5
Material to landfill	1683.92	1657.31	1657.31	1657.31	1683.92	1683.92	1683.92	1653.89	1653.89	1653.89	1680.5	1680.5	1680.5

Table A4. Life cycle inventory of material transportation for the replacement stage of case study buildings.

Carriage (tkm)	Building-50	1. TF-CB-CC	2. TF-CB-FAGC	3. TF-CB-GGBFS	4. TF-DB-CC	5. TF-DB-FAGC	6. TF-DB-GGBFS	7. SF-CB-CC	8. SF-CB-FAGC	9. SF-CB-GGBFS	10. SF-DB-CC	11. SF-DB-FAGC	12. SF-DB-GGBFS
Material to site		2442.0	2442.0	2442.0	1194.6	1194.6	1194.6	2442.0	3087.6	3087.6	1194.6	1424.4	1424.4
Construction waste		122.1	122.1	122.1	59.7	59.7	59.7	122.1	154.4	154.4	59.7	71.2	71.2
Material to landfill		4070.0	1341.5	1341.5	1991.0	1991.0	1991.0	4070.0	5146.0	5146.0	1991.0	2374.0	2374.0
Total material to landfill		4192.1	1463.6	1463.6	2050.7	2050.7	2050.7	4192.1	5300.4	5300.4	2050.7	2445.2	2445.2

Table A5. Life cycle inventory of energy for the use stage of case study buildings.

Energy	Energy Consumption GJ/year
Non-thermal energy (home appliances, lighting)	18.834
Thermal energy (heating/cooling)	10.430
Natural gas consumption (hot water)	22.610
Total energy (use stage)	51.874

Table A6. Estimated number of replacements of building components in active service life of buildings.

Replacements of Building Components	Building-50	1. TF-CB-CC	2. TF-CB-FAGC	3. TF-CB-GGBFS	4. TF-DB-CC	5. TF-DB-FAGC	6. TF-DB-GGBFS	7. SF-CB-CC	8. SF-CB-FAGC	9. SF-CB-GGBFS	10. SF-DB-CC	11. SF-DB-FAGC	12. SF-DB-GGBFS
Building	50	57	57	57	57	57	57	57	65	65	57	66	69
Terracotta tiles	0	1	1	1	1	1	1	1	1	1	1	1	1
Gypsum board	0	1	1	1	1	1	1	1	2	2	1	2	2
Plaster	0	2	2	2	2	2	2	2	2	2	2	2	2
Render	0	3	3	3	n/a	n/a	n/a	3	4	4	n/a	n/a	n/a
Ceramic tiles	0	0	0	0	0	0	0	0	1	1	0	1	1

Table A7. Estimated Remaining life of building components at post-use stage of buildings.

Remaining Life (years)	Building-50	1. TF-CB-CC	2. TF-CB-FAGC	3. TF-CB-GGBFS	4. TF-DB-CC	5. TF-DB-FAGC	6. TF-DB-GGBFS	7. SF-CB-CC	8. SF-CB-FAGC	9. SF-CB -GGBFS	10. SF-DB-CC	11. SF-DB-FAGC	12. SF-DB-GGBFS
Building	50	57	57	57	57	57	57	57	65	65	57	66	69
Terracotta tiles		41	41	41	41	41	41	41	32	29	41	32	29
Gypsum board		3	3	3	3	3	3	3	25	25	3	24	21
Roof frame		0	0	0	0	0	0	29	21	21	29	20	17
Plaster		21	21	21	21	21	21	21	13	13	21	12	9
Render		3	3	3	n/a	n/a	n/a	3	10	10	n/a	n/a	n/a
Wall frame		8	8	8	25	25	25	8	0	0	25	16	13
Ceramic tiles		0	0	0	0	0	0	0	39	39	0	38	35
Slab footing		0	9	12	0	9	12	0	9	12	0	9	12

References

1. Grant, A.; Ries, R.; Kibert, C. Life cycle assessment and service life prediction: A case study of building envelope materials. *J. Ind. Ecol.* **2014**, *18*, 187–200. [CrossRef]
2. Nunen, H.V. Assessment of the Sustainability of Flexible Building: The Improved Factor Method: Service Life Prediction of Buildings in The Netherlands, Applied to Life Cycle Assessment. Ph.D. Thesis, Universiteit Eindhoven, Eindhoven, The Netherlands, 2010.
3. Leticia, O.M.; Begoña, S.L. Proposed method of estimating the service life of building envelopes. *J. Constr.* **2015**, *14*, 60–88.
4. Rauf, A.; Crawford, R.H. The relationship between material service life and the life cycle energy of contemporary residential buildings in Australia. *Artich. Sci. Rev.* **2013**, *56*, 252–261. [CrossRef]
5. Robert, H.; Crawford, R.J.F. Energy and greenhouse gas emissions implications of alternative housing types for Australia. In Proceedings of the State of Australian Cities National Conference Australian Sustainable Cities and Regions Network (ASCRN), Melbourne, VIC, Australia, 29 November–2 December 2011; pp. 1–12.
6. Robert, H.; Crawford, I.C.; Robert, J.F. A comprehensive framework for assessing the life-cycle energy of building construction assemblies. *Archit. Sci. Rev.* **2010**, *53*, 288.
7. Silvestre, J.D.; Silva, A.; de Brito, J. Uncertainty modelling of service life and environmental performance to reduce risk in building design decisions. *J. Civ. Eng. Manag.* **2015**, *21*, 308–322. [CrossRef]
8. Ramesh, T.; Prakash, R.; Shukla, K.K. Life cycle energy analysis of a residential building with different envelopes and climates in Indian context. *Appl. Energy* **2012**, *89*, 193–202. [CrossRef]
9. Allacker, K. Sustainable Building: The Developement of an Evaluation Method. Ph.D. Thesis, Katholieke Universiteit Leuven, Leuven, Belgium, 2010.
10. Audenaert, A.; de Cleyn, S.H.; Buyle, M. Lca of low energy flats using the eco-indicator 99 method: Impact of insulation materials. *Energy Build.* **2012**, *47*, 68–73. [CrossRef]
11. Carre, A. *A Comparative Life Cycle Assessment of Alternative Constructions of a Typical Australian House Design*; PNA147-0809; Forest & Wood Products Australia: Melbourne, Australia, 2011.
12. Iyer-Raniga, U.; ChewWong, J.P. Evaluation of whole life cycle assessment for heritage buildings in australia. *Build. Environ.* **2012**, *47*, 138–149. [CrossRef]
13. Rouwette, R. *Lca of Brick Products*; Think Brick Australia: Melbourne, Australia, 2010.
14. Cuellar-Franca, R.M.; Azapagic, A. Environmental impacts of the UK residential sector: Life cycle assessment of houses. *Build. Environ.* **2012**, *54*, 86–99. [CrossRef]
15. Nemry, F.; Uihlein, A.; Colodel, C.M.; Wetzel, C.; Braune, A.; Wittstock, B.; Hasan, I.; Kreißig, J.; Gallon, N.; Niemeier, S.; et al. Options to reduce the environmental impacts of residential buildings in the european union-potential and costs. *Energy Build.* **2010**, *42*, 976–984. [CrossRef]
16. Ortiz-Rodríguez, O.; Castells, F.; Sonnemann, G. Life cycle assessment of two dwellings: One in spain, a developed country, and one in colombia, a country under development. *Sci. Total Environ.* **2010**, *408*, 2435–2443. [CrossRef] [PubMed]
17. Cabeza, L.F.; Rincón, L.; Vilariño, V.; Pérez, G.; Castell, A. Life cycle assessment (lca) and life cycle energy analysis (lcea) of buildings and the building sector: A review. *Renew. Sustain. Energy Rev.* **2014**, *29*, 394–416. [CrossRef]
18. Biswas, W.K. Carbon footprint and embodied energy assessment of a civil works program in a residential estate of western australia. *Int. J. Life Cycle Assess.* **2014**, *19*, 732–744. [CrossRef]
19. Islam, H.; Jollands, M.; Setunge, S. Life cycle assessment and life cycle cost implication of residential buildings—A review. *Renew. Sustain. Energy Rev.* **2015**, *42*, 129–140. [CrossRef]
20. Atmaca, A. Life-cycle assessment and cost analysis of residential buildings in south east of turkey: Part 2—A case study. *Int. J. Life Cycle Assess.* **2016**, *21*, 925–942. [CrossRef]
21. Lawania, K.K.; Biswas, W.K. Achieving environmentally friendly building envelope for western australia's housing sector: A life cycle assessment approach. *Int. J. Sustain. Built Environ.* **2016**, *5*, 210–224. [CrossRef]
22. Manish, K.; Dixit, J.L.F.; Lavy, S.; Charles, H.C. Need for an embodied energy measurement protocol for buildings: A review paper. *Renew. Sustain. Energy Rev.* **2012**, *16*, 3730–3743.
23. Vitale Pierluca, A.N.; Fabrizio, D.G.; Umberto, A. Life cycle assessment of the end of life phase of a residential building. *Waste Manag.* **2017**, *60*, 311–321. [CrossRef]

24. Pierluca Vitale, U.A. An attributional life cycle assessment for an italian residential multifamily building. *Environ. Technol.* **2018**, *39*, 3033–3045. [CrossRef]
25. Balasbaneh, A.T.; Marsono, A.K.B.; Khaleghi, S.J. Sustainability choice of different hybrid timber structure for low medium cost single-story residential building: Environmental, economic and social assessment. *J. Build. Eng.* **2018**, *20*, 235–247. [CrossRef]
26. Buildings and constructed assets—Service life planning. *Part-1: General Principles and Framework*; ISO 15686-1; International Standards Organisation: Geneva, Switzerland, 2011.
27. Hed, G. Service life planning of building components. *Durab. Build. Mater. Compnt.* **1999**, *8*, 1543–1551.
28. Hovde, P.J.; Moser, K. *Performance Based Methods for Service Life Prediction*; In-House Publishing: Rotterdam, The Netherlands, 2004.
29. Masters, L.W.; Brandt, E. Prediction of service life of building materials and components. *Mater. Struct.* **1987**, *20*, 55–77. [CrossRef]
30. Building and constructed assets—Service planning. *Part 1: General Principals and Framework*; ISO 15686-1:2000; International Standards Organisation: Geneva, Switzerland, 2000.
31. Buildings and constructed assets—Service life planning. *Part-2: Service Life Prediction Procedures*; ISO 15686-2; International Standards Organisation: Geneva, Switzerland, 2012.
32. Cecconi, F. Performance lead the way to service life prediction. In Proceedings of the 9th International Conference on Durability of Building Materials and Components, Brisbane, Australia, 17–20 March 2002.
33. Ivan Cole, P.C. Predicting the service life of buildings and components. *Constr. Mater.* **2011**, *164*, 305–314. [CrossRef]
34. Chen, C.-J.; Juan, Y.-K.; Hsu, Y.-H. Developing a systematic approach to evaluate and predict building service life. *J. Civ. Eng. Manag.* **2017**, *23*, 890–901. [CrossRef]
35. Hovde, P.J. Performance Based Methods for Service Life Prediction, Part A: Factor Methods for Service Life Prediction. Available online: http://site.cibworld.nl/dl/publications/Pub294.pdf (accessed on 1 January 2019).
36. Hovde, P.J. *Evaluation of the Factor Method to Estimate the Service Life of Building Components*; Department of Building and Construction Engineering, The Norwegian University of Science and Technology: Trondheim, Norway, 1998.
37. Buildings and constructed assets—Service-life planning. *Part 8: Reference Service Life and Service-Life Estimation*; ISO 15686-8; International Standards Organisation: Geneva, Switzerland, 2008.
38. *Environmental Management—Life Cycle Assessment, Principles and Framework*; ISO 14040; International Standards Organization: Geneva, Switzerland, 2006.
39. Guinee, J.B.; Udo de Haes, H.A.; Huppes, G. Quantitative life cycle assessment of products: 1: Goal definition and inventory. *J. Clean. Prod.* **1993**, *1*, 3–13. [CrossRef]
40. Consoli, F.A.; Boustead, I.; Fava, J.; Franklin, W.; Jensen, A.; Oude, N.; Parrish, R.; Perriman, R.; Postlethwaite, D.; Quay, B.; et al. *Guide Lines for Life-Cycle Assessment: A Code of Practice*; Society of Environmental Toxicology and Chemistry: Pensacola, FL, USA, 1993.
41. Bekker, P.C.F. A life cycle approach in building. *Build. Environ.* **1982**, *17*, 3–13. [CrossRef]
42. Ortiz, O.; Castells, F.; Sonnemann, G. Sustainability in the construction industry: A review of recent developments based on lca. *Constr. Build. Mater.* **2009**, *23*, 28–39. [CrossRef]
43. Lotteau, M.; Loubet, P.; Pousse, M.; Dufrasnes, E.; Sonnemann, G. Critical review of life cycle assessment (lca) for the built environment at the neighborhood scale. *Build. Environ.* **2015**, *93*, 165–178. [CrossRef]
44. Anderson, J.E.; Wulfhorst, G.; Lang, W. Expanding the use of life-cycle assessment to capture induced impacts in the built environment. *Build. Environ.* **2015**, *94*, 403–416. [CrossRef]
45. Passer, A.; Ouellet-Plamondon, C.; Keneally, P.; John, V.; Habert, G. Impact of future scenarios on building renovation strategies towards plus energy buildings. *SBE16 Zurich* **2016**, *124*, 2.
46. Norman, J.; MacLean, H.; Kennedy, C.A. Comparing high and low residential density: Life-cycle analysis of energy use and greenhouse gas emissions. *J. Urban Plan. Dev.* **2006**, *132*, 10–21. [CrossRef]
47. Wahidul, K.; Biswasa, Y.A.; Krishna, K.L.; Prabir, K.S.; Elsarrag, E. Life cycle assessment for environmental product declaration of concrete in the gulf states. *Sustain. Cities Soc.* **2017**, *35*, 36–46.
48. Lawania, K.K.; Biswas, W.K. Cost-effective ghg mitigation strategies for western australia's housing sector: A life cycle management approach. *Clean Technol. Environ. Policy* **2016**, *18*, 2419–2428. [CrossRef]

49. Werner, F.; Richter, K. Wooden building products in comparative lca. *Int. J. Life Cycle Assess.* **2007**, *12*, 470–479.
50. Wahidul, K.; Biswas, L.B.; Carter, D. Global warming potential of wheat production western australia: A life cycle assessment. *Water Environ. J.* **2008**, *22*, 206–216.
51. Mirabella, N.; Röck, M.; Saade, M.R.M.; Spirinckx, C.; Bosmans, M.; Allacker, K.; Passer, A. Strategies to improve the energy performance of buildings: A review of their life cycle impact. *Buildings* **2018**, *8*, 105. [CrossRef]
52. Vincenzo Franzitta, M.L.; Gennusa, G.P.; Rizzo, G.; Scaccianoce, G. Toward a european eco-label brand for residential buildings: Holistic or by-components approaches? *Energy* **2011**, *36*, 1884–1892. [CrossRef]
53. Francesco Barreca, P.P. Post-occupancy evaluation of buildings for sustainable agri-food production—A method applied to an olive oil mill. *Buidings* **2018**, *8*, 83. [CrossRef]
54. Samaratunga, M.; Ding, L.; Bishop, K.; Prasad, D.; Yee, K.W.K. Modelling and Analysis of Post-Occupancy Behaviour in Residential Buildings to Inform Basix Sustainability Assessments in Nsw. *Procedia Eng.* **2017**, *180*, 343–355. [CrossRef]
55. Aneurin Grant, R.R. Impact of building service life models on life cycle assessment. *Build. Res. Inf.* **2013**, *41*, 168–186. [CrossRef]
56. Galle, W.; de Temmerman, N.; de Meyer, R. Integrating scenarios into life cycle assessment: Understanding the value and financial feasibility of a demountable building. *Buidings* **2017**, *7*, 64. [CrossRef]
57. TCI. *Australia's Emissions: What Do the Numbers Really Mean?* The Climate Institute: Sydney Australia, 2015.
58. Argent, R.M. *Australia State of the Environment 2016: Inland Water*; Australian Government Department of the Environment and Energy: Canberra, Australia, 2017.
59. BOM. *Average Anual, Seasonal and Monthly Rainfall*; BOM: Melbourne, Australia, 2018.
60. Pfister, S.; Koehler, A.; Hellweg, S. Assessing the environmental impacts of freshwater consumption in lca. *Environ. Sci. Technol.* **2009**, *43*, 4098–4104. [CrossRef] [PubMed]
61. Fenton, G.G. Desalination in australia-past and future. *Desalination* **1981**, *39*, 399–411. [CrossRef]
62. Radcliffe, J.C. The water energy nexus in australia-the outcome of two crises. *Water-Energy Nexus* **2018**, *1*, 66–85. [CrossRef]
63. Joe Pickin, P.R. *Australian National Waste Report 2016*; Department of the Environment and Energy & Blue Environment Pty Ltd.: Docklands, Australia, 2017.
64. PRe'-Consultants. *Simapro 8.4 Lca Software*; Pre' Consultants: Amersfoort, The Netherlands, 2016.
65. Shahana, Y.; Janjua, P.K.S.; Wahidul, K.B. Sustainability implication of a residential building using a lifecycle assessment approach. In Proceedings of the 4th International Conference on Low Carbon Asia and Beyond (ICLCA), Johor Bahru, Malaysia, 24–26 October 2018; Chemical Engineering Transactions: Johor Bahru, Malaysia, 2018; Volume 73, pp. 1–6.
66. AccuRate. *Accurate Sustainability Tool Version 2.3.3.13 sp3*; CSIRO Land & Water Flagship and Distributed by Energy Inspection Pty Ltd.: Canberra, Austrailia, 2015.
67. Crowther, P. Design for disassembly to recover embodied energy. In Proceedings of the 16th International conference on Passive and Low Energy Architecture, Melbourne/Brisbane/Cairns, Australia, 22–24 September 1999.
68. Lawania, K.K. *Improving the Sustainability Performance of Western Australian House Construction: A Life Cycle Management Approach*; Curtin University: Perth, Australia, 2016.
69. Treloar, G.; Fay, R.; Love, P.E.D.; Iyer-Raniga, U. Analysing the life-cycle energy of an australian residential building and its householders. *Build. Res. Inf.* **2000**, *28*, 184–195. [CrossRef]
70. Fay, R.; Treloar, G.; Iyer-Raniga, U. Life-cycle energy analysis of buildings: A case study. *Build. Res. Inf.* **2000**, *28*, 31–41. [CrossRef]
71. Vass, N. Western Australia Energy Fact Sheet. Available online: http://www.australianpowerproject.com.au/western-australia-energy-fact-sheet/ (accessed on 19 December 2018).
72. Lawania, K.; Biswas, W.K. Application of life cycle assessment approach to deliver low carbon houses at regional level in western australia. *Int. J. Life Cycle Assess.* **2018**, *23*, 204–224. [CrossRef]
73. Allacker, K.; Maia de Souza, D.; Sala, S. Land use impact assessment in the construction sector: An analysis of lcia models and case study application. *Int. J. Life Cycle Assess.* **2014**, *19*, 1799–1809. [CrossRef]
74. Spang, E.S.; Moomaw, W.R.; Gallagher, K.S.; Kirshen, P.H.; Marks, D.H. The water consumption of energy production: An international comparison. *Environ. Res. Lett.* **2014**, *9*, 1–14. [CrossRef]

75. Bardhan, S. Assessment of water resource consumption in building construction in India. *WIT Trans. Ecol. Environ.* **2011**, *144*, 93–101.
76. Robert, H.G.; Treloar, J.T. An assessment of the energy and water embodied in commercial building construction. In Proceedings of the 4th Australian Life Cycle Assessment Conference, Sydney, NSW, Australia, 23–25 February 2005; pp. 1–10.

6

Mechanical and Durability Properties of Green Star Concretes

Faiz Shaikh

School of Civil and Mechanical Engineering, Curtin University, Perth, WA 6120, Australia;
s.ahmed@curtin.edu.au

Abstract: This paper presents mechanical and durability properties of green star concretes. Four series of concretes are considered. The first series is control concrete containing 100% ordinary Portland cement, 100% natural aggregates and fresh water. The other three series of concretes are green star concretes according to Green Building Council Australia (GBCA), which contain blast furnace slag, recycled coarse aggregates and concrete wash water. In all above concretes compressive strength, indirect tensile strength, elastic modulus, water absorption, sorptivity and chloride permeability are measured at 7 and 28 days. Results show that mechanical properties of green star concretes are lower than the control concrete at both ages with significant improvement at 28 days. Similar results are also observed in water absorption, sorptivity and chloride permeability where all measured durability properties are lower in green star concretes compared to control concrete except the higher water absorption in some green star concretes.

Keywords: green star concrete; slag; recycled aggregate; wash water; sustainability

---

1. Introduction

Concrete is the most widely used construction materials in the world. Ordinary Portland cement (OPC) and aggregates are the important ingredients for concrete, where the former when mixed with water forms matrix which bind the aggregates and the latter contributes to the volume of the concrete. Ordinary concrete is not environmental friendly due to energy intensive manufacturing of OPC. It is reported that manufacturing of OPC contributes between 5 and 7% of global CO_2 emission and about one ton of CO_2 is released to produce the equal amount of OPC [1–3]. On the other hand, the aggregates which are sourced from various natural sources (e.g., river beds, rocks, sand dunes, etc.) also affects the natural eco-system. As a result, significant efforts are being directed to reduce the amount of OPC and natural aggregates in concrete to reduce the adverse impact on the environment. Significant research are also conducted to reduce the carbon footprint of concrete by partially replacing OPC and natural aggregates using various supplementary cementitious materials (SCMs) and recycled aggregates, respectively [4–9]. Among various SCMs fly ash, slag and silica fume, which are industrial by-products, are most widely used in concrete as partial replacement OPC. Recycled aggregates are sourced from crushed demolished concrete structures and are used as partial replacement of natural aggregates in concrete. A very good level of understanding exists on various properties of concretes containing SCMs and recycled aggregates either individually or combined through significant amount of research [10–16].

On the other hand, huge amount of tap water is used for washing activities in ready mix concrete plants. It is reported that about 500–1500 litres of tap water is used to wash one concrete mixer truck in the ready mix concrete plant [17]. Therefore, huge amount of clean water is being used in the concrete mixing plants. The reusing of this wash water in concrete mixing is the most sustainable way of reducing the use of clean tap water in concrete mixing facilities. Generally, the wash water from concrete mixing plants contains mostly cementitious fines and few chemicals from the use of superplasticizers, which are not different from the ingredients of concrete [17]. In a number of studies the effect of wash water and recycled water on the properties of concretes are studied and all are on concrete containing natural aggregates and OPC as the main binder [17–19].

Energy efficiency of buildings is another important requirement for asset owners, builders as well as the building regulators in many countries. Generally, the reduction of heating and cooling energy of buildings are the main objective in the design of energy efficient building. In one estimate it is found that built environment is world's single largest contributor of greenhouse gas and consumes about a third of water and generates 40% of wastes (Green Building Council Australia (GBCA)). To promote the environmental efficiency of buildings "green star" rating is introduced in many countries including in Australia, which is an internationally-recognized sustainability rating system for buildings. Among various green star points, up to three points are allocated for concrete. According to green building council of Australia [20] up to "two green star" points can be awarded to the concrete containing 40% SCMs as partial replacement of OPC, while up to "three green star" points can be awarded to concrete containing 40% SCMs as partial replacement of OPC, 40% recycled materials as partial replacement of natural aggregates and 50% reclaimed water as partial replacement of tap/drinking water. The above incentive encourages the promotion of environmentally friendly concrete in the construction of buildings.

Significant research has been devoted to study the properties of concrete containing partial replacement of OPC using various SCMs of various quantities and partial replacement of natural aggregates using various amounts of recycled concrete aggregates separately as well as their combined used. No study so far reported the properties of green star concretes where industrial by-products, recycled materials and reclaimed wash water are used together. This research presents the first study of this kind where mechanical and durability properties of green star concretes are studied and compared with control concrete containing 100% OPC, natural aggregates and fresh water.

2. Materials and Methods

In this study four series of mixes were considered. The first series was control concrete consisting of 100% OPC, 100% natural aggregates and 100% tap water. The second series was "two green star concrete" contained 40% slag (a by-product of steel industry) as partial replacement of OPC, while the rest of the materials were similar to the first series. The third series was similar to the second series except where 40% natural coarse aggregate (NCA) was replaced by recycled coarse aggregates (RCA) that were sourced from construction and demolition wastes. The fourth series was "three green star concrete" which was similar to the third series except where 50% fresh water was replaced by reclaimed wash water collected from the concrete laboratory. The fourth series was between two and three green star concrete and was considered to study the effect of 50% reclaimed wash water on the properties of three green star concrete. The mix proportions of all concretes are shown in Table 1. The percentage replacements of NCA by RCA and OPC by slag were on the basis of weight. The water/binder ratio was kept constant at 0.45 in all mixes. In all three types of green star concretes as well as control concrete mechanical properties in terms of compressive strength, elastic modulus and indirect tensile strength and durability properties in terms of sorptivity, water absorption and chloride permeability were measured at 7 and 28 days.

2.1. Materials

Ordinary Portland cement (OPC) was used in all concrete mixes. The blast furnace slag used in this study was obtained from a local supplier. The properties and chemical compositions of OPC and slag are shown in Table 2. The recycled coarse aggregate was obtained from a local construction and demolition (C&D) waste recycling plant in Perth, Western Australia. Figure 1 shows the analysis of 5 kg sample of the C&D waste used as RCA in this study. The percentages are based on mass. It can be seen that approximately 69% are from concrete and the rest consisted of limestone, masonry, etc. Table 3 shows the properties of recycled and natural aggregates. Sieve analysis of RCA is also conducted and about 23% were 20 mm size and 75% were 10 mm size. The natural coarse aggregates (NCA) used in this study were mixture of 10 mm and 20 mm sizes coarse aggregates at a ratio of 1:2. The NCA and RCA used in this study were in saturated and surface dry condition before used in the mixing. Chemical analysis of tap water and reclaimed wash water in concrete laboratory is shown in Table 4. It can be seen that the pH level of wash water is higher than that of tap water, while the turbidity of the wash water is much higher than the recommended limit for drinking water (5 NTU (Nephelometric Turbidity Unit)) by World health organization [21]. The higher turbidity in wash water can be attributed to various cementitious materials, superplasticizers and alkali chemicals used in the laboratory for the production of concrete and other cementitious materials.

Table 1. Mix proportion of concretes.

Series	Mix Proportions in Kg/m^3						
	OPC	Slag	Fine Aggregate	Natural Coarse Aggregate	Recycled Coarse Aggregate	Tap Water	Wash Water
Mix 1 (Control)	413	-	512	1254	-	190	-
Mix 2 (2 Green star)	248	165	512	1254	-	190	-
Mix 3	248	165	512	752	502	190	-
Mix 4 (3 Green star)	248	165	512	752	502	95	95

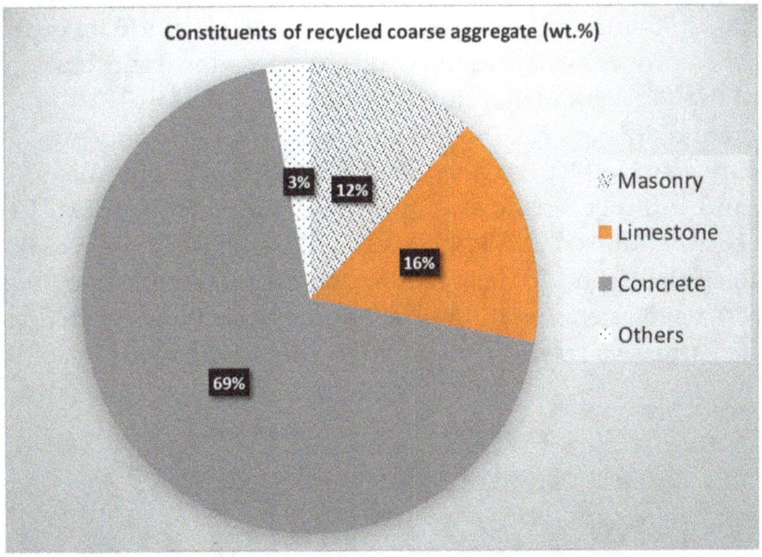

Figure 1. Analysis of construction and demolition wastes used as recycled coarse aggregates in this study.

Table 2. Chemical and physical properties of Ordinary Portland cement (OPC) and slag.

Chemical Analysis	OPC (wt %)	Slag (wt %)
SiO_2	21.1	32.45
Al_2O_3	5.24	13.56
Fe_2O_3	3.1	0.82
CaO	64.39	41.22
MgO	1.1	5.1
K_2O	0.57	0.35
Na_2O	0.23	0.27
SO_3	2.52	3.2
LOI	1.22	1.11
Specific gravity	3.17	3.00

Table 3. Properties of natural coarse aggregate (NCA) and recycled coarse aggregate (RCA).

Properties	NCA	RCA	NFA
Un-compacted bulk density (kg/m^3)	1547	1301	1498
Water absorption (%)	1.6	7.1	1.16

Table 4. Chemical analysis of tap water and wash water.

Chemical Properties	Tap Water	Wash Water
pH value	8.95	12.26
Chloride content (mg/L)	60	61
Turbidity (NTU)	-	41

2.2. Methods

The compressive strength, indirect tensile strength, elastic modulus, water absorption, water sorptivity and chloride ion permeability were measured at 7 and 28 days for all mixes. At least three specimens were cast and tested in each series for each property measured in this study. All specimens were water cured until the day before the test date. The compressive strength and elastic modulus tests were carried out on 100ø × 200 mm cylinders and the indirect tensile strength was determined on 150ø × 300 mm cylinders. The water absorption, water sorptivity and chloride ion permeability tests were conducted on 100ø × 50 mm thick specimens, which were made by cutting the standard 100ø × 200 mm cylinders. The compressive strength, indirect tensile strength and elastic modulus were measured according to the Australian standards AS1012.9 [22], AS1012.10 [23] and AS1012.17 [24], respectively. The concrete cylinders for the compressive strength and modulus of elasticity were sulphur capped to ensure a smooth surface. A MCC8 3000kN capacity machine was used to test the compressive strength and indirect tensile strength of all concrete samples. In the determination of modulus of elasticity a DMG/Rubicon 2500kN Universal Testing Machine was used to apply a constant load rate up to 40% of the ultimate load of respective concrete mix, while two linear variable differential transducers (LVDT) were used as shown in Figure 2 to measure the axial deformation of the cylinder. The slope of the recorded stress vs strain curve yielded the elastic modulus of the concrete.

Figure 2. Test setup to measure the elastic modulus of concrete cylinder.

The rate of water absorption (sorptivity) of concrete samples was measured according to ASTM C1585 [25]. The principle of the method is that a specimen had one surface in free contact with water (no more than 5 mm above the base of the specimen) while the other sides were sealed. This test determined the rate of absorption of water by hydraulic cement concrete by measuring the increase in the mass of a specimen resulting from absorption of water as a function of time. In this study the mass of the concrete specimen was measured regularly to determine the initial absorption from 1 min to the first 6 hours. The absorption *I* was the change in mass divided by the product of the cross-sectional area of the test specimen and the density of water. The initial rate of water absorption value (mm/sec$^{1/2}$) was calculated as the slope of the line that is the best fit to *I* plotted against the square root of time (sec$^{1/2}$). The chloride ion penetration resistance of concrete, popularly called the rapid chloride permeability test (RCPT), was conducted according to ASTM C1202 [26], details of which can be found in the standard. Water absorption of all concretes were measured according to the Australian standard AS1012.21 [27].

3. Results and Discussion

The workability of green star concretes in terms of measured slump values are shown in Figure 3 and is bench marked with respect to control concrete. It can be seen that the slump value of two green star concrete is higher than that of control concrete and can be attributed to the use of 40% slag as partial replacement of OPC. The relatively lower specific gravity of slag than that of cement increases the paste volume in concrete containing 40% slag, which cause the improvement in workability. The slump of mix 2 is decreased from 175 mm to 160 mm when 40% RCA is used as partial replacement of NCA in mix 3. This reduction can be attributed to the presence of 40% RCA whose water absorption is higher than NCA and contain more 10 mm size aggregates than the NCA. The workability is slightly affected in three green star concrete (mix 4) due to addition of 50% reclaimed wash water as partial replacement of tap water. High turbidity and pH of wash water are attributed to this slight reduction in workability. Nevertheless the workability of all green star concretes is better than the control concrete.

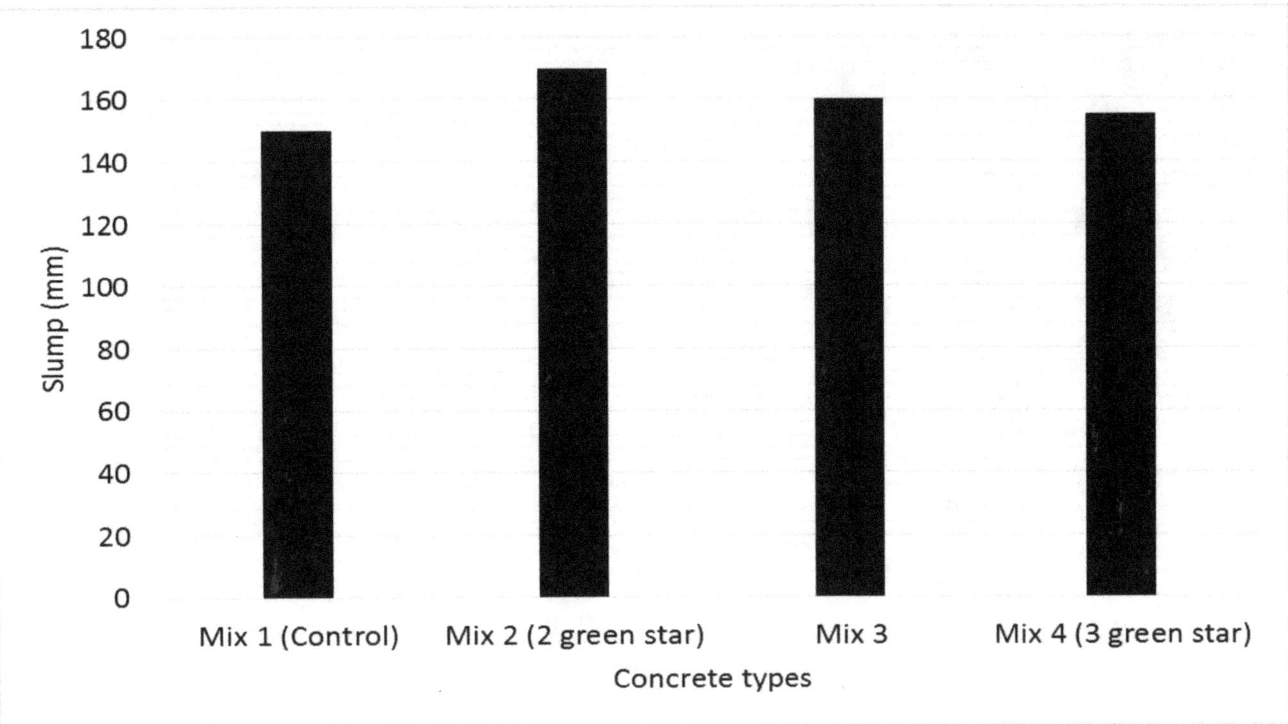

Figure 3. Measured slump values of green star concretes and control concrete.

The measured compressive strength of all concretes at 7 and 28 days are shown in Figure 4. It can be seen that the compressive strength of green star concretes is lower than the control concrete at both ages especially at 7 days. The partial replacement of OPC by 40% slag in two green star concrete exhibited about 31% and 12% reduction, respectively at 7 and 28 days compared to control concrete. The slow pozzalonic reaction of slag is the reason for higher compressive strength loss at 7 days in this concrete than at 28 days.

However, when 40% RCA is used as partial replacement of NCA in mix 3 a slight reduction in both 7 and 28 days compressive strength is observed compared to mix 2. A slight reduction in both 7 and 28 days compressive strength is also observed in three green star concrete (Mix 4) when 50% wash water is used as partial replacement of tap water compared to Mix 3. The presence of weak interfacial transition zone between old mortar of RCA with new matrix and the higher turbidity of wash water could be the reason of the observed slight reduction in compressive strength in mixes 3 and 4 than mix 2.

Nevertheless the three green star concrete containing 40% less OPC, 40% less NCA and 50% less fresh water exhibited a 28 days compressive strength of 38 MPa, which is 18% lower than the control concrete but still adequate enough for structural application. Similar reduction trend in indirect tensile strength and elastic modulus is also observed in all green star concretes compared to control concrete. It can be seen in Figures 5 and 6 that the reduction of indirect tensile strength of all green star concretes is slightly lower than that of compressive strength at both ages, however, in the case of elastic modulus the reduction was higher than the compressive strength. It is also interesting to observe that indirect tensile strength and elastic modulus of three green star concrete at both ages are similar and slightly higher, respectively than those of mix 3 concrete.

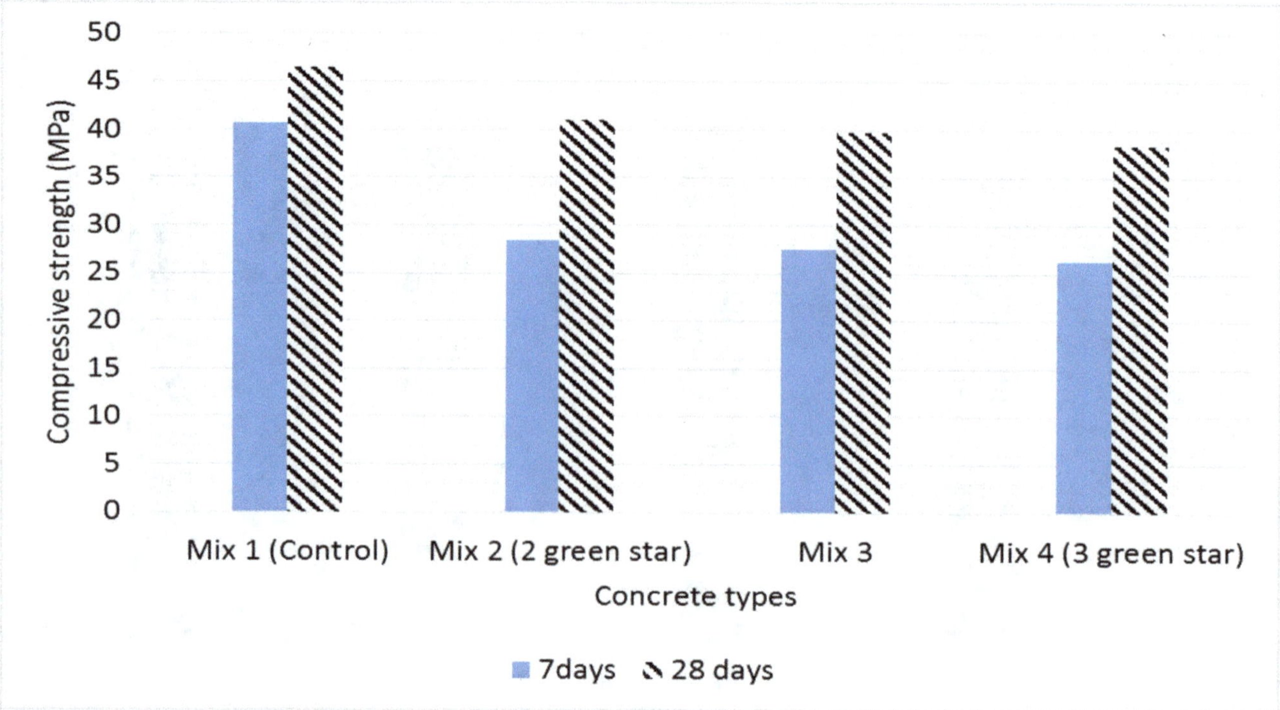

Figure 4. Measured compressive strength of green star concretes and control concrete.

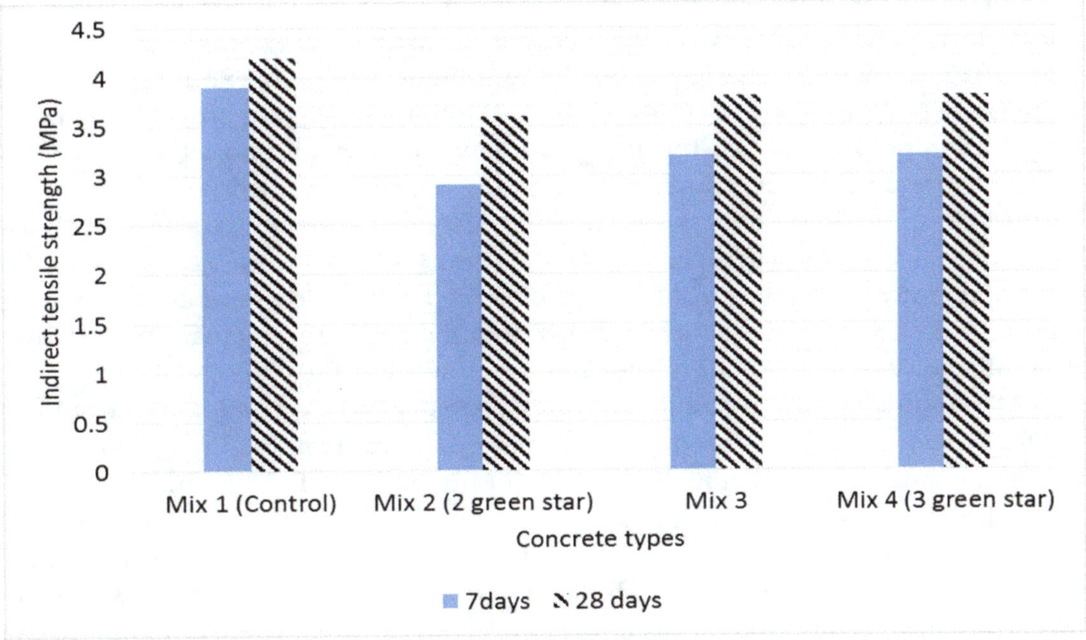

Figure 5. Measured indirect tensile strength of green star concretes and control concrete.

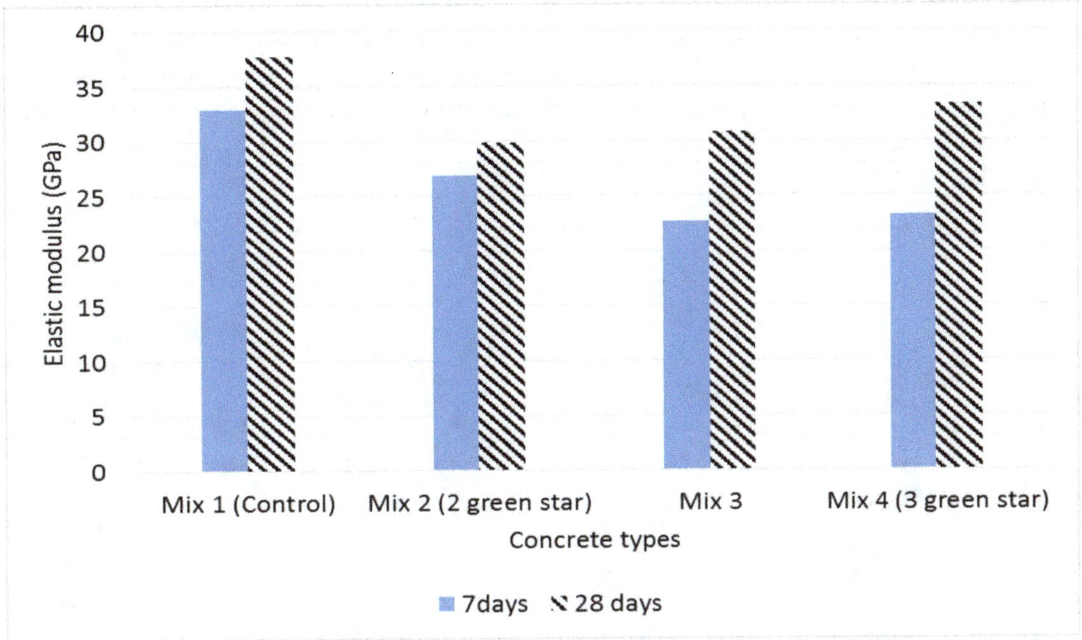

Figure 6. Measured elastic modulus of green star concretes and control concrete.

The measured durability properties of green star concretes are shown in Figures 7–10. Figure 7 shows the water absorption of green star concretes. No changes in water absorption of two green star concrete compared to control concrete after 7 days of wet curing can be seen, however, about 15% reduction is observed after 28 days curing. This can be attributed to the pozzolaanic reaction of SiO_2 and Al_2O_3 of slag with $Ca(OH)_2$ of hydration reaction which densified the matrix through formation of additional hydration products and pore filling by the slag particles. The three green star concrete and mix 3, as expected, showed much higher water absorption than control and two green star concretes due to the presence of 40% RCA, whose water absorption capacity is much higher than NCA as shown in Table 3. The effect of wash water however is not affected the water absorption of three green star concrete than of mix 3 concrete. Unlike water absorption, the rate of water absorption of all green star concretes is lower than that of control concrete at both ages. Among green star concretes, the two green star concrete containing 40% slag as partial replacement of OPC exhibited the lowest rate of water absorption at both ages and is believed to be due to the pore refinement of its matrix due to additional hydration product formed through pozzolanic reaction and pore filling. After 28 days of curing this rate further reduced. A summary of water sorptivity, which is the slope of the rate of water absorption lines in Figure 8, is shown in Figure 9. It can be seen that the sorptivity of mix 3 and mix 4 concretes is very similar, where no adverse effect of wash water on sorptivity is observed in three green star concrete. Similar result is also observed in chloride ion penetration in green star concretes in RCPT test. It can be seen in Figure 10 that the chloride permeability in two green star concrete is significantly reduced by 62% after 28 days of wet curing. This can be contributed by the pore refinement of matrix in two green star concrete due to pozzolanic reaction of slag and pore filling. However, after 7 days of curing the reduction in chloride permeability is much lower than that at 28 days due to slow pozzolanic reaction of slag. The addition of 40% RCA in mix 3 concrete, however, increased the chloride permeability possibly due to higher porosity of RCA than NCA and the presence of more interfacial transition zone with matrix and RCA than with matrix and NCA. Interestingly, the three green star concrete shows lower chloride permeability than the other two green concrete mixes, which is not observed in water absorption and sorptivity tests, where a slight increase in water absorption and sorptivity is observed in three green star concrete than mix 3. While all three measured durability properties are affected by the pores in the concrete, the water absorption and sorptivity are mostly depend on overall porosity of the concrete as the specimens in those tests are kept in water or in contact with water for long time. However, in RCPT test Cl^{-} is forced to pass from one side of the concrete specimen to other side. Therefore, the observed lower chloride ion penetration in three green star concrete than other two green star concretes indicate that the pores in that concrete are not interconnected rather disperse.

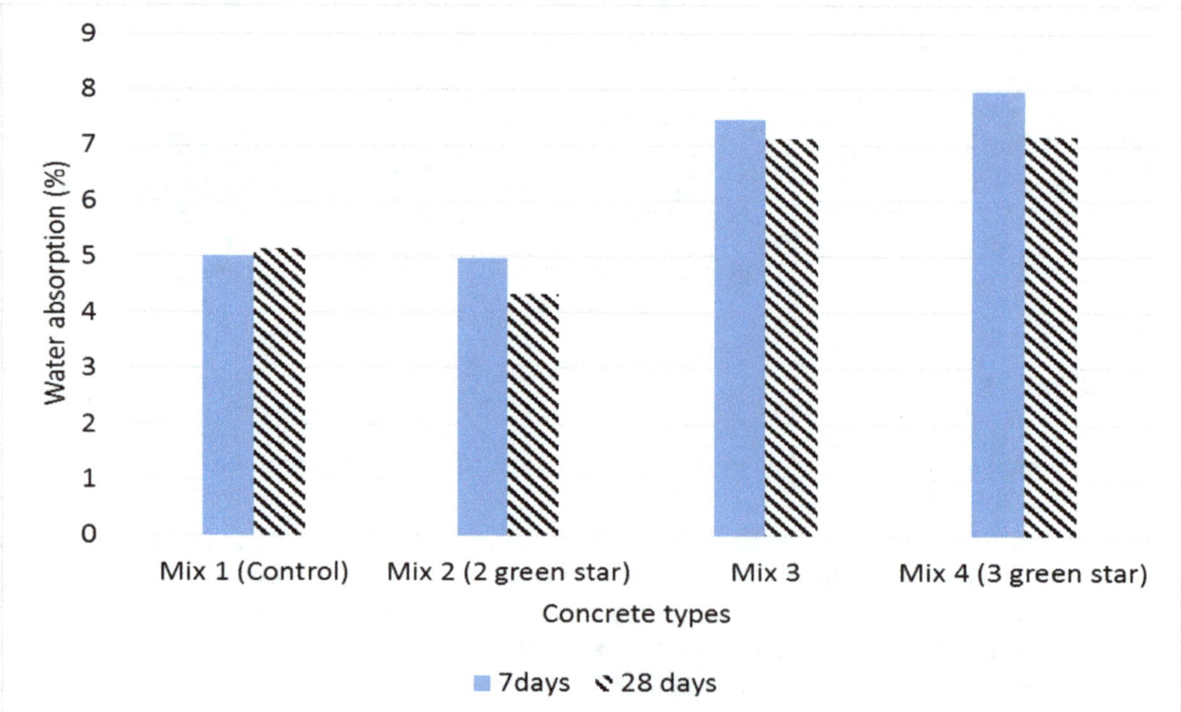

Figure 7. Measured water absorption of green star concretes and control concrete.

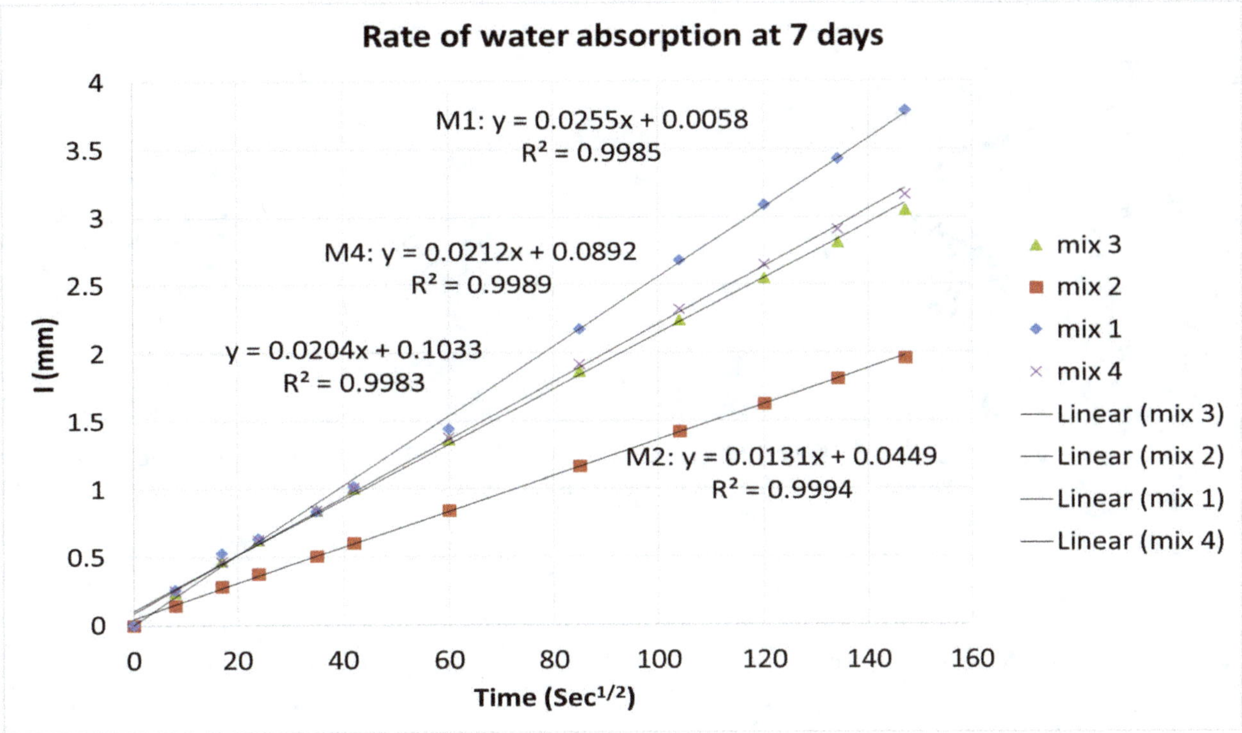

(a)

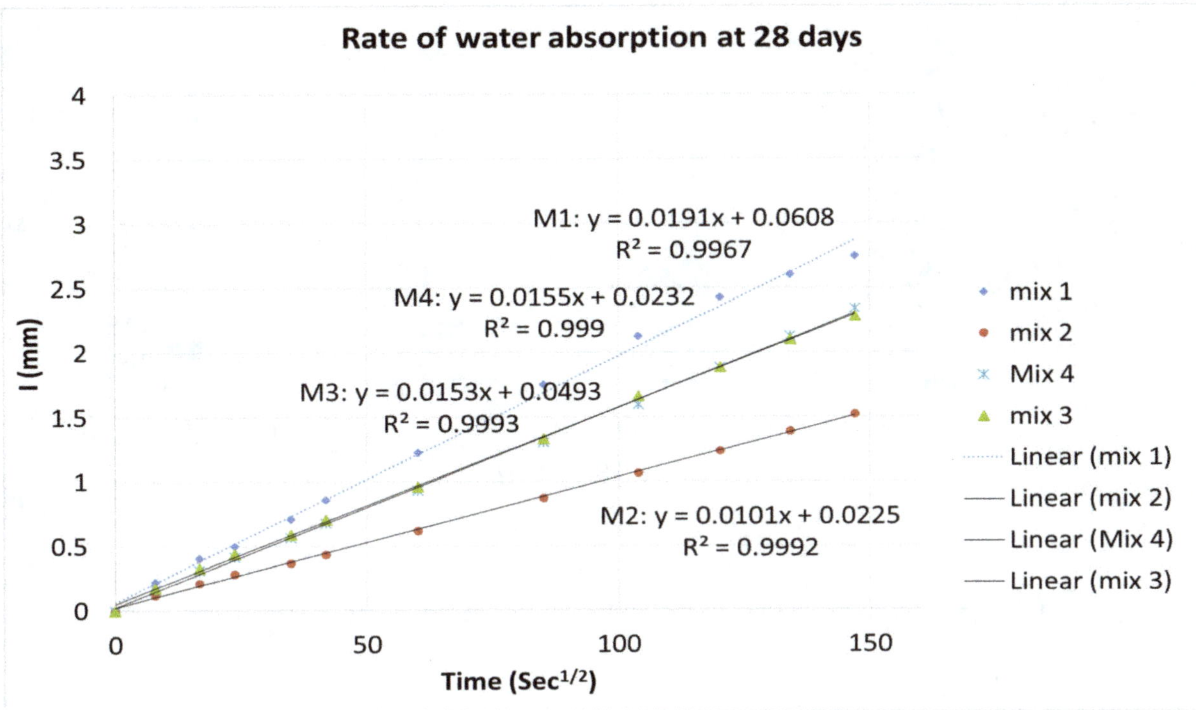

(b)

Figure 8. Rate of water absorption of green star concretes and control concrete at (**a**) 7 days and (**b**) 28 days.

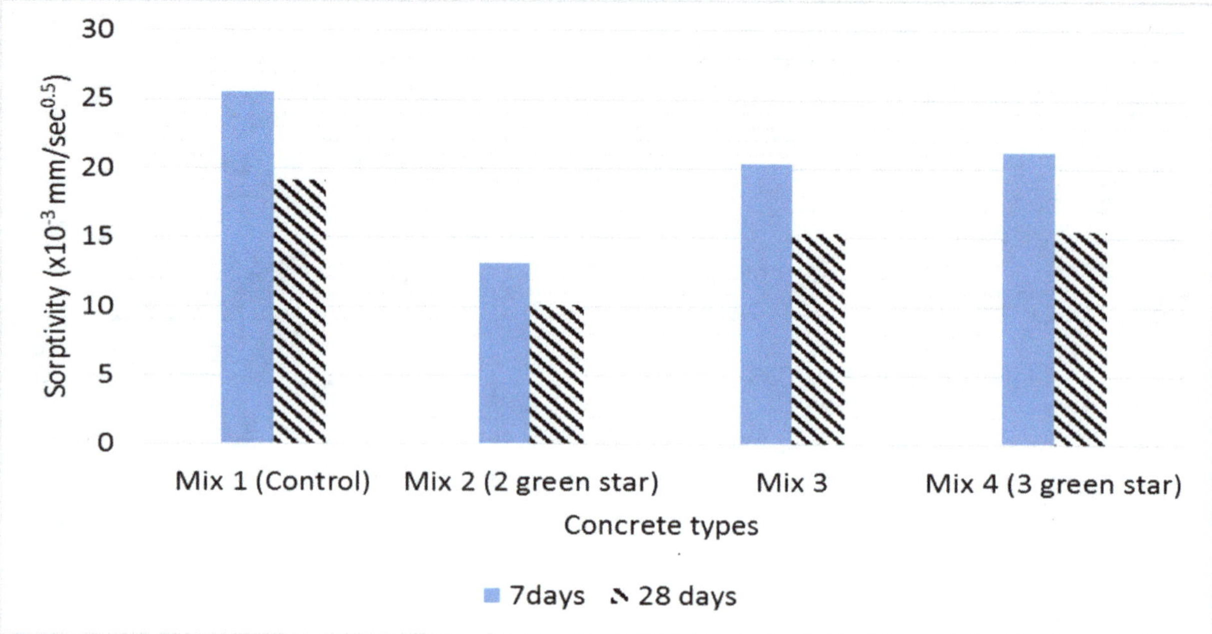

Figure 9. Summary of sorptivity of green star concretes and control concrete.

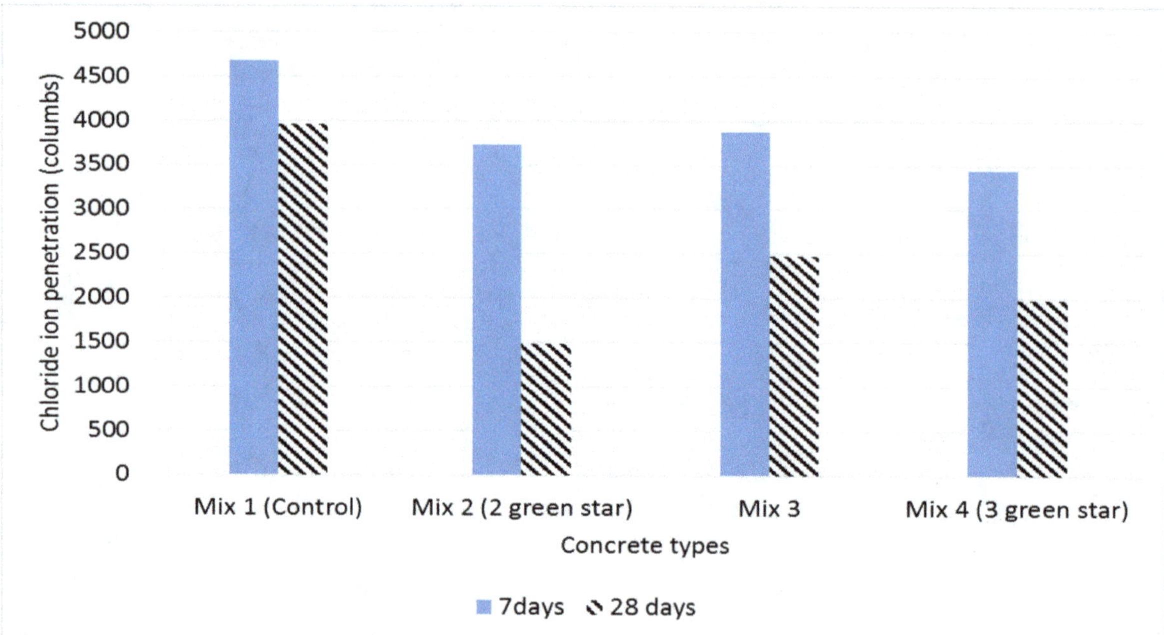

Figure 10. Measured chloride ion prenetration of green star concretes and control concrete.

4. Conclusions

This study evaluated mechanical and durability properties of green star concretes according to the definition of green building council Australia and bench marked with control concrete. Within limited studies the following conclusions can be drawn:

- Green star concretes exhibited reduction in compressive strength, indirect tensile strength and elastic modulus at both 7 and 28 days compared to control concrete. However, the reduction at 7 days of green star concrete is much higher than at 28 days. Formation of additional hydration products and particle packing due to pozzolianic reaction and small particle size of slag are contributed to the increase in above mechanical properties at 28 days. Three green star concrete exhibited about 18%, 10% and 12% reduction in compressive strength, indirect tensile strength and elastic modulus, respectively at 28 days compared to control concrete while containing 40% less OPC, 40% less NCA and 50% less fresh water.

- The two green star concrete containing 40% slag exhibited lower water absorption, sorptivity and chloride permeability than control concrete at both ages, while three green star concrete containing 40% slag, 40% RCA and 50% concrete wash water exhibited better durability properties at both ages except the water absorption which is about 25% higher than two green star concrete at 28 days. The presence of 40% RCA, whose water absorption capacity is much higher than the NCA, is the reason for such increase in water absorption.
- Both two and three green star concretes exhibited 28 days compressive strength of about 40 MPa and much lower chloride permeability than control concrete. Therefore, these green concretes can be used in structural application with good resistance against reinforcement corrosion.

While only slag and concrete wash water are used to partially replace the OPC and tap water, respectively in the green star concretes in this study, other SCMs. E.g., fly ash, silica fume, combination of fly ash/slag/silica fume and recycled water from other sources can be used in green star concretes and studied their properties to establish their use.

Acknowledgments: Author acknowledge to two final year project students Brett Milman and Miss Aije for their assistance in casting and testing of all concrete specimens in this study.

References

1. Moretti, L.; Caro, S. Critical analysis of the life cycle assessment of the Italian cement industry. *J. Cleaner Prod.* **2017**, *152*, 198–210. [CrossRef]
2. Limbachiya, M.; Bostanci, S.C.; Kew, H. Suitability of BS EN197 CEM II and CEM V cement for production of low carbon concrete. *Constr. Build. Mater.* **2014**, *71*, 397–405. [CrossRef]
3. Malhotra, V.M. Sustainable development and concrete technology. *Concr. Int.* **2002**, *24*, 1–22.
4. Malhotra, V.M.; Mehta, P.K. *High-Performance, High-Volume Fly Ash Concrete: Materials, Mixture Proportioning, Properties, Construction Practice, and Case Histories*; HVFA High-Performance-Inc.: Ottawa, ON, Canada, 2002.
5. Siddique, R. Performance characteristics of high-volume class F fly ash concrete. *Cem. Concr. Res.* **2004**, *34*, 487–493. [CrossRef]
6. Kou, S.C.; Poon, C.S.; Chan, D. Influence of fly ash as a cement addition on the hardened properties of recycled aggregate concrete. *Mater. Struct.* **2007**, *41*, 1191–1201. [CrossRef]
7. Shaikh, F.U.A.; Supit, S.W.M. Compressive strength and durability properties of high volume fly ash concretes containing ultrafine fly ash. *Constr. Build. Mater.* **2015**, *82*, 192–205. [CrossRef]
8. Berndt, M.L. Properties of sustainable concrete containing fly ash, slag and recycled concrete aggregate. *Constr. Build Mater.* **2009**, *23*, 2606–2613. [CrossRef]
9. Corinaldesi, V.; Moriconi, G. Influence of mineral additions on the performance of 100% recycled aggregate concrete. *Constr. Build. Mater.* **2009**, *23*, 2869–2876. [CrossRef]
10. Poon, C.S.; Shui, Z.H.; Lam, L. Effect of microstructure of ITZ on compressive strength of concrete prepared with recycled aggregates. *Constr. Build. Mater.* **2004**, *18*, 461–468. [CrossRef]
11. Sagoe-Crentsil, K.K.; Brown, T.; Taylor, A.H. Performance of concrete made with commercially produced coarse recycled concrete aggregate. *Cement Concr. Res.* **2001**, *31*, 707–712. [CrossRef]
12. Ahmed, S.F.U. Properties of concrete containing construction and demolition wastes and fly ash. *J. Mater. Civ. Eng.* **2013**, *25*, 1864–1870. [CrossRef]
13. Ahmed, S.F.U. Properties of concrete containing recycled fine aggregate and fly ash. *J. Solid Waste Technol. Manage.* **2014**, *40*, 70–78. [CrossRef]
14. Shaikh, F.U.A.; Odoh, H.; Than, A.B. Effect of nano silica on properties of concrete containing recycled coarse aggregates. *Constr. Mater.* **2014**, *168*, 68–76. [CrossRef]
15. Shaikh, F.U.A. Effect of ultrafine fly ash on properties of concretes containing construction and demolition wastes as coarse aggregates. *Struct. Concr.* **2016**, *17*, 116–122. [CrossRef]

16. Zhang, W.; Ingham, J.M. Using recycled concrete aggregates in New Zealand ready-mix concrete production. *J. Mater. Civ. Eng.* **2010**, *22*, 443–450. [CrossRef]
17. Kadir, A.A.; Shahidan, S.; Yee, L.H.; Hassan, M.I.H.; Abdullah, M.A. The effect on slurry water as a fresh water replacement in concrete properties. *IOP Conf. Ser. Mater. Sci. Eng.* **2016**, *133*, 012041. [CrossRef]
18. Al-jabri, K.S.; Al-saidy, A.H.; Taha, R.; Al-kemyani, A.J. Effect of using waste water on the properties of high strength concrete. *Procedia Eng.* **2011**, *14*, 370–376. [CrossRef]
19. Lobo, C.; Mullings, G.M. Recycled water in ready mixed concrete operations. *Concr. Focus* **2003**, *2*, 1–10.
20. Revised Green Star Concrete Credit, Green Building Council Australia. Available online: https://www.gbca.org.au/green-star/materials-category/revised-green-star-concrete-credit/34008.htm (accessed on 30 October 2017).
21. Water Research Centre National Secondary Drinking Water Standards. Available online: https://www.water-research.net/index.php/standards/secondary-standards (accessed on 9 July 2018).
22. AS 1012.9. *Determination of Compressive Strength of Concrete*; Australia Standards: Sydney, Australia, 2010.
23. AS 1012.10. *Determination of Indirect Tensile Strength of Concrete*; Australia Standards: Sydney, Australia, 2010.
24. AS 1012.17. *Determination of Elastic Modulus of Concrete*; Australia Standards: Sydney, Australia, 2010.
25. ASTM C1585-13. *Standard Test Method for Measurement of Rate of Absorption of Water by Hydraulic Cement Concretes*; ASTM International: West Conshohocken, PA, USA, 2013.
26. ASTM C1202. *Standard Test Method for Electrical Indication of Concrete's Ability to Resist Chloride ion Penetration*; American Society for Testing and Materials: Philadelphia, PA, USA, 2012.
27. AS 1012.21. *Determination of Water Absorption and Apparent Volume of Permeable Voids in Hardened Concrete*; Australia Standards: Sydney, Australia, 1999.

7

Comprehensive Evaluation of Carbon Emissions for the Development of High-Rise Residential Building

Stephen Y. C. Yim [1], S. Thomas Ng [2,*], M. U. Hossain [2] and James M. W. Wong [3]

1 Development and Construction Division, Housing Department, Housing Authority Headquarters, Ho Man Tin, Hong Kong, China; stephen.yim@housingauthority.gov.hk
2 Department of Civil Engineering, The University of Hong Kong, Pokfulam, Hong Kong, China; uzzal@hku.hk
3 Research & Development, Construction Industry Council, 38/F, COS Centre, 56 Tsun Yip Street, Kwun Tong, Kowloon, Hong Kong, China; jameswong@cic.hk
* Correspondence: tstng@hku.hk

Abstract: Despite the fact that many novel initiatives have been put forward to reduce the carbon emissions of buildings, there is still a lack of comprehensive investigation in analyzing a buildings' life cycle greenhouse gas (GHG) emissions, especially in high-density cities. In addition, no studies have made attempt to evaluate GHG emissions by considering the whole life cycle of buildings in Hong Kong. Knowledge of localized emission at different stages is critical, as the emission varies greatly in different regions. Without a reliable emission level of buildings, it is difficult to determine which aspects can reduce the life cycle GHG emissions. Therefore, this study aims to evaluate the life cycle GHG emissions of buildings by considering "cradle-to-grave" system boundary, with a case-specific high-rise residential housing block as a representative public housing development in Hong Kong. The results demonstrated that the life cycle GHG emission of the case residential building was 4980 kg CO_2e/m^2. The analysis showed that the majority (over 86%) of the emission resulted from the use phase of the building including renovation. The results and analysis presented in this study can help the relevant parties in designing low carbon and sustainable residential development in the future.

Keywords: greenhouse gases; residential building; life cycle assessment

---

1. Introduction

Climate change has become an unprecedented challenge for humanity. The annual greenhouse gas (GHG) emissions grew on average by 1.0 giga ton carbon dioxide equivalent ($GtCO_2e$) per year from 2000 to 2010 compared to 0.4 $GtCO_2e$ per year from 1970 to 2000, and total anthropogenic GHG emissions were the highest in human history reaching 49.0 $GtCO_2e/y$ in 2010 [1]. These phenomena are primarily due to various human activities, in particular the use of fossil fuels, deforestation, and change in land use [2]. Any delay in stabilizing and reducing the atmospheric CO_2e concentration would only exacerbate the global warming crisis and increase the difficulty to tackle the disastrous consequences in the future [3].

Currently, the building sector represents the single largest contributor to GHG emissions [4,5]. To help reduce GHG emissions, the building sector has an undeniable role to play as buildings worldwide account for up to one-third of the GHG emissions [6]. For subtropical countries and cities like Hong Kong, buildings can contribute to almost 60% of final energy consumption [7]. Of this, residential buildings take up a significant portion of total energy consumption and hence the GHG emissions, resulting from energy used for construction, operation, and demolition of buildings.

With a continuous growth in population, a preference for smaller family sizes, and the desire for a more comfortable living environment, the energy demands and GHG emitted from residential buildings in Hong Kong are expected to escalate even further [8].

Establishing pragmatic policy to encourage the building sector to cut down on GHG emissions is clearly an important goal for governments around the world. This is particularly the case for high-density cities not only because there are lots of high-rise buildings but also due to the rather limited opportunities to adopt emerging renewable energy solutions like photovoltaic panels or wind turbines. For Hong Kong, having committed to reducing the energy intensity by at least 25% by 2030 compared with the 2005 levels, its government has begun to examine the overall life cycle environmental burdens of buildings under their jurisdiction, and the public housing developments would be an ideal starting point as they share around one-third of the entire residential stock in Hong Kong, which is equivalent to 700,000 flats [9]. In 2008, public housing in Hong Kong consumed 6988 million kWh of electricity or five million tons (Mt) of CO_2e [7].

Various studies have been conducted to gauge the environmental impacts of buildings. For example, Chen and Ng [10] proposed factoring in the embodied GHG emissions when assessing the environmental performance of buildings. De Wolf et al. [11] investigated the GHG emissions from 200 recently completed buildings based on the quantities of structural materials (data were based on different design firms) in the United States, without considering the whole life cycle of buildings. A life cycle assessment model was developed to evaluate the environmental impacts of building construction [12]. Peuportier [13] compared the environmental performance of three types of houses located in France, and a sensitivity analysis was performed based on the choice of alternative construction materials, types of heating energy, and transportation using an EQUER tool. Similar studies were conducted in France [14], the Netherlands [15], Japan [16], the United Kingdom [17], and China [18]. Some of the reviews were conducted in assessment of GHG emissions, energy consumption, and other environmental impacts of buildings [19–22].

Focusing on the assessment of GHG emissions generated from the building sector, Suzuki and Oka [23] proposed quantifying the energy consumed and carbon emitted due to the construction, operation, and renovation of office buildings in Japan using input/output tables. On the other hand, Seo and Hwang [24] estimated the life cycle CO_2 emissions of different types of residential buildings. Similarly, Bastosa et al. [25] presented a life cycle energy and GHG analysis of three residential building types in Lisbon. Some studies have also focused on the specific stage of the buildings, such as the material level [26,27], building construction [28,29], renovation [30,31], demolition, and end-of-life treatment [32]. The collection of a large variety of data to model a comprehensive assessment is not only time consuming but also, is often impossible. However, a few studies focused on assessing the GHG emissions of the whole building by considering different stages, but excluding the renovation and end-of-life waste treatment [33–37]. Recent reviews also concluded that the occupancy and end-of-life phases are overlooked in most of the life cycle assessment (LCA) studies of building assessment [19,38–40].

In addition, environmental impacts of buildings can significantly vary among the studies depending upon the regions or countries [38,39]. A few studies were conducted on environmental assessment, including GHG emissions of buildings in Hong Kong [12,41,42]. However, these studies have excluded some important aspects in their assessment, e.g., considerations of use, renovation, and end-of-life phases of the building. The aim of this research therefore, is to evaluate the life cycle GHG emissions of high-rise residential building comprehensively by including the construction, use and renovation, and end-of-life phases as a case in Hong Kong. The results of the study can be used as a benchmark for comparing and setting up mitigation measures for new building construction.

2. Methodology

The life cycle assessment (LCA) method has been used for assessing the GHG emissions of high-rise public housing blocks in this study. LCA enables the quantification and evaluation of environmental impacts of a building [41]. Governed by the ISO 14040 standard [43], an analytical skeleton is applied in this study which consists of four main phases; goal and scope definition, life cycle inventory analysis, life cycle impact assessment, and interpretation.

2.1. Goal and Scope of Study

This study aimed to evaluate the GHG emissions (in terms of CO_2e) from cradle-to-grave of a public housing block as shown in Figure 1. The GHG emissions are calculated by assessing the GHGs as defined in the Kyoto Protocol of the United Nations Framework Convention on Climate Change (UNFCC), including carbon dioxide (CO_2), methane (CH_4), nitrous oxide (N_2O), hydro fluorocarbons (HFCs), perfluorocarbons (PFCs), and sulphurhexafluoride (SF_6) [44,45]. These GHG emissions are converted into kg or tCO_2e emissions using the Intergovernmental Panel on Climate Change (IPCC) 100-year global warming potential (GWP) coefficients [46]. In this study, the functional unit was the unit of flat and gross floor area (GFA) of the building, i.e., m^2.

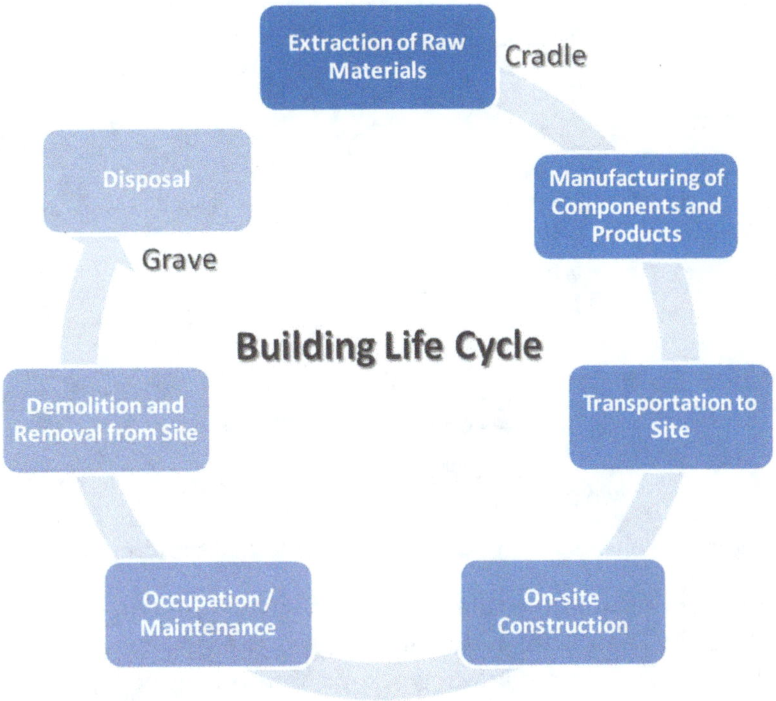

Figure 1. Life cycle process of a building.

A standard housing block design named "New Harmony One" (NH1)—Option 6, as shown in Figure 2, was selected for the analysis and used for setting a benchmark for the life cycle GHG emissions of public residential buildings. The Housing Authority adopted a site specific design approach and the internal floor area since 2004, by applying micro-climate studies at the early planning stage [47]. NH1 is selected as a basis of this study, as such a design can be applied to various sites in Hong Kong on a repetitive basis. Typically, a NH1 block is a reinforced concrete tower of 40 domestic levels which contains 799 flats with a gross floor area of 33,078 m^2. The ground floor is used for non-domestic purpose to accommodate the necessary ancillary facilities. There are 16–20 modular flats per floor which are arranged in four groups in a cruciform configuration attached to the central core where building services, lifts, and staircases are located. The compact form of NH1 makes it suitable for use in smaller urban area sites in Hong Kong.

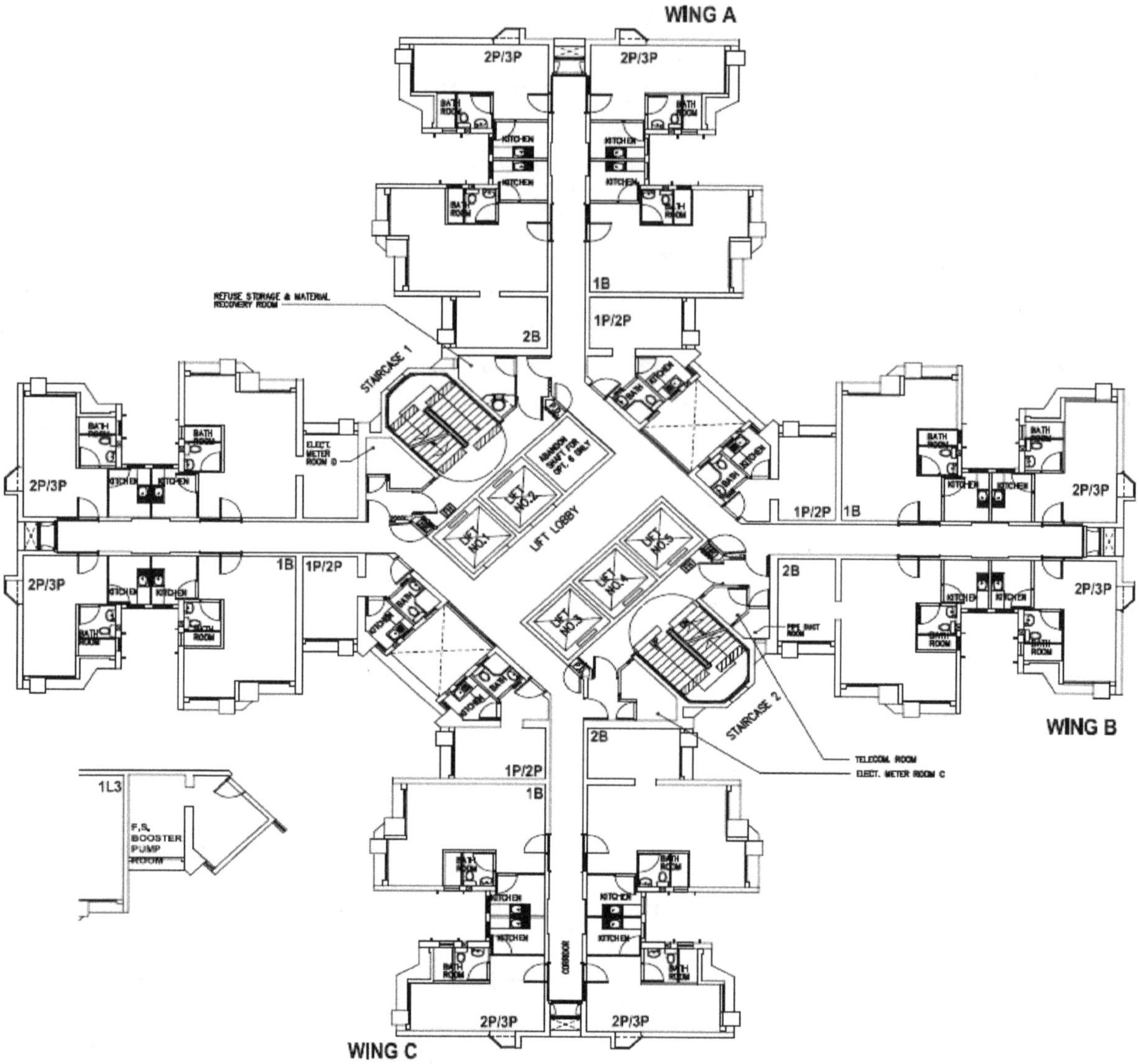

Figure 2. New Harmony One (NH1) residential building design.

The GHG emissions over the building's life cycle are assessed based on their sources and components, so as to evaluate the carbon footprint of the building meaningfully. Based on ISO 21931-1 [48] "Sustainability in Building Construction—Framework for Methods of Assessment of the Environmental Performance of Construction Works—Part 1: Buildings", the scope of a carbon audit study includes eight aspects associated with the following three distinct but interlinked stages: (i) production and construction; (ii) occupation (both energy consumed by tenants and communal installations) and renovation; and (iii) demolition as shown in Table 1. The guidelines provided by this ISO standard were used to derive the equations (Equations (1)–(7)) in this study. In addition, the said method aligns well with the carbon emission estimation model developed by the Hong Kong Housing Department [9], and Equations (1)–(7) were therefore used for the assessment of each individual stage accordingly. These aspects cover the major sources of GHG emissions of a building's life cycle as suggested by Seo and Hwang [24] and Fieldson et al. [49] which form the system boundary of this study. As a result, a "cradle-to-grave" system boundary with the functional unit of 1 m^2 of building floor area was considered in this study. The system boundary covers the production and transportation of principal construction materials; the use stage of buildings including the energy consumed by building services equipment and utilities; renovations including the material's production and transportation; waste transportation and disposal; and the end-of-life stage of buildings including the dismantling of buildings, and transportation of waste materials to the disposal sites (Table 1). However, the energy and fuel used on site during construction were excluded, as their emissions are minimal compared with the emissions of the entire building's life cycle. For example, the construction processes contribute to only about 2–5% of the total emissions (except refurbishments, demolition, and waste treatment) [28,50]. Taking into account the quantity and environmental profile, this study initially focused on three major materials, namely concrete, steel, and timber as they are the dominant contributors of embodied carbon of a housing block [51]. An inventory of materials and energy consumed over the sampled building's life cycle is assessed to calculate the associated GHG emissions. However, the inventory given in Table 1 omits non-structural materials and associated emissions over the building's life cycle.

Table 1. Study scope and system boundary of the greenhouse gas (GHG) evaluation.

Stage	Aspect	Sources of GHG Emissions	
Production and Construction stage	I	Materials consumed during construction	Steel formwork for superstructure Timber formwork for superstructure Steel formwork for substructure Timber formwork for substructure
	II	Materials for structure	Steel for superstructure Concrete for superstructure Steel for substructure Concrete for substructure
	III	Transportation of materials (in Aspects I and II)—from factory gate to site	-
Occupation stage	IV	Energy consumption by communal building services	Lighting Lift Security TV A/C and ventilation Fire services Water supply Electrical distribution
	V	Energy consumption by tenants	Cooking Space conditioning Hot water Lighting Refrigeration Others (laundry, audiovisual and miscellaneous equipment)
	VI	GHG removals	Planting trees (taller than 5 m)
-	VII	Renovation	Materials replacement (production and transport) Waste transport and disposal
Demolition stage	VIII	Disposal	Dismantling of building Transportation of building debris from site to disposal sites

2.2. *Inventory Analysis and Analytical Framework*

2.2.1. Emissions in the Construction Stage (Aspects I, II, and III)

The construction process of building follows the NH1 design of construction. As indicated, the GHG emissions of the construction process is excluded from this study, as it contributes to a negligible amount of emissions compared to the total emission associated with building. While the energy used and the consequential GHG emissions over the occupation of a building contribute to the majority of its carbon footprint, a considerable amount of GHG is emitted during the manufacturing and transportation of building materials [6]. Equation (1) calculates the embodied GHG emissions of key building materials consumed during construction and for structure (i.e., Aspects I and II, respectively), including concrete, steel, and formwork. This accounts for 84–95% of the total materials (structural and non-structural) related GHG emissions for a reinforced concrete framed building [26,52].

This measures the GHG emitted from the extraction, processing and manufacturing of building materials [53].

$$GHG_m = \sum_{i=0}^{n} Q_i \times F_i^m, \quad (1)$$

where GHG_m is the total embodied GHG emissions of concrete, steel, and formwork (in kg CO_2e); Q_i is the amount of building material i (in m^3); and F_i^m is the GHG emission factor for building material i (in kg CO_2e/m^3).

The quantities of concrete, steel, and timber employed during the construction of the NH1 housing block were obtained from the tender documents as well as the drawings. Local, regional, and international sources and databases were used to retrieve the GHG emission factors for the selected building materials (Table 2). For example, the GHG emission factor for steel production was extracted from the Inventory of Carbon and Energy (ICE) compiled by Hammond and Jones [54], which is within the range of steel production in China [55]; local concrete production was according to Zhang et al. [56]; and regional (Southern China) timber production was from Zhang [57]. The use of local or regional GHG emission factors for the principal building materials is important for achieving representative results. Therefore, local or regional GHG emission factors for such materials were used in this study.

Table 2. Embodied carbon of materials (unit: kg CO_2e/m^3).

References	Timber	Steel	Concrete
Hammond and Jones [54]	468	15,210 [b]	317
Morris [58]	450	14,287	326
Eaton and Amato [59]	-	15,313	-
Zhang et al. [56]	-	-	426 [c]
Zhang [57]	962 [a]	-	-
Alcorn [60]	-	10,441	376

Note: [a] plywood, embodied carbon: 1.78 kg CO_2e/kg in China with the density as of 540 kg/m^3 [57]; [b] steel bar and rod, embodied carbon: 1.95 kg CO_2e/kg which is within the range of steel production in China (1.72–1.96 kg CO_2e/kg steel) according to Jing et al. [55]; non-EU average recycled content: 35.5%; density is assumed as 7800 kg/m^3 [61]; [c] concrete grade is assumed as 32/40 MPa, embodied carbon: 0.177 kg CO_2e/kg, density is assumed as 2400 kg/m^3 [61].

Transportation emissions are also an integral part of the LCA study, generated from the transportation of construction materials from cradle-to-gate and from gate-to-site. Aspect III focuses on the latter stage, i.e., the transportation distances from the manufacturing plant to the construction site. The former stage has already been included in the embodied carbon emission factors as shown in Table 2, which embraces all energy required for extraction, manufacturing, and transportation until the materials leave the factory gate. In the case of the NH1 block, 95% of the building materials can be sourced from Hong Kong or South China, they are therefore transported through the land routes using diesel trucks and by sea [51]. Equation (2) calculates the GHG emissions generated from the transportation of building materials.

$$GHG_t = \sum_{i=0}^{n} \frac{(Q_i^l \times E_l \times D_i^l \times F_l^t)}{5}, \quad (2)$$

where GHG_t is the total GHG emissions from fuel combustion of transportation of the key building materials (in kg CO_2e); Q_i^l is the amount of building material i transported by land (in m^3), assuming the loading limit per truck is 5 m^3; E_l is the diesel consumption (in liter/km/truck), which is 0.325 L/km; D_i^l is the total distances of transporting building materials i by land (in km), and the distances between the site and the manufacturing plant in Hong Kong and South China are assumed as 20 km [52] and 250 km [62], respectively; and F_l^t is the emission factors of transporting by diesel truck, which is 2.62 kg CO_2e/liter [52].

2.2.2. Emissions in the Occupation Stage (Aspects IV, V, VI and VII)

During the occupation phase, heating and electricity account for the major portion of GHG emissions [63]. For this study, the GHG emissions at the occupation stage are quantified and classified into three aspects, namely: Aspect IV—Energy consumption by communal building services; Aspect V—Energy consumption by tenants; Aspect VI—GHG removals; and Aspect VII—Renovation. Equation (3) is used to simulate the GHG emissions from the energy consumption by communal building services and tenants.

$$GHG_o = \sum_{i=0}^{n} \left(E_i^e \times F_e + E_i^g \times F_g \right) \times 50, \tag{3}$$

where GHG_o is the total GHG emissions due to the energy used over the 50-year building life cycle (in kg CO_2e); E_i^e and E_i^g are the annual quantity of electricity and gas consumption for building services system i, in kWh or in gas unit (i.e., 1 unit as registered by the gas meter = 48 mega joules consumed), respectively; F_e and F_g are the emission factors of the energy consumed by electricity and gas, respectively, with the territory-wide default values of F_e and F_g being 0.7 kg CO_2e/kWh and 0.59 kg CO_2e/unit of gas purchased, respectively [64].

The electricity consumption data for communal building services installations was calculated from the sampled NH1 housing block. However, the energy consumed by tenants of the sampled housing block was not accessible. Therefore, the energy end-use data of the public housing group (in terajoules) as published by the Electrical and Mechanical Services Department [7] was used for the energy use estimation in this study. It was assumed that 80% of the tenants used electric water heaters while the rest used gas water heaters [65] and 90% of tenants used gas for cooking [66].

GHG removals are calculated by assessing the respective GHG absorption by the assimilation of CO_2 by plants as shown in Equation (4). According to EPD [64], 23 kg CO_2 can be removed by each tree based on Hong Kong's location, woodland types, and estimated density of trees. The figure is applicable to all trees commonly found in Hong Kong which are able to reach at least 5 m in height. Since this figure is derived as an annual average based on an extended period of time corresponding to the life cycle of the trees, the figure is applicable to trees at all ages.

$$GHG_r = (T \times F_t) \times 50, \tag{4}$$

where GHG_r is the total GHG absorption over the 50-year building life cycle by tree planting (in kg CO_2e); T is the number of newly planted trees within the building's physical boundary (e.g., within building premises, associated with the surroundings that are used for multipurpose activities including planting trees for a particular housing estate) after the beginning stage of construction which are able to reach at least 5 m in height; and F_t is the GHG removal factor, which is taken as 23 kg CO_2/tree per annum [64].

In this study, a 50-year service life of a residential block in Hong Kong was considered. As a complex system, buildings would often undergo various changes by means of renovation. Considering the building's service life, typical replacement of principal elements with their number of replacements over the entire life of a building in Hong Kong are shown in Table 3 (adjusted based on Chiang et al. [67]). The production and transport of these materials/elements were included in the LCA. Ecoinvent databases were used for collecting their upstream data, for instance, ceramic tile, emulsion paint, sealing materials, and hardwood doors production. Based on the renovation of typical flats, average per unit (m^2) was calculated according to Chiang et al. [67]. In addition, the transportation and disposal (in landfills) of materials generated during renovation were also considered in this assessment.

Table 3. Typical replacement of building elements during renovation in Hong Kong.

Element/Material	Service Life (years)	Number of Replacements over the Service Life of Building
Ceramic tiles	20	2
Emulsion paint	5	9
Silicone seal	10	4
Hardwood solid-core doors	20	2

Therefore, the total GHG emissions due to the renovation of the building during its service life can be estimated by Equation (5).

$$GHG_R = \sum\nolimits_{service\ life=50}(M_R \times N_R) + T_M + D_L, \tag{5}$$

where GHG_R is the total GHG emissions over the 50-year service life of building (in kg CO_2e); M_R is the materials/elements replacement during renovation; N_R is the number of replacements of the respective elements/materials; T_M is the transport of the materials; and D_L is the disposal into landfill.

2.2.3. Emissions in the Demolition Stage (Aspects VIII)

According to ISO [68], recycling of steel and concrete should be assessed in the subsequent loop of a building life cycle. Typically, inert wastes generated from buildings are disposed at public fill sites, whereas non-inert waste are dumped into landfills, in Hong Kong. After transporting to off-site sorting facilities, inert materials are crushed and screened to recycle materials. In addition, a certain amount of concrete is recycled [69]. After screening, the remaining inert materials are sent to public fills, while the non-inert materials are disposed at landfills. However, the impacts of recycling and disposal were excluded due to the complexity of different management strategies and the lack of data in Hong Kong. In this study, the GHG emitted during the demolition stage is mainly due to the energy consumption for the machinery operation at the demolition site and the transportation of building debris from the site to disposal sites [23]. It is also assumed that the saving of GHG emissions due to steel and concrete recovery, and the induced GHG emissions of other materials' disposal (whether to public fills or landfills) would be similar, and thus they were excluded from this analysis. Equation (6) serves as the basis to assess the emissions.

$$GHG_d = Q_d \times F_d + \frac{Q_d \times E_d \times D_i^l \times F_l^t}{5}, \tag{6}$$

where GHG_d is the total GHG emissions in the demolition stage; Q_d is the amount of building materials to be dismantled or building debris (in m^3), for transportation of building debris, 5 m^3 load per truck is assumed; F_d is the emission factor for dismantling a building, which is 17 kg CO_2e/m^3 according to Nielsen [70]; E_d is the diesel consumption (in liter/km/truck), which is 0.325 L/km [71]; D_i^l is the total distance for transporting building materials i by land (in km) and the distance is taken as 26 km [52]; and F_l^t is the emission factors for transportation by diesel truck, which is 2.62 kg CO_2e/liter [52].

Therefore, the total GHG emissions over the NH1's building life cycle can be estimated by Equation (7):

$$GHG_{NH1} = GHG_m + GHG_t + (GHG_o + GHG_r + GHG_R) + GHG_d, \tag{7}$$

2.3. Limitations

Conducting an environmental assessment of the whole building is complicated due to the differences in materials, associated transportation, diverse considerations at the use stage, lifespan of the building, and different considerations of the end-of-life of the building. Although this study has attempted to include all these in the assessment, several limitations cannot be avoided. For example, this study did not consider the contributions of non-structural materials. Waste material (generated during the construction and end-of-life stages) treatments were not considered in this assessment. During the use stage of building, actual energy consumed by tenants is not accessible. Thus, average energy consumption data was collected from the relevant department [7]. Due to the lack of local/regional data, Ecoinvent databases were used for carbon emissions of the replacement materials for the renovation of building. This study only focused on carbon assessment and excluded other impact categories. However, some of the limitations were further discussed and justified in the Discussion section.

3. Results

The result of GHG emissions throughout the life cycle of the standard NH1 public housing block, based on Equation (7), is presented in Table 4. Using the data collected and assumptions, the estimated total life cycle GHG emissions of the block are 186,150 tCO_2e for the NH1 2000 Edition. The GHG emissions intensity of the sampled building is 232.98 tCO_2e/flat or 5.38 tCO_2e/m^2 of GFA. Figure 3 shows the distribution of the emissions in various life cycle phases. The operating energy consumption by communal building services and tenants is clearly impacting the environment the most, accounting for about 85.82% of the emissions. The materials consumed during construction, though emit considerably less GHGs than that in the operation stage, are taking up about 12.69% of the life cycle emissions. The remaining aspects, including the renovation, transportation of materials, and the disposal of the block are accounted to 1.14%, 0.07%, and 0.28%, respectively, of the building's carbon footprint.

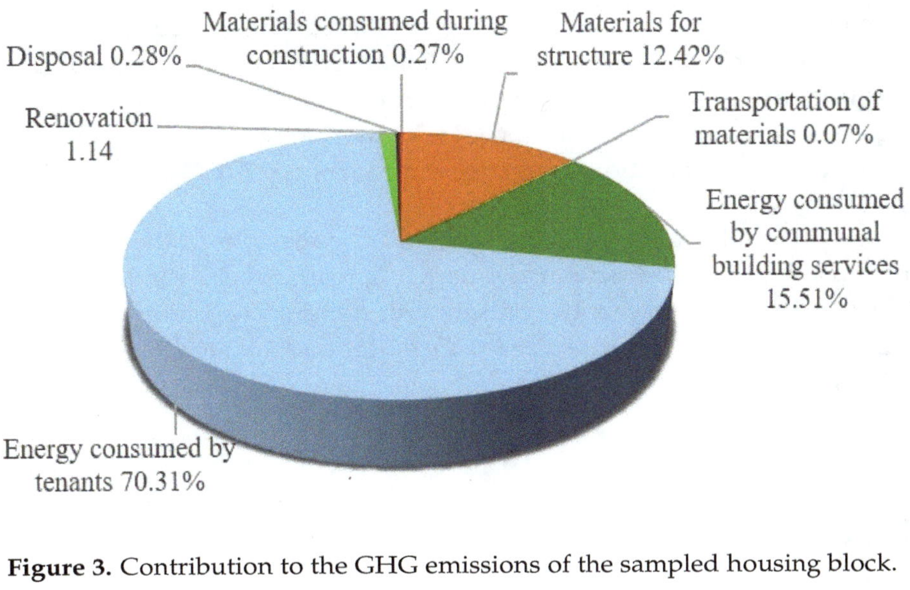

Figure 3. Contribution to the GHG emissions of the sampled housing block.

Hotspots for GHG emissions of building have been highlighted in Figure 3 (i.e., the contribution to GHG emissions). It can be seen that over 85% of the total GHG emissions is associated with energy consumption of tenants and building services equipment. This supports the results from previous studies [16,24,72]. The elements emitting the most significant amount of GHGs are found to be communal lighting and lifts, as well as the energy consumed by tenants for hot water, space conditioning, and refrigeration. This reflects that GHG emitted from a public housing block are strongly dependent not only on the building and occupancy factors such as ventilation and efficiency of appliances, but also on the source of energy. Therefore, it is important to install energy efficient building service equipment, and encourage tenants to use energy efficient appliances to reduce energy consumption and GHG emissions. Apart from reducing the energy consumption and embodied energy in buildings, switching to low carbon fuels and utilizing renewable energy are considered effective in tackling the climate change problem [46]. Materials (including their production and transportation) emit about 13% of the total emissions. However, it is possible to reduce GHG emissions by sourcing sustainable materials and using low carbon materials.

During the construction stage, concrete is the dominant building material for the NH1 housing block, not only in terms of quantities but also the embodied carbon. The NH1 housing block studied consumed over 28,600 m^3 of concrete, producing over 12 million kg of CO_2e. Reducing the carbon content of concrete through the manufacturing process is therefore influential. A saving in embodied carbon can be achieved by increasing the proportion of off-site manufacturing of components and/or adopting recycled materials or materials with lower environmental impact [6,52]. For instance, by replacing cement with alternative binding materials (e.g., pulverized fuel ash, ground granulated blast furnace slag, and silica fume) in the concrete mixes can save significant amount of cement and the associated CO_2 emissions [51,73]. In addition, the use of alternative or low carbon cement, i.e., eco-glass cement or Portland fly-ash cement, can also considerably reduce the carbon footprint of concrete [74].

The study also estimated the GHG emissions of the residential block according to the latest "Model Client Brief 2010" as presented in Table 4. According to the Hong Kong Energy End-Use Data 2010 provided by the Electrical and Mechanical Services Department of Hong Kong SAR [7], this brief has incorporated the latest development of various communal building services installations. These include the employment of electronic ballasts and two illumination levels in the lighting system, adjusting the capacity and weight of lifts, adopting a variable speed drive system in booster pumps, as well as using more energy efficient motors. As a result, a significant reduction in annual electricity consumption is achieved from 1032 kWh/flat in the Client Brief January 2000 Edition to 596 kWh/flat in 2008 [7]. In addition, the brief requires the planting of one tree for every 15 flats in a newly built public housing estate. Consequently, the annual GHG emissions caused by electricity consumption of communal building services installations have decreased from 577 tCO_2e to 337 tCO_2e (Table 4).

According to the "Model Client Brief 2010", the GHG emissions of the NH1 housing block were 215.69 tCO_2e/flat and 4.98 tCO_2e/m^2 per flat and per GFA, respectively. While the energy consumption of tenant areas is beyond the management's control, with Aspect V—"Energy consumption by tenants" being excluded, the GHG emissions were 51.88 tCO_2e/flat and 1.20 tCO_2e/m^2 per flat and per GFA, respectively (Table 4).

Table 4. Evaluation of GHG emissions of the studied case in Hong Kong.

Stage Aspect	Category	Category	Benchmarking Block NH1 Standard Block (2000 Edition)			Benchmarking Block NH1 Standard Block (Model Client Brief 2010)		
		Type	Quantity (m^3)	Per Unit CO_2e Emission (kg/m^3)	Total Emission (kg CO_2e)	Quantity (m^3)	Per Unit CO_2e Emission (kg/m^3)	Total Emission (kg CO_2e)
Production and Construction stage	I	Materials consumed during construction (GHG_m)						
		Steel formwork for superstructure	23.3	15,210	354,393	22.5	15,210	342,073
		Timber formwork for superstructure	149.8	962	144,108	144.6	962	139,105
		Steel formwork for substructure	0.0	15,210	0	0.0	15,210	0
		Timber formwork for substructure	3.4	962	3271	3.4	962	3271
		Sub Total	-	-	501,772	-	-	484,449
	II	Materials for structure (GHG_m)						
		Steel for superstructure	548.2	15,210	8,338,122	433.0	15,210	6,585,930
		Concrete for superstructure	20,419.0	426	8,698,494	20,460.4	426	8,716,130
		Steel for substructure	170.0	15,210	2,585,700	170.0	15,210	2,585,700
		Concrete for substructure	8212.0	426	3,498,312	8212.0	426	3,498,312
		Sub Total	-	-	23,120,628	-	-	21,386,072
	III	Transportation of materials (GHG_t)						
		Type	Quantity (m^3)	Distance from site (km)	Total Emission (kg CO_2e)	Quantity (m^3)	Distance from site (km)	Total Emission (kg CO_2e)
		Steel	741.5	250	31,531	625.5	250	26,598
		Timber	153.2	250	6515	148.0	250	6291
		Concrete	28,631.0	20	97,398	28,672.4	20	97,539
	-	-	Sub Total	-	-	135,443	-	-	130,428

Table 4. *Cont.*

Stage Aspect		Category	Category	Benchmarking Block NH1 Standard Block (2000 Edition)			Benchmarking Block NH1 Standard Block (Model Client Brief 2010)		
Operation stage	IV	Energy consumption by communal building services (GHG_o)	System	Yearly Energy Consumption (kWh)	Annual Emission (kg CO_2e)	Total Emission in 50 Years (kg CO_2e)	Yearly Energy Consumption (kWh)	Annual Emission (kg CO_2e)	Total Emission in 50 Years (kg CO_2e)
			Lighting	538,188	376,732	18,836,580	264,799	185,359	9,267,965
			Lift	165,783	116,048	5,802,405	103,175	72,223	3,611,125
			Security	3793	2655	132,755	3793	2655	132,755
			TV	4205	2944	147,175	4205	2944	147,175
			A/C and ventilation	17,800	12,460	623,000	17,800	12,460	623,000
			Fire services	1764	1235	61,740	1764	1235	61,740
			Water supply	76,972	53,880	2,694,020	76,972	53,880	2,694,020
			Electrical distribution (2% of above items)	16,170	11,319	565,954	9450	6615	330,756
	-	-	Sub Total	824,675	577,273	28,863,629	481,958	337,371	16,868,536
	V	Energy consumption by tenants (GHG_o)	Energy End Use	Yearly Energy Consumption (Tera joule)	Annual Emission (kg CO_2e)	Total Emission in 50 Years (kg CO_2e)	Yearly Energy Consumption (Tera joule)	Annual Emission (kg CO_2e)	Total Emission in 50 Years (kg CO_2e)
			Cooking (90% gas; 10% electricity)	4.2	158,997	7,949,854	4.2	158,997	7,949,854
			Space conditioning	2.8	551,754	27,587,695	2.8	551,754	27,587,695
			Hot water (20% gas; 80% electricity)	4	634,835	31,741,738	4	634,835	31,741,738
			Lighting	1.1	219,947	10,997,347	1.1	219,947	10,997,347
			Refrigeration	2.2	421,694	21,084,722	2.2	421,694	21,084,722
			Others (laundry, audiovisual and miscellaneous equipment)	3.2	630,544	31,527,209	3.2	630,544	31,527,209
	-	-	Sub Total	17.5	2,617,771	130,888,564	17.5	2,617,771	130,888,564

Table 4. *Cont.*

Stage Aspect	Category		Category	Benchmarking Block NH1 Standard Block (2000 Edition)			Benchmarking Block NH1 Standard Block (Model Client Brief 2010)		
Operation stage	VI	Renovation	-	-	-	Total Emission in 50 Years	-	-	Total Emission in 50 Years
	-	-	Materials/components replacement and transport	-	-	2,127,938	-	-	2,127,938
	-	-	Sub Total	-	-	2,127,938	-	-	2,127,938
	VII	GHG removals (GHG_r)	Tree	Quantity	Annual Emission Absorption (kg CO_2e)	Total Absorption in 50 Years (kg CO_2e)	Quantity	Annual Emission Absorption (kg CO_2e)	Total Absorption in 50 Years (kg CO_2e)
			Tree (Taller than 5 m)	0	0	0	53	1219	60,950
	-	-	Sub Total	-	-	0	-	-	−60,950
Demolition stage	VIII	Disposal (GHG_d)	Demolition	Quantity (m^3)	Per Unit CO_2e Emission (kg/m^3)	Total Emission (kg CO_2e)	Quantity (m^3)	Per Unit CO_2e Emission (kg/m^3)	Total Emission (kg CO_2e)
			Dismantling of building	23,919	17	406,623	23,960	17	407,320
			Transportation	Quantity (m^3)	Distance from site (km)	Total Emission (kg CO_2e)	Quantity (m^3)	Distance from site (km)	Total Emission (kg CO_2e)
			Transportation of building debris from site to disposal sites	23,919	26	105,779	23,960	26	105,961
	-	-	Sub Total	-	-	512,402	-	-	513,281
	Overall results			-	Total GHG Emissions		-	Total GHG Emissions	
	-			-	(tons CO_2e)	(kg CO_2e)	-	(tons CO_2e)	(kg CO_2e)
Sum of above	Grand total (I + II + III + IV + V + VI + VII + VIII)			-	186,150	186,150,376	-	172,338	172,338,318
	Grand total discounting tenant areas (I + II + III + IV + VI + VII + VIII)			-	55,262	55,261,812	-	41,450	41,449,754
Project data	Total no. of flat			-	799	-	-	799	-
	Gross floor area (GFA) (m^2)			-	34,609	-	-	33,078	-
Results	CO_2e emission per flat = a/c			-	232.98	232,979	-	215.69	215,693
–	CO_2e emission per flat (discounting tenant areas) = b/c			-	69.16	69,164	-	51.88	51,877
-	CO_2e emission per GFA (m^2) = a/d			-	5.38	5379	-	4.98	4980
-	CO_2e emission per GFA (m^2) (discounting tenant areas) = b/d			-	1.60	1597	-	1.20	1198

4. Discussion

This study has comprehensively evaluated the GHG emissions of a concrete reinforced high-rise residential building in Hong Kong. In addition to the structural materials, the study also considered the carbon emitted from communal building services, tenants due to energy end use, renovation, building demolition, and transportation of waste materials. It can be seen that the GHG emissions of the studied case ranged from 4980 kg CO_2e/m^2 to 5379 kg CO_2e/m^2 (based on design). The comparison of GHG emissions among different studies in different regions per functional unit is given in Figure 4. The variation of GHG emissions is relatively high (which ranges from 1657–6276 kg CO_2e/m^2) among different studies due to the use of different structural materials (concrete, steel, wood, composite, and so forth), heating and cooling requirements for different regions based on the climate, as well as other considerations. However, the GHG evaluated in this study is in the upper range of the emissions (Figure 4). This may be due to the higher GHG emission factors for different structural materials used in Hong Kong including the long transport distance, as Hong Kong has sourced most of the construction materials from China, which have higher emission factors (Table 2) which is also supported by the previous studies. For instance, De Wolf et al. [11] estimated the GHG emissions of 200 completed buildings based on structural materials quantities in the US, and calculated the GHG emissions range from 150–600 kg CO_2e/m^2. However, the GHG emissions are even higher than the upper range (for structural materials) found in this study (about 686 kg CO_2e/m^2, Table 4).

Based on the collected data and assumptions for renovation works in Hong Kong, it is estimated that renovation contributes to 61.50 kg CO_2e/m^2 of the building during its considered service life (e.g., 5 years). The value is considerably higher than 45 kg CO_2e/m^2 estimated by Ortiz-Rodríguez et al. [75] and 38 CO_2e/m^2 by Kumanayake and Luo [76]. However, energy efficient and low carbon refurbishments and replacement of building services can significantly help reduce the total embodied CO_2 emissions of buildings [30,77].

Although the evaluation of GHG emissions was based on a single case study in this study, the sampled building is a typical design of housing blocks in Hong Kong. Comparison on the emissions of new housing development can be conducted by making references at different building life cycle stages [36]. GHG emissions can also be minimized by using environmentally-friendly materials or energy efficient appliances, lighting, heating, and cooling equipment [78]. While the tenants of public rental housing estates represent almost 28% of Hong Kong population, their behavior might have a substantial impact on energy use, especially when the building services equipment is controlled manually. Reducing material use as well as specifying the use of localized materials, recycled materials, and/or alternative low carbon material are the options available for implementation during the design stage for reducing the embodied carbon of buildings [79–86].

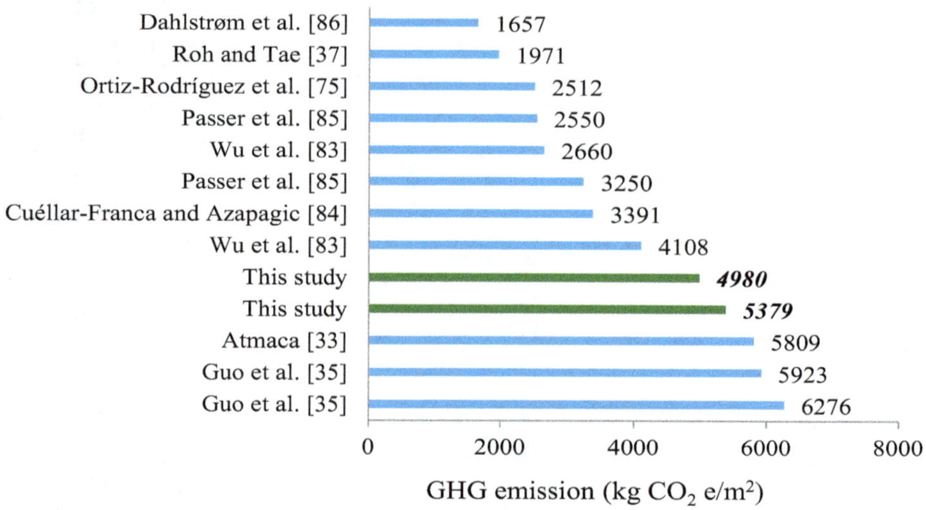

Figure 4. Comparison of GHG emissions of residential buildings.

The contribution of cladding and other non-structural materials, including windows, doors and roof coverings, internal partitions, and internal cladding was not considered in this study. Although some of the materials used during the construction process are negligible in terms of weight, their impacts can be significant to the total impacts. For example, polyamide safety nets and aluminum are used <0.1% by weight, but would contribute 2–3% towards the total GHG emissions [28]. These aspects were also not considered due to data unavailability despite the fact they can affect the emission inventory. In addition, carbon emissions can be expected with an anticipated growth in building activity and higher performance of buildings due to greater material use [80].

Although demolition and waste transportation were included in this study, the assessment of waste material treatments was not considered due to the different end-of-life considerations and the lack of available data. The demolition phase of buildings including demolition, waste transportation, and waste treatment contributed to about 2–5% of the total GHG emissions depending on the types of waste treatment [34,36]. However, according to Coelho and de Brito [32], GHG emissions of waste disposal could be about 65–283 kg CO_2e/m^2 of building (depending on the types of waste treatment). Therefore, the amount is insignificant compared to the total emissions estimated in this study (4980 kg CO_2e/m^2). More comprehensive investigations on the overall environmental performance of buildings by including other impact categories are desirable.

5. Conclusions

Environmental impacts associated with building construction, use, and end-of-life are greatly dependent on the region, climate, and type of buildings. Therefore, a case-specific assessment is important to benchmark the evaluation, as well as to reduce and mitigate the impacts from buildings. In this study, the GHG emissions from a typical high-rise residential building in Hong Kong was comprehensively evaluated using a case-specific analysis with a "cradle-to-grave" system boundary. Through this analytical regime, the GHG emissions were estimated to about 213.03 tCO_2e/flat and 4980 kg CO_2e/m^2, respectively. Considering the GHG emissions over the service life of the sampled residential building, the operating energy causes over 85.82% of the emissions, whereas 12.69% for materials, 1.14% for renovation, 0.28% for end-of-life of the building, and 0.07% for other factors. Therefore, various carbon reduction measures should be attempted and evaluated such as the use of energy efficient equipment, renewable energy, recycled/recyclable materials, and eco-design by utilizing natural lighting and ventilation. Policy and decision makers should explore different low carbon construction initiatives to maximize the opportunity for emission reduction. For future work, the residential buildings including all kinds of public and private buildings should be assessed by considering the limitations of this study on their carbon emissions as well as other environmental impact indicators. Tremendous effort is required to advocate low carbon construction at various levels including building materials, building components, and the entire building through effective incentive and reward schemes. For a more sustainable future, there is an urgent call for immediate, community-wide actions to reduce GHG emissions to help combat climate change.

Author Contributions: The authors contributed equally to this manuscript. S.Y.C.M. and the Housing Department initiated and designed the Carbon and Emission Estimation model, S.T.N. and J.M.W.W. provided suggestions to improve the reliability of the model; and M.U.H. conducted further analysis based on the local data and the developed model.

Acknowledgments: The authors would like to acknowledge the Hong Kong Housing Department as the life cycle building carbon estimation method described in this paper originated from the carbon emission estimation model of the Hong Kong Housing Department.

References

1. IPCC. *Climate Change 2014: Synthesis Report. Contribution of Working Groups I, II and III to the Fifth Assessment Report of the Intergovernmental Panel on Climate Change (IPCC)*; IPCC: Geneva, Switzerland, 2014.
2. Rogner, H.H.; Zhou, D.; Bradley, R.; Crabbé, P.; Edenhofer, O.; Hare, B.; Kuijpers, L.; Yamaguchi, M. Introduction. In *Climate Change 2007: Mitigation. Contribution of Working Group III to the Fourth Assessment Report of the Intergovernmental Panel on Climate Change*; Cambridge University Press: Cambridge, UK, 2007.
3. Stern, N.H. *Stern Review on the Economics of Climate Change*; HM Treasury: London, UK, 2006.
4. Ibn-Mohammed, T. Application of mixed-mode research paradigms to the building sector: A review and case study towards decarbonising the built and natural environment. *Sustain. Cities Soc.* **2017**, *35*, 692–714. [CrossRef]
5. Emami, N.; Marteinsson, B.; Heinonen, J. Environmental impact assessment of a School building in Iceland using LCA-including the effect of long distance transport of materials. *Buildings* **2016**, *6*, 46. [CrossRef]
6. Monahan, J.; Powell, J.C. An embodied carbon and energy analysis of modern methods of construction in housing: A case study using a lifecycle assessment framework. *Energy Build.* **2011**, *43*, 179–188. [CrossRef]
7. EMSD. *Hong Kong Energy End-Use Data 2010*; Electrical and Mechanical Services Department, Government of HKSAR: Hong Kong, China, 2010.
8. Wang, X.; Chen, D.; Ren, Z. Global warming and its implication to emission reduction strategies for residential buildings. *Build. Environ.* **2011**, *46*, 871–883. [CrossRef]
9. HKHA. *Housing in Figures 2010*; The Hong Kong Housing Authority, Government of HKSAR: Hong Kong, China, 2010.
10. Chen, Y.; Ng, S.T. Factoring in embodied GHG emissions when assessing the environmental performance of building. *Sustain. Cities Soc.* **2016**, *27*, 244–252. [CrossRef]
11. De Wolf, C.; Yang, F.; Cox, D.; Charlson, A.; Hattan, A.S.; Ochsendorf, J. Material quantities and embodied carbon dioxide in structures. *Proc. Inst. Civ. Eng. Eng. Sustain.* **2016**, *169*, 150–161. [CrossRef]
12. Dong, Y.H.; Ng, S.T. A life cycle assessment model for evaluating the environmental impacts of building construction in Hong Kong. *Build. Environ.* **2015**, *89*, 183–191. [CrossRef]
13. Peuportier, B.L.P. Life cycle assessment applied to the comparative evaluation of single family houses in the French context. *Energy Build.* **2001**, *33*, 443–450. [CrossRef]
14. Hoxha, E.; Habert, G.; Lasvaux, S.; Chevalier, J.; Roy, R.L. Influence of construction material uncertainties on residential building LCA reliability. *J. Clean. Prod.* **2017**, *144*, 33–47. [CrossRef]
15. Meijer, A.; Huijbregts, M.; Reijnders, L. Human health damages due to indoor sources of organic compounds and radioactivity in life cycle impact assessment of dwellings. *Int. J. Life Cycle Assess.* **2005**, *10*, 383–392. [CrossRef]
16. Gerilla, G.P.; Teknomo, K.; Hokao, K. An environmental assessment of wood and steel reinforced concrete housing construction. *Build. Environ.* **2007**, *42*, 2278–2784. [CrossRef]
17. Hacker, J.N.; De Saulles, T.P.; Minson, A.J.; Holmes, M.J. Embodied and operational carbon dioxide from housing: A case study on the effects of thermal mass and climate change. *Energy Build.* **2008**, *40*, 375–384. [CrossRef]
18. Li, D.Z.; Chen, H.X.; Hui, E.C.M.; Zhang, J.B.; Li, Q.M. A methodology for estimating the life-cycle carbon efficiency of a residential building. *Build. Environ.* **2013**, *59*, 448–455. [CrossRef]
19. Anand, C.K.; Amor, B. Recent developments, future challenges and new research directions in LCA of buildings: A critical review. *Renew. Sustain. Energy Rev.* **2017**, *67*, 408–416. [CrossRef]
20. Dixit, M.; Fernández-Solís, J. Need for an embodied energy measurement protocol for buildings: A review paper. *Renew. Sustain. Energy Rev.* **2012**, *16*, 3730–3743. [CrossRef]
21. Pomponi, F.; Moncaster, A. Scrutinising embodied carbon in buildings: The next performance gap made manifest. *Renew. Sustain. Energy Rev.* **2018**, *81*, 2431–2442. [CrossRef]
22. Teng, Y.; Li, K.; Pan, W.; Ng, S.T. Reducing building life cycle carbon emissions through prefabrication: Evidence from and gaps in empirical studies. *Build. Environ.* **2018**, *132*, 125–136. [CrossRef]
23. Suzuki, M.; Oka, T. Estimation of life cycle energy consumption and CO_2 emission of office buildings in Japan. *Energy Build.* **1998**, *28*, 33–41. [CrossRef]
24. Seo, S.; Hwang, Y. Estimation of CO_2 emissions in life cycle of residential buildings. *J. Constr. Eng. Manag.* **2001**, *127*, 414–418. [CrossRef]

25. Bastosa, J.; Battermanb, S.A.; Freirea, F. Life-cycle energy and greenhouse gas analysis of three building types in a residential area in Lisbon. *Energy Build.* **2014**, *69*, 344–353. [CrossRef]
26. Cattarinussi, L.; Hofstetter, K.; Ryffel, R.; Zumstein, K.; Ioannidou, D.; Klippel, M. Life cycle assessment of a post-tensioned timber frame in comparison to a reinforced concrete frame for tall buildings. In *Expanding Boundaries: Systems Thinking for the Built Environment, Proceedings of the Sustainable Built Environment (SBE) Regional Conference, Zürich, Switzerland, 15–17 June 2016*; vdf Hochschulverlag AG an der ETH Zürich: Zürich, Switzerland, 2016; pp. 656–661.
27. Hildebrandt, J.; Hagemann, N.; Thrän, D. The contribution of wood-based construction materials for leveraging a low carbon building sector in Europe. *Sustain. Cities Soc.* **2017**, *34*, 405–418. [CrossRef]
28. Hong, J.; Shen, G.P.; Feng, Y.; Lau, W.S.; Ma, C. Greenhouse gas emissions during the construction phase of a building: A case study in China. *J. Clean. Prod.* **2015**, *103*, 249–259. [CrossRef]
29. Sandanayake, M.; Lokuge, W.; Zhang, G.; Setunge, S.; Thushar, Q. Greenhouse gas emissions during timber and concrete building construction—A scenario based comparative case study. *Sustain. Cities Soc.* **2018**, *38*, 91–97. [CrossRef]
30. Almeida, M.; Ferreira, M. Cost effective energy and carbon emissions optimization in building renovation (Annex 56). *Energy Build.* **2017**, *152*, 718–738. [CrossRef]
31. Assefa, G.; Ambler, C. To demolish or not to demolish: Life cycle consideration of repurposing buildings. *Sustain. Cities Soc.* **2017**, *28*, 146–153. [CrossRef]
32. Coelho, A.; de Brito, J. Influence of construction and demolition waste management on the environmental impact of buildings. *Waste Manag.* **2012**, *32*, 532–541. [CrossRef] [PubMed]
33. Atmaca, A. Life-cycle assessment and cost analysis of residential buildings in South East of Turkey: Part 2—A case study. *Int. J. Life Cycle Assess.* **2016**, *21*, 925–942. [CrossRef]
34. Cho, S.-H.; Chae, C.-U. A study on life cycle CO_2 emissions of low-carbon building in South Korea. *Sustainability* **2016**, *8*, 579. [CrossRef]
35. Guo, H.; Liu, Y.; Meng, Y.; Huang, H.; Sun, C.; Shao, Y. A comparison of the energy saving and carbon reduction performance between reinforced concrete and cross-laminated timber structures in residential buildings in the severe cold region of China. *Sustainability* **2017**, *9*, 1426. [CrossRef]
36. Peng, C. Calculation of a building's life cycle carbon emissions based on Ecotect and building information modeling. *J. Clean. Prod.* **2016**, *112*, 453–465. [CrossRef]
37. Roh, S.; Tae, S. An integrated assessment system for managing life cycle CO_2 emissions of a building. *Renew. Sustain. Energy Rev.* **2017**, *73*, 265–275. [CrossRef]
38. Cabeza, L.F.; Rincón, L.; Vilarino, V.; Pérez, G.; Castell, A. Life cycle assessment (LCA) and life cycle energy analysis (LCEA) of buildings and the building sector: A review. *Renew. Sustain. Energy Rev.* **2014**, *29*, 394–416. [CrossRef]
39. Pomponi, F.; Moncaster, A.M. Embodied carbon in the built environment: Management, mitigation, and reduction—What does the evidence say? *J. Environ. Manag.* **2016**, *181*, 687–700. [CrossRef] [PubMed]
40. De Wolf, C.; Pomponi, F.; Moncaster, A. Measuring embodied carbon dioxide equivalent of buildings: Areview and critique of current industry practice. *Energy Build.* **2017**, *140*, 68–80. [CrossRef]
41. Dong, Y.H.; Ng, S.T. Comparing the midpoint and endpoint approaches based on ReCiPe—A study of commercial buildings in Hong Kong. *Int. J. Life Cycle Assess.* **2014**, *19*, 1409–1423. [CrossRef]
42. Gan, V.J.L.; Cheng, J.C.P.; Lo, I.M.C.; Chan, C.M. Developing a CO_2-e accounting method for quantification and analysis of embodied carbon in high-rise buildings. *J. Clean. Prod.* **2017**, *141*, 825–836. [CrossRef]
43. ISO. *ISO 14040—Environmental Management: Life Cycle Assessment Principles and Framework*; The International Organisation for Standardization: Paris, France, 2006.
44. ISO. *ISO 14064—International Standard on Greenhouse Gases—Part 1: Specification with Guidance at the Organization Level for Quantification and Reporting of Greenhouse Gas Emissions and Removals*; The International Organisation for Standardization: Geneva, Switzerland, 2006.
45. WRI/WBCSD. *The Greenhouse Gas Protocol: A Corporate Accounting and Reporting Standard*, Revised Edition; World Resources Institute and World Business Council for Sustainable Development: Washington, DC, USA, 2010.
46. IPCC. *Climate Change 2007: The Physical Science Basis, Summary for Policy Makers*; Intergovernmental Panel on Climate Change: Geneva, Switzerland, 2007.

47. LC. Design of the New Public Housing Flats by the Hong Kong Housing Authority. Legislative Council (LC) Panel on Housing, 2015. Available online: http://www.legco.gov.hk/yr14-15/english/panels/hg/papers/hg20150706cb1-1037-1-e.pdf (accessed on 22 October 2018).
48. ISO. *ISO 21931-1:2010—Sustainability in Building Construction—Framework for Methods of Assessment of the Environmental Performance of Construction Works—Part 1: Buildings*; The International Organisation for Standardization: Geneva, Switzerland, 2010.
49. Fieldson, R.; Rai, D.; Sodagar, B. Towards a framework for early estimation of lifecycle carbon footprinting of buildings in the UK. *Constr. Inf. Q.* **2009**, *11*, 66–75.
50. Takano, A.; Pittau, F.; Hafner, A.; Ott, S.; Hughes, M.; De Angelis, E. Greenhouse gas emission from construction stage of wooden buildings. *Int. Wood Prod. J.* **2014**, *5*, 217–223. [CrossRef]
51. HKHA. *Life Cycle Assessment (LCA) and Life Cycle Costing (LCC) Study of Building Materials and Components*; The Hong Kong Housing Authority, Government of HKSAR: Hong Kong, China, 2005.
52. Yan, H.; Shen, Q.P.; Fan, L.C.H.; Wang, Y.; Zhang, L. Greenhouse gas emission in building construction: A case study of One Peking in Hong Kong. *Build. Environ.* **2010**, *45*, 949–995. [CrossRef]
53. Ng, S.T.; To, C.; Li, G. Unveiling the embodied carbon of construction materials through a product-based carbon labeling scheme. *Int. J. Clim. Chang. Impacts Responses* **2015**, *7*, 1–9. [CrossRef]
54. Hammond, G.; Jones, C. *Inventory of Carbon and Energy (ICE)*; Version 2.0; Sustainable Energy Research Team, Department of Mechanical Engineering, University of Bath: Bath, UK, 2011.
55. Jing, R.; Cheng, J.C.P.; Gan, V.J.L.; Woon, K.S.; Lo, I.M.C. Comparison of greenhouse gas emission accounting methods for steel production in China. *J. Clean. Prod.* **2014**, *83*, 165–172. [CrossRef]
56. Zhang, J.; Cheng, J.C.P.; Lo, I.M.C. Life cycle carbon footprint measurement of Portland cement and ready mix concrete for a city with local scarcity of resources like Hong Kong. *Int. J. Life Cycle Assess.* **2014**, *19*, 745–757. [CrossRef]
57. Zhang, J. Life Cycle Carbon Measurement of Hong Kong Construction Materials: Cement, Concrete and Plywood. Master's Thesis, The Hong Kong University of Science and Technology University, Hong Kong, China, 2013.
58. Morris, J. The ethics and evaluation of embodied carbon in buildings. *Struct. Eng.* **2008**, *86*, 30–34.
59. Eaton, K.J.; Amato, A. *A Comparative Environmental Life Cycle Assessment of Modern Office Buildings*; Steel Construction Institute: Ascot, UK, 1998.
60. Alcorn, A. Embodied energy and CO_2 coefficients for New Zealand building materials. In *Report Series: Centre for Building Performance Research Report*; Victoria University of Wellington: Wellington, New Zealand, 2003.
61. Catalonia Institute of Construction Technology (CICT). BEDEC PR/PCT ITEC Materials Database. 2011. Available online: https://en.itec.cat/database/ (accessed on 22 October 2018).
62. EMSD. *Life Cycle Energy Analysis of Building Construction*; Final Report of Consultancy Study; Electrical and Mechanical Services Department, Government of HKSAR: Hong Kong, China, 2006.
63. Junnila, S. Life cycle assessment of environmentally significant aspects of an office building. *Nord. J. Surv. Real Estate Res. Spec. Ser.* **2004**, *2*, 81–97.
64. EPD. *Guidelines to Account for and Report on Greenhouse Gas Emissions and Removals for Buildings (Commercial, Residential or Institutional Purposes) in Hong Kong*, 2010 ed.; Environmental Protection Department and the Electrical and Mechanical Services Department, Government of HKSAR: Hong Kong, China, 2010.
65. Tso, G.K.F.; Yau, K.K.W. Predicting electricity energy consumption: A comparison of regression analysis, decision tree and neural networks. *Energy* **2007**, *32*, 1761–1768. [CrossRef]
66. Wan, K.S.Y.; Yik, F.W.H. Building design and energy end-use characteristics of high-rise residential buildings in Hong Kong. *Appl. Energy* **2004**, *78*, 19–36. [CrossRef]
67. Chiang, Y.H.; Li, V.J.; Zhou, L.; Wong, F.; Lam, P. Evaluating sustainable building-maintenance projects: Balancing economic, social, and environmental impacts in the case of Hong Kong. *J. Constr. Eng. Manag.* **2016**, *142*, 06015003. [CrossRef]
68. ISO. *ISO/TS 14067:2013—Greenhouse Gases—Carbon Footprint of Products—Requirements and Guidelines for Quantification and Communication*; The International Organisation for Standardization: Geneva, Switzerland, 2013.
69. Hossain, M.U.; Wu, Z.; Poon, C.S. Comparative environmental evaluation of construction waste management through different waste sorting systems in Hong Kong. *Waste Manag.* **2017**, *69*, 325–335. [CrossRef] [PubMed]
70. Nielsen, C.V. *Carbon Footprint of Concrete Buildings Seen in the Life Cycle Perspective*; Claus Vestergaard Nielsen of the Danish Technological Institute, Concrete Centre: Taastrup, Denmark, 2008.

71. Schlitter, M. State-of-the-art and emerging truck engine technologies for optimized performance, emissions, and life-cycle costing. In Proceedings of the 9th Diesel Engine Emissions Reduction Conference, Newport, RI, USA, 24–28 August 2003.
72. Ortiz, O.; Castells, F.; Sonnemann, G. Sustainability in the construction industry: A review of recent developments based on LCA. *Constr. Build. Mater.* **2009**, *23*, 28–39. [CrossRef]
73. Hossain, M.U.; Poon, C.S.; Dong, Y.H.; Xuan, D. Environmental impact distribution methods for supplementary cementitious materials. *Renew. Sustain. Energy Rev.* **2018**, *82*, 597–608. [CrossRef]
74. Hossain, M.U.; Poon, C.S.; Lo, I.M.C.; Cheng, J.C.P. Comparative LCA on using waste materials in the cement industry: A Hong Kong case study. *Resour. Conserv. Recycl.* **2017**, *120*, 199–208. [CrossRef]
75. Ortiz-Rodríguez, O.; Castells, F.; Sonnemann, G. Life cycle assessment of two dwellings: One in Spain, a developed country, and one in Colombia, a country under development. *Sci. Total. Environ.* **2010**, *408*, 2435–2443. [CrossRef] [PubMed]
76. Kumanayake, R.; Luo, H. A tool for assessing life cycle CO_2 emissions of buildings in Sri Lanka. *Build. Environ.* **2018**, *128*, 272–286. [CrossRef]
77. Saynajoki, A.; Heinonen, J.; Junnila, S. A scenario analysis of the life cycle greenhouse gas emissions of a new residential area. *Environ. Res. Lett.* **2012**, *7*, 034037. [CrossRef]
78. Dernie, D.; Gaspari, J. Building Envelope Over-Cladding: Impact on Energy Balance and Microclimate. *Buildings* **2015**, *5*, 715–735. [CrossRef]
79. Chau, C.K.; Hui, W.K.; Ng, W.Y.; Powell, G. Assessment of CO_2 emissions reduction in high-rise concrete office buildings using different material use options. *Resour. Conserv. Recycl.* **2012**, *61*, 22–34. [CrossRef]
80. Giesekam, J.; Barrett, J.R.; Taylor, P. Construction sector views on low carbon building materials. *Build. Res. Inf.* **2016**, *44*, 423–444. [CrossRef]
81. Ng, T.S.K.; Yau, R.M.H.; Lam, T.N.T.; Cheng, V.S.Y. Design and commission a zero-carbon building for hot and humid climate. *Int. J. Low-Carbon Technol.* **2016**, *11*, 222–234. [CrossRef]
82. Colangelo, F.; Forcina, A.; Farina, I.; Petrillo, A. Life cycle assessment (LCA) of different kinds of concrete containing waste for sustainable construction. *Buildings* **2018**, *8*, 70. [CrossRef]
83. Wu, X.; Peng, B.; Lin, B. A dynamic life cycle carbon emission assessment on green and non-green buildings in China. *Energy Build.* **2017**, *149*, 272–281. [CrossRef]
84. Cuéllar-Franca, R.M.; Azapagic, A. Environmental impacts of the UK residential sector: Life cycle assessment of houses. *Build. Environ.* **2012**, *54*, 86–99. [CrossRef]
85. Passer, A.; Kreiner, H.; Maydl, P. Assessment of the environmental performance of buildings: A critical evaluation of the influence of technical building equipment on residential buildings. *Int. J. Life Cycle Assess.* **2012**, *17*, 1116–1130. [CrossRef]
86. Dahlstrøm, O.; Sørnes, K.; Eriksen, S.T.; Hertwich, E.G. Life cycle assessment of a single-family residence built to either conventional- or passive house standard. *Energy Build.* **2012**, *54*, 470–479. [CrossRef]

8

Industrial Data-Based Life Cycle Assessment of Architecturally Integrated Glass-Glass Photovoltaics

Jeeyoung Park [1,†], Dirk Hengevoss [2,†] and Stephen Wittkopf [1,*]

1 School of Engineering and Architecture, Lucerne University of Applied Sciences and Arts, Technikumstrasse 21, CH-6048 Horw, Switzerland; jeeyoung.park@hslu.ch
2 Institute for Ecopreneurship, School for Life Sciences, University of Applied Sciences Northwestern Switzerland, Hofackerstrasse 30, 4132 Muttenz, Switzerland; dirk.hengevoss@fhnw.ch
* Correspondence: stephen.wittkopf@hslu.ch
† These authors contributed equally to this work.

Abstract: Worldwide, an increasing number of new buildings have photovoltaics (PV) integrated in the building envelope. In Switzerland, the use of coloured PV façades has become popular due to improved visual acceptance. At the same time, life cycle assessment of buildings becomes increasingly important. While a life cycle inventory for conventional glass-film PV laminates is available, this is not the case for glass-glass laminates, and in particular, coloured front glasses. Only conventional glass-film PV laminates are considered in databases, some of which are partly outdated. Our paper addresses this disparity, by presenting life cycle inventory data gathered from industries producing coloured front glass by digital ceramic printing and manufacturing glass-glass PV laminates. In addition, we applied this data to a hypothetical façade made of multi-coloured glass-glass laminates and its electricity generation in terms of Swiss eco-points, global warming potential, and cumulative energy demand as impact indicators. The results of the latter show that the effect of the digital ceramic printing is negligible (increase of 0.1%), but the additional glass (4% increase) and reduction of electricity yield (20%) are significant in eco-points. The energy pay-back time for a multi-coloured PV façade is 8.1 years, which decreases by 35% to 5.3 years when replacing the glass rain cladding in an existing façade, leaving 25 years for surplus electricity generation.

Keywords: coloured glass; life cycle assessment; building integrated photovoltaic; rain cladding; LCA; LCI; BIPV

---

1. Introduction

1.1. Building Integrated Photovoltaics

Photovoltaics (PV) offer great opportunities in densely populated regions such as Switzerland, particularly when installed on the surfaces of buildings. This is primarily due to their ability to generate electricity without noise emissions, and deployment in a large range of sizes and forms. Ground mounted photovoltaics are impractical due to the scarcity and consequently high cost of land in Switzerland. However, the Swiss energy strategy 2050 aims to increase the national electricity supply from PVs from about 2% in 2016 [1] up to 20% by 2050. Global warming (climate change) is a critical issue drawing wordwide attention, and Switzerland should reduce green house gas emission by 20% in comparison to their 1990 level by 2020 according to Swiss law (the CO_2 Act). In addition, nuclear power, which is CO_2-free and generates about 40% of Swiss electricity, will phase out [2].

Building integrated photovoltaics (BIPV) contribute to achieving this goal by fully utilising building surfaces, such as the roof or façade, to maximise electricity generation. This goal is difficult to achieve with monochromatic BIPV, as they compromise the architectural aesthetic of the building, leading to low visual acceptance on behalf of the architect or the building's owner. Various types of technologies for aesthetically appealing PV systems in buildings have recently been developed, such as thin films or special foils [3]. Such systems are referred to as *architecturally integrated*. However, the conservative nature of the construction sector, as well as vague or even conflicting building regulations regarding façade elements have thus far delayed the application of these technologies [4].

Multi-coloured glass-glass (MCGG) and crystalline silicon cell (c-Si) PV laminates are an approach to overcome some of these aforementioned issues and achieve aesthetically pleasing, yet technically and economically viable building integrated PV systems, by utilising market-proven technologies which also retain high market potential in the future. This solution combines translucent digital ceramic print technology [5–7] with existing mainstream technologies, notably crystalline silicon cells and glass-glass sheets (see Figure 1).

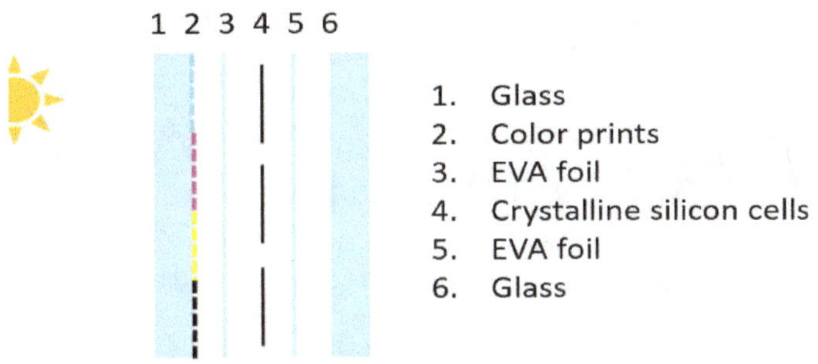

Figure 1. Layers comprising a multi-coloured glass-glass photovoltaic laminate. A translucent multi-coloured motif (layer 2) is printed on the inner surface of the front glass sheet [3].

C-Si is the mainstream technology in commercial PV markets accounting for 94% of the market share, while thin film accounts for 6% [8]. Despite the emergence of many new technologies, the former is most likely to remain dominant for next decade, driving substantial market growth by ongoing cost reduction and improvement in efficiency [9].

The most common configuration of PV laminates for grid connected systems comprises 60 cells, which are typically arranged as a 6 by 10 matrix, occupying an area of approximately $1.00 \times 1.615\ m^2$. The cells are wired in series, thus metallic tab wires attached to a front side of a cell are connected to the backside of a neighbouring cell. This interconnecting wiring technology is subject to ongoing improvements in cell efficiency, with a triple busbar arrangement currently dominating the market [10]. Deployed without a frame, this system is referred to as a photovoltaic laminate, while an additional surrounding frame designates it as a photovoltaic panel. A junction box and two ca. 1m cables are attached to the laminate or the panel for connection to adjacent units.

Interconnected cells are brittle and easily corroded by moisture. To protect against impact and moisture from hail and rain in outdoor environments, the solar cell strings are encapsulated by two sheets of polymer foils (e.g., Ethylene-vinyl acetate, or EVA), which are in turn encapsulated by two sheets of glass. Finally, these layers are laminated under heat and pressure in a vacuum to form a PV laminate. For a non-BIPV solution, a Tedlar® polymer film is also commonly used instead of a glass back-sheet. However, a BIPV laminate should resist fire hazards and wind loads to comply with architectural regulations on a building façade. Glass sheets are a typical solution to achieve this, as glass is a commonly used façade component under the established building codes. Moreover, even for a non-BIPV solution, glass back-sheets are increasingly popular as they prevent moisture intrusion.

MCGG PV laminates can readily replace façade panels in a rainscreen wall. A rainscreen wall system [11] prevents penetration of rainwater into a building wall and is widely used since the 1970s (see Figure 2). It consists of a rainscreen cladding, ventilated and drained air cavity, and an air barrier system from the outermost layer of a building. Common materials for rainscreen cladding include fibre cement, metal, and timber, which are selected based on various criteria. Glass is frequently chosen when an aesthetically pleasing façade finish is preferred. The multi-coloured glass-glass PV laminates have the potential to satisfy this aesthetic criterion and be accepted as rainscreen cladding, making them economically and environmentally beneficial as the existing rainscreen mounting system can accommodate the PV (see Figure 2b).

(a) (b)

Figure 2. Rainscreen wall system: Glass cladding installed on the outside of a building (**a**) and a rainscreen cladding sub-construction (**b**) [12].

1.2. Life Cycle Assessment

Environmental Life cycle Assessment (LCA) is a methodology to assess all the environmental impacts of a product or a service during the whole life cycle from the raw material extraction to final disposal [13]. Alternative products have different eco-profiles over their lifetime; one product's environmental impact contributes mainly to its production phase, while another's dominates during its use or operating phase. A classical example are disposable diapers, which create 90 times more solid waste than reusable cloth diapers. The latter, however, produce tenfold water pollution due to detergents etc., and consume triple the energy [14]. LCA does not provide a simple answer, but it enables rational judgements with trade offs. For this reason, LCA is becoming an increasingly important and popular tool in policy and industry [13].

The International Organisation for Standardisation (ISO) standardised an LCA methodological framework and terminology (ISO 14040:2006) [15] and provided general guidelines and requirements (ISO 14044:2006) [16]. According to ISO, the LCA procedure consists of the following steps:

1. Goal and scope: Definition of the System boundary & functional unit.
2. Life Cycle Inventory (LCI): Elaboration of a mass balance for a process with all inputs and outputs.
3. Life Cycle Impact Assessment (LCIA): Assessment of environmental consequences of the LCI such as climate change, natural resource depletion, ozone depletion, ecotoxicity etc. with specific indicators. A sensitivity analysis considers the individual effects of the choices made, i.e., flows and indicators.
4. Interpretation: Identification of processes and flows with main environmental impact and recommendation of measures for improvement.

BIPV affects the heating and cooling load in buildings, as it generates heat as well as electricity and replaces building elements which may have a different thermal resistance [17]. The current reviews covering the state-of-the-art of LCA on BIPV assess the influence of BIPV on the energy performance of buildings [18,19]. Adaptive BIPV deployed on windows provides shading and admits daylight, while also reducing the energy consumption for lighting, heating, and cooling [20]. These energy reduction factors can be considered in two ways:

1. Place the factors into a system boundary. For example, Jayathissa [20] estimated the energy load (heating, cooling, and lighting energy) in an office with windows fitted with dynamic BIPV, static shading, and no shading at all. The study compares the environmental impact of the German grid electricity to that generated by different BIPV technologies (thin film and crystalline), and the resulting reduction in energy loads. These results are further discussed in the context of this study in Section 3.5.
2. Alternatively, these factors can be excluded from a system boundary along with building materials serving a similar purpose, (e.g., when replaced by BIPV in a façade). For example, Ng [21] estimated the lifetime performance of semi-transparent BIPV glazing when it replaces double glazing windows, which similarly impacts building energy performance.

In this study, the latter approach was adopted by replacing the rainscreen cladding in a façade with BIPV. This better caters to the focus on electricity generation within the scope of the study, the purpose of which is explained in the following section.

1.3. Purpose of the Study

Glass is a commonly used material in both photovoltaic and building industries, and its proportional weight in a glass-glass PV laminate is over 90%. Moreover, glass-glass c-Si PV laminates are the de facto standard for a building integrated solution. However, its environmental impact assessment (LCA) is not available, thus the resulting approximated environmental impact based on life cycle inventory databases may not reflect reality.

To realistically assess the environmental impact of the entire life cycle of a MCGG c-Si PV laminate, this study presents up-to-date data on c-Si production gathered from literature, as well as material and energy flow data from a solar glass and PV laminate supplier. Moreover, a façade case study considers a realistic BIPV use case, in which the PV laminate replaces the rainscreen cladding.

Unlike ground mounted photovoltaics, building integrated photovoltaic systems need to satisfy multiple goals, i.e., aesthetic appeal, fail-safe installation and operation, and electricity generation with lower environmental impact than that of the existing Swiss low-voltage grid. As a consequence, multiple parties contribute towards achieving these goals, including the glass and photovoltaic manufacturer, architect, building owner, planner, and installer. These goals need to be achieved collectively as each party has only a limited scope of influence, thus the relevant information must be consolidated. However, existing PV life cycle inventory data and assessment results are aggregated and averaged, thus individual data relevant to the building context cannot be often identified.

The purpose of this study is therefore to provide life cycle perspectives of BIPV in general, and glass-glass crystalline silicon cell photovoltaics in particular, by presenting life cycle inventory data and the assessment results in an itemised and organised manner, such that they are relevant and comprehensible to the involved parties. Furthermore, the influence of façade components on the environmental impact of BIPV are analysed, as well as the additional benefits of BIPV for a building. In the long term, these results support the parties in making future decisions in their respective areas of influence.

The rest of the paper is organised as follows: In Section 2, we elaborate our LCA methodology of a PV façade, and the impact indicators used; Section 3 presents results obtained from the LCA of the façade components, and the electricity generated in terms of energy pay-back time; in Section 4, we draw our conclusions. A glossary of abbreviated terminology used in this paper can be found in Table 1.

Table 1. Abbreviations and terminology used in this paper.

Term	Definition
AC	Alternate current
BIPV	Building integrated photovoltaics
BOS	Balance of system
CED	Cumulative energy demand
c-Si	Crystalline silicon cell
EPBT	Energy pay-back time
EVA	Ethylene-vinyl acetate
GWP	Global warming potential
ISO	International Organization for Standardisation
IPCC	Intergovernmental Panel of Climate Change
LCA	Life cycle assessment
LCI	Life cycle inventory
MCGG	Multi coloured glass-glass
m-Si	Multi crystalline silicon cell
PV	Photovoltaics
s-Si	Single crystalline silicon cell

2. Methodology

The following section specifies the PV laminate and façade installation used in this study. In accordance with the sequence of the four LCA phases in Section 1.2, the goal and scope definition, inventory analysis, and impact indicators used in this study are presented. The impact assessment and interpretations is presented as results in Section 3.

2.1. PV Laminate Specifications for Life Cycle Assessment

Table 2 lists detailed specifications of a PV laminate used in this study, which is representative of one of the most common configurations in the market. The basic configuration of a laminate consists of multi-crystalline silicone (m-Si) cells and a multi-coloured front glass sheet. Common variants on the market replace the multi-coloured front glass sheet with a clear glass sheet, and m-Si by s-Si.

Table 2. Typical specifications of a photovoltaic laminate considered in this study.

Parameter	Basic Configuration	Variants
Dimension of PV laminate	1.00 m × 1.615 m	
Weight of PV laminate	35.47 kg	
Glass thickness	4 mm × 2 = 8 mm	
Glass weight	16.15 kg × 2 = 32.3 kg	
Glass type	multi-coloured (80% performance)	clear glass
Type of cells	60 m-Si cells (220 Wp)	60 s-Si cells (250 Wp)
Thickness of cell	0.2 mm	
Wiring technology for cells	3 busbar tap wirings	36 active(smart) wirings [10]

To consider the entire life cycle of a photovoltaic laminate, the study considers a hypothetical façade installation consisting of 140 laminates (see Table 3). The system capacity using the m-Si clear glass PV laminates (250 Wp) is 30.8 kWp. When using multi-coloured glass sheets, electricity production decreases by 20%, reducing the capacity to 24.6 kWp. The system with clear glass s-Si PV laminates has 35 kWp capacity. The annual electricity yield expressed in kWh/kWp for each orientation is measured in Dübendorf, Switzerland, after the inverter and before being fed into the grid; this takes into account losses due to AC conversion, while transmission losses in the grid are disregarded. The calculation of electricity production assumes a projected PV lifetime of 30 years. To consider various degradation modes such as discolouration and corrosion [22], the annual yield is linearly derated by 0.69% as recommended by IEA PVPS [23].

Table 3. Façade installation considered in this study as use case for photovoltaic (PV) laminates.

Parameter	Basic Configuration	Variants
PV type	m-Si (15% efficiency) multi-coloured	s-Si (17% efficiency)
Laminate capacity (System capacity)	176 Wp (24.6 kWp)	m-Si clear glass: 220 Wp (30.8 kWP) s-Si clear glass 250 Wp (35 kWp) s-Si coloured glass 200 Wp (28 kWP)
Number of laminates	140	
Laminates used over lifetime	144.23 (1% rejected in construction of façade, 2% replaced due to early end of life)	
Façade dimension	21 m × 12 m = 252 m^2	
Projected lifetime	30 years	
Annual yield/kWp installed before degradation according to façade orientation	Facing south: 700 kWh/kWp Facing east/west: 530 kWh/kWp Facing north: 200 kWh/kWp	
Degradation of PV	0.69 % per year from the first year, total 20 % for 30 years	
Lifetime yield according to façade orientation	Facing south: 582 MWh Facing east/west: 440 MWh Facing north:166 MWh	

2.2. *Goal and Scope of Life Cycle Assessment*

In accordance with the purpose of the study stated in Section 1.3, the scope of the LCA covers:

- up-to-date crystalline silicon cell production
- clear and multi-coloured glass production by a specific manufacturer
- glass-glass laminate production with various configurations by a specific manufacturer
- a hypothetical but realistic PV façade installation
- electricity generated from the façade facing south, east/west, and north
- and a comparison of the generated electricity to that of the Swiss low voltage electricity grid.

Figure 3 presents functional units of each product along the supply chain of the BIPV system, leading to the generated electricity as final product. The laminate is assessed by applying the "cradle to gate" approach. The system boundary includes the material and energy flows of the upstream processes for the laminate. Nevertheless take back and recycling of a laminate is already included. The electricity is assessed by applying the "cradle to cradle" approach. The system boundary includes all processes for laminate production, including production of the Balance of System (BOS) [24], such as mounting system, cabling, and an inverter. Furthermore, this considers the transportation, installation, cleaning, and maintenance of the PV installation, and its end of life. To obtain the environmental impact per 1kWh electricity generated using the model in this study, the total environmental impact of the façade (including water and waste water from cleaning) is simply divided by the total electricity generated during its lifetime.

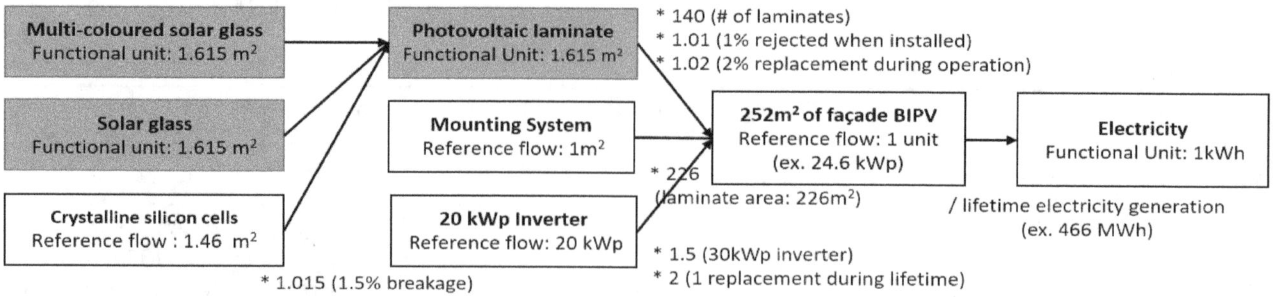

Figure 3. Configuration of an architecturally integrated 24.6 kWp photovoltaic (PV) façade facing south consisting of 140 multi coloured glass-glass (MCGG) laminates generating 466 MWh over 30 years. While a 30 kWp inverter is installed, its data is derived from that of a 20 kWp inverter scaled by a factor of 1.5. Product systems with grey background were specified from industrial data, while those with white background were derived from literature values and estimates.

2.3. *Life Cycle Inventory*

A main challenge in LCA is the collection of data for the Life Cycle Inventory (LCI). This can be mitigated by using the Ecoinvent database [25], which contains basic data on materials, energy, waste, transportation, and other factors. This data is collected from averages of material and energy flows provided by industrial sectors, expert estimates, and literature values. This study uses the APOS [26] data set, which is one of the three sets of data available in Ecoinvent 3.4. The calculations were performed in software mainly using Simapro 6.4, supplemented by MS Excel.

Ecoinvent data is updated frequently, but a literature review has shown that updated LCI data on the requisite photovoltaic components for this study are not yet integrated in the latest Ecoinvent version 3.4. Thus, this study carried out an LCI for system components based on recent literature and current industry data, which are stated in the sections below.

2.3.1. LCI of Crystalline Silicon Cells

The LCI for m-Si and s-Si cells in Ecoinvent 3.4 were updated with an LCA of photovoltaics dated 2011 [27] (Part 1: Data Collection, Table 37). A main difference compared to the LCI in Ecoinvent 3.4 is the an increase in electricity consumption for the m-Si wafer process from 4.7 kWh to 20.8 kWh. For further differences, see the cited reference.

2.3.2. LCI of Solar Glass

Solar glass is made from float glass, i.e., floated on molten tin to obtain a flat and polished surface [28]. The LCI for float glass in Ecoinvent 3.4 was itemised with material flow data from two European float glass manufacturers [29,30].

The energy consumption for the final solar glass production was obtained from the Swiss manufacturer (personal communication P. Schaad, Glas Trösch AG, 25.05.2018). The resulting material and energy flows for the processes to produce 1.615 m^2 of clear and multi-coloured solar glass sheets are shown in Figure 4 and Table S1 (supplementary material). The company produces multiple glass types of different sizes, which run through different process steps. Consequently, the production is not optimised for solar glass. This implies higher energy stand-by losses in continuous production flows as well as material losses. Energy flows (electricity and compressed air) are expressed in kWh, while those for the material are expressed in kg. The thickness of an arrow corresponds to the quantity of material flow.

The production of clear solar glass requires four main processes: Cutting the raw material (i.e., float glass) to the desired size; bevelling (grinding the edges); washing off grinding sludge and drying; and finally tempering to increase the strength of the solar glass. Losses and breakages from the raw glass at cutting are about 20%. The cullet is recycled directly in float glass production, which reduces the quantity of the primary raw material. Breakages at tempering and quenching of the solar glass are about 5%. To produce 32.30 kg of two final glass sheets, 43.24 kg of raw float glass are consumed. Coloured solar glass needs two additional passes before tempering: A second washing and drying step immediately following the first (on the same washing machine), and digital printing with ceramic ink on the clear glass, followed by tempering. The total electricity consumption in production adds up to 19 kWh.

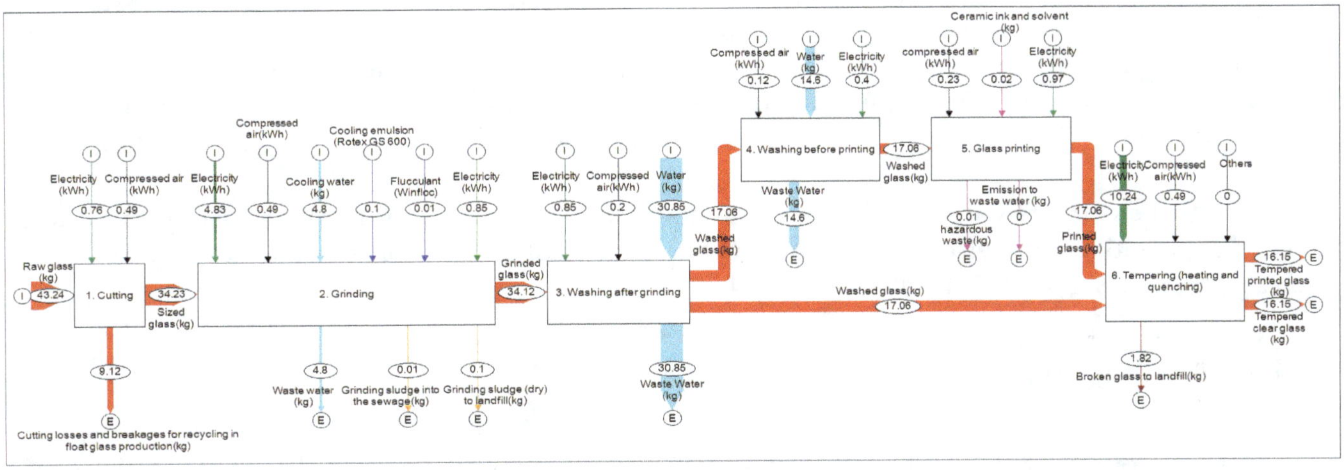

Figure 4. Material (kg) and energy (kWh) flows to produce 1.615 m^2 of a multi-coloured and a clear solar glass sheet with 4 mm thickness at a Swiss glass supplier. Arrowheads labelled *I* denote import flow, indicating an input material or energy, while those labelled *E* denote export flow, indicating a product, waste, or loss. The data was derived from the material and energy balance as well as the measured electricity consumption of the production steps and support processes. These were obtained from the Swiss glass manufacturer (courtesy of Glas Trösch AG, 2017).

2.3.3. LCI of PV Laminate

The front and back glass are transported from the Swiss glass manufacturer to a PV laminate manufacturer in Germany. The production the glass-glass PV laminate (Figure 5) requires four main processes: Cleaning the glasses in a washing machine (closed water cycle); cutting the EVA foil; brazing crystalline silicon cell strings (by an external supplier) and connecting tap wires; laminating under pressure and heat; cutting EVA leftovers; and final mounting and test.

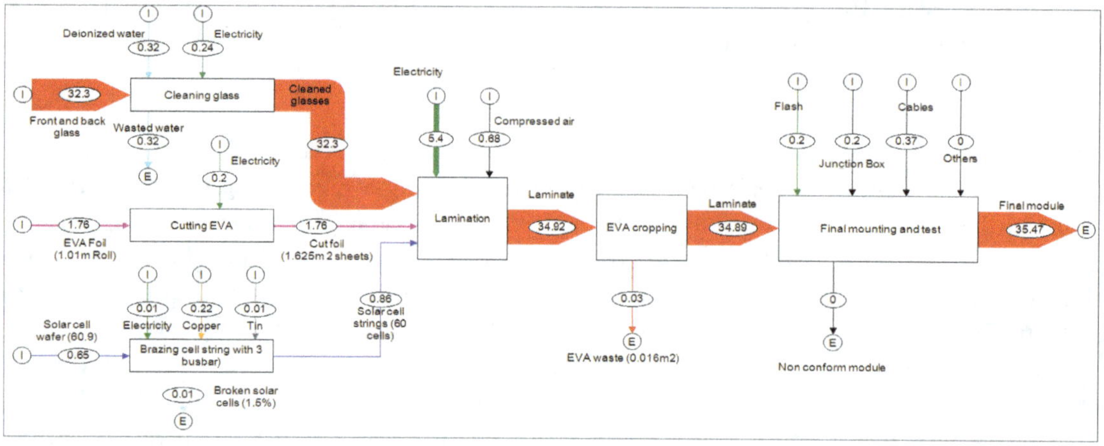

Figure 5. Material (kg) and energy (kWh) flows to produce 1.615 m^2 of a multi-coloured glass-glass PV laminate with 60 crystalline silicon cells, a junction box, and connecting cable. Arrowheads labelled *I* denote and import flow, indicating an input material or energy, while those labelled *E* denote an export flow, indicating a waste or loss. Data was derived from literature [27] and updated with material balance data and measured electricity consumption of the production steps and supporting processes provided by a German PV module manufacturer (courtesy of GES Gebäude- und Energiesysteme GmbH, 2018).

The total electricity consumption in production adds up to 6.05 kWh. When using crystalline silicone cells with active wires instead of 3 busbars, the total amount of copper in the PV laminate increases from 219 g to 229 g by 4%.

2.3.4. LCI of Further Processes and Components

The data sources for the LCI of other processes and components considered in the LCA are listed in Table 4.

Table 4. Data sources for the elaboration of the life cycle inventory (LCI) for other processes and components.

Processes and Components	Remark, Source/Reference
30 kWp inverter	LCA of low power solar inverters (2.5 to 20 kW) [31]
Mounting and electric system	Ecoinvent 3.4 [25]
Installation	Transportation and electricity for mounting, Ecoinvent 3.4
Cleaning and maintenance	Water consumption and waste water treatment, Ecoinvent 3.4
Take-back & recycling of laminate with one glass sheet	Energy consumption for shredding was adapted to the higher glass quantity in a glass-glass laminate. It is assumed the quality of the glass cullet is too low for recycling in float glass production [32].

2.4. Life Cycle Impact Indicators Selected for This Study

Impact categories and indicators commonly used in LCA of photovoltaic system are presented in [23] (Table 3.2). For this study, the five indicators in Table 5 were selected.

Table 5. Indicators selected for the life cycle impact assessment (LCIA) in this study.

Name	Unit	Remark
Global warming potential (GWP)	Grams CO_2-equivalents [g CO_2 eq]	Contains Intergovernmental Panel of Climate Change (IPCC) climate change factors for a timespan of 100 years [33]
Cumulative energy demand (CED)	MJ-equivalents [MJ eq]	Contains the energy content of renewable and non-renewable primary energy [23]. In this study, only non-renewable primary energy is considered.
Ecological Scarcity 2013	Eco-points [EP]	Eco-points reflect both the actual emission situation and the national or international emission targets pursued by Switzerland [34].
Energy Payback Time (EBPT)	Years	Time required to generated enough electricity, so that Non-renewable CED/kWh becomes the same as one of the reference electricity [23]. Refers to the efficiency of the Swiss low voltage electricity grid at the consumer end (9.43 MJ/kWh acc. to Ecoinvent 3.4)

In this study, the LCIA of solar glass, photovoltaic laminate, and façade system mainly use the ecological scarcity as indicator (expressed in eco-points), while the electricity generated by the PV façade is assessed with CED and EPBT, which are commonly used indicators to compare energy generation systems, along with GWP.

3. Results

This chapter presents the results of the Life Cycle Impact Assessment (LCIA) according to the sequence of the supply chain described in Figure 3. First, the environmental impact of the production of the front and back glass for a PV laminate is presented in terms of eco-points. Subsequently, the eco-points of the production of PV cells and laminates and of an architecturally integrated PV façade installation are detailed. Finally, the EBPT of the system will be presented along with the eco-points and GWP of 1 kWh of generated electricity at different façade orientations compared to the current impact of the Swiss low voltage electricity grid.

3.1. Production of Multi-Coloured and Clear Glass

Figure 6 shows the eco-points of a multi-coloured front glass sheet (1.615 m^2) by components and processes, which adds up to a total of 34.2 kPt. The eco-point of an unprinted back sheet adds up to 33.2 kPt. Consequently, the total eco-points (Figures S1 and S2) for the front and back glass of a PV laminate is 67.4 kPt. The eco-points for the clear solar glass (33.2 kPt) in this study are 47% higher than those from Ecoinvent (22.6 kPt). This is because the reference data for float glass used in this study [29] has different input materials from that of Ecoinvent. As mentioned in Section 2.3.2, this is further attributed to the high glass losses and breakages in cutting and tempering.

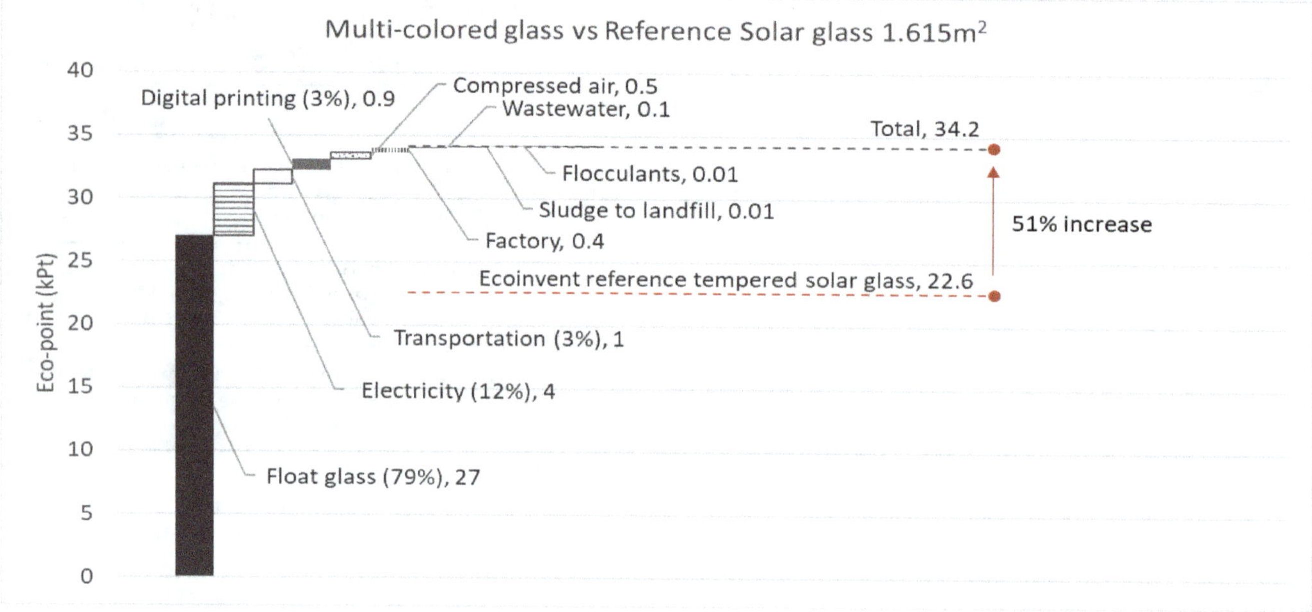

Figure 6. Per-process and component breakdown of eco-points for 1.615 m^2 of multi-coloured glass for a PV laminate modelled in this study. The total (34.2 kPt) is 51% higher than that of tempered solar glass (22.6 kPt) in Ecoinvent. The printing step to obtain the multi-coloured glass only contributes 3% to the total impact.

The raw float glass input incurs a major contribution (79%) to the total eco-points of the multi-coloured glass sheet production. This is attributed to the energy and CO_2 intensive, which has a strong weighing in eco-points, float glass production process and the losses and breakages in the solar glass production. The electricity and digital printing account for only 12% and 3%, respectively. The European electricity mix contributes to not just GWP and eco-points but also the category nuclear waste.

3.2. *Production of Crystalline Silicon Cells*

The LCIA results for the production of the 60 m-Si and s-Si cells for a PV-laminate (1.615 m^2) using the latest published data (see Section 2.3.1) that deviate from those in Ecoinvent 3.4 for the same cell types. The eco-points for the m-Si cells increase by 22% from 311 to 380 kPt, while those for s-Si cells are reduced by 5% from 524 to 496 kPt.

3.3. *Production of PV Laminate*

The front and back glass from the Swiss glass manufacturer are transported by road to the German PV laminate manufacturer. The crystalline silicon cells, EVA foil for lamination and the electric components (bus wire, tap wires, junction box and the connecting cable), as well as packaging material, are purchased on the international market. When smart wires [10] are used instead of triple busbars as wiring technology for the c-Si cells, the total amount of copper in the PV laminate increases by 4%, which has a negligible effect on its total eco-points.

Figure 7 shows the eco-points for three different types of PV laminate. In the production of an MCGG m-Si PV laminate (Figure S3), the multi-coloured front and clear back glass accounts only 15% of the total eco-points (460 kPt). For s-Si MCGG PV, the glass contributes 10% to the total eco-points (673 kPt). The impact of printing is negligible. By contrast, the impact of the PV cells is significant; PV glass laminates using s-Si cells have a 46% higher impact in eco-points compared to laminates with m-Si cells, while the electric components contribute 7% and 10%, respectively. The impact of these components stems from the copper. The reference (m-Si glass-foil PV laminate) has lower eco-points than an m-Si glass-glass PV laminate due to the lower contribution from the backsheet foil compared to the glass sheet. The proportion of eco-points for take-back and recycling at the PV laminate's end of life is 3% and included in the *Others* category.

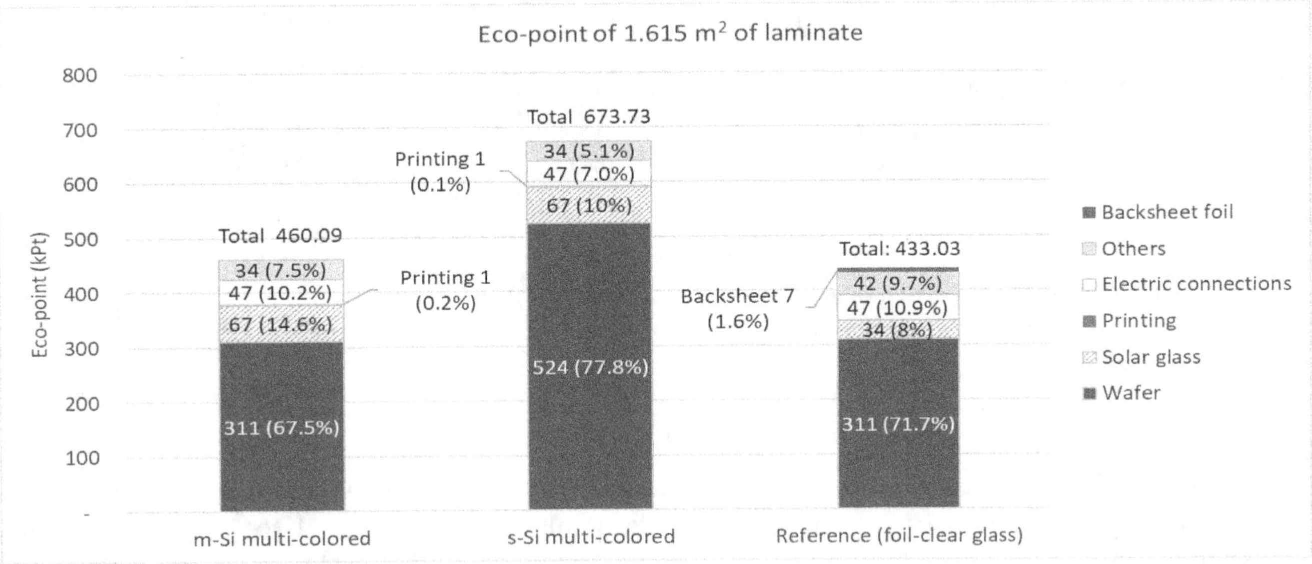

Figure 7. Eco-points for the production for various glass-glass PV laminate configurations. The reference data is for a 1.615 m^2 glass-foil.

3.4. Installation of an Architecturally Integrated Photovoltaic Façade

For the installation of the façade in Switzerland, 140 modules are delivered by road. The BOS (inverter, mounting system, electric components) are purchased from the international market. The LCIA of the installation of the architecturally integrated photovoltaics façade considers these system components: The transportation, 1% of rejected PV laminates at the construction site, replacement of 2% of PV laminates due to early end of life during operation, a one-time replacement of the inverter during the operating lifespan, and furthermore the electric energy for mounting.

The eco-points for the façade with s-Si cells and clear glass is 141,000 kPt. For m-Si cells, it is 22% lower (110,000 kPt). The solar glass and the mounting system can be excluded from the system boundary and allocated to the building, e.g., if an existing rain cladding glass façade approaches the end of its useful life and needs to be replaced, or the multi-coloured solar glass assumes a function of the building. In this case, the eco-points decrease by 25% to 83,000 kPt for the m-Si configuration, and by 19% to 114,500 kPt for the s-Si configuration.

Figure 8 shows the breakdown of the eco-points for façade configurations consisting of 24.6 kWp m-Si and 28 kWp s-Si MCGG PV laminates, and 30.8 kWp conventional clear glass-foil PV laminates, containing 140 modules of each component in ratio. The major contributors are the wafers in all three façades, making up 41% (glass-glass) and 42% (glass-foil) for m-Si, and 54% for s-Si. Front and back glass contributes 9% (m-Si) and 7% (s-Si) for the glass-glass PV façade. By contrast, the foil-glass façade contributes only 5% for the front glass, and 1% for the foil. Lastly, the mounting system contributes 13% resp. 10%, while the inverter contributes 12% resp. 9% for the glass-glass PV façade.

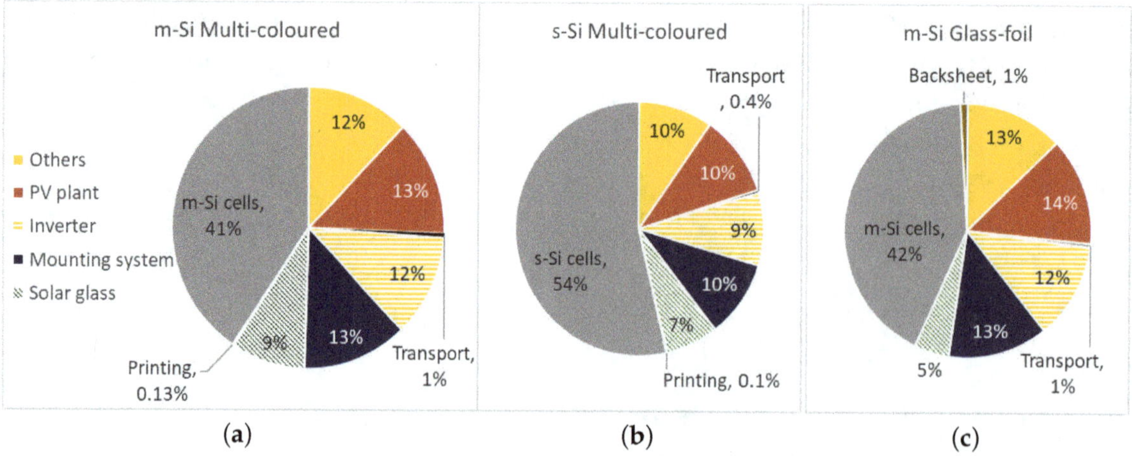

Figure 8. Percentile breakdown of eco-points for various façade configurations consisting of 140 PV modules: 24.6 kWp m-Si (**a**) and 28 kWp s-Si (**b**) multi-coloured glass-glass, and 30.8 kWp traditional foil-glass (**c**).

3.5. Electricity Generation Based on Façade Orientation

All environmental impact of the façades are allocated to electricity generated, so the percentage of eco-points contributions of the façades (Figure 8) also can be applied to 1 kWh electricity. A recommended functional unit for the LCA of the electricity generating system is 1 kWh, facilitating comparisons irrespective of size and capacity. The environmental impact of 1 kWh electricity can be decreased if that of any contributors are decrease. For example, to decrease the impact by 10%, the following options are viable:

- Increase electricity production by 3% (for example, by improving c-Si cell efficiency)
- Reduce impact during façade installation by 3% (for example, mounting system with lower eco-points)
- Reduce impact during lamination processes by 4% ($0.04 \times 0.6 = 2.4\%$, since 140 laminates contribute 60% of the total eco-points) (for example, reducing electricity consumption)
- Decrease solar glass impact by 20% ($0.2 \times 0.09 = 1.8\%$, since glass contributes 9% to the total) (for example, reducing loss and breakage)

In the same way, magnitude of the uncertainty of data in each contributor affecting to 1 kWh of electricity also calculated. For example, 20% of uncertainty of solar glass data affects 1.8% of impact (eco-points) of 1 kWh electricity. The float glass contributes 79% of the solar glass, so 20% of float glass reduction reduces the total impact of electricity by 1.4%.

Depending on the façade orientation, the amount of electricity generated by a PV system varies, along with its environmental impact per kWh. Therefore its orientation is considered when comparing the façade to the Swiss low voltage electricity grid as reference. The reference electricity has an impact of 0.273 kPt per kWh. In this study, this is derated by 6% to 0.257 kPt to account for the grid loss of electricity generated by the façade. Thus, the environmental impact of the reference electricity is decreased instead of reducing the amount of electricity generated by the façade.

The GWP and eco-points per kWh generated according to façade orientation are shown in Figure 9. In terms of eco-points, all configurations facing south are superior to the reference. In the façade facing east/west, only the m-Si configurations are viable. In general, GWP indicates less favourable results than eco-points compared to the Swiss low voltage electricity grid. Consequently, the discussion that follows focuses on the former indicator.

The Swiss grid electricity mix (GWP of 0.13 kg CO_2 eq), used as reference in this study, has a lower share of carbon-based electricity, but a higher share of nuclear and hydro power than the European electricity mix. This contrasts with the German grid electricity mix (GWP of 0.61 kg CO_2 eq) used as reference in [20], which is dominated by high-impact coal sources. Almost all the façade configurations in Figure 9, except s-Si multi-coloured facing north, outperform the German electricity mix. Furthermore, the figure reveals that:

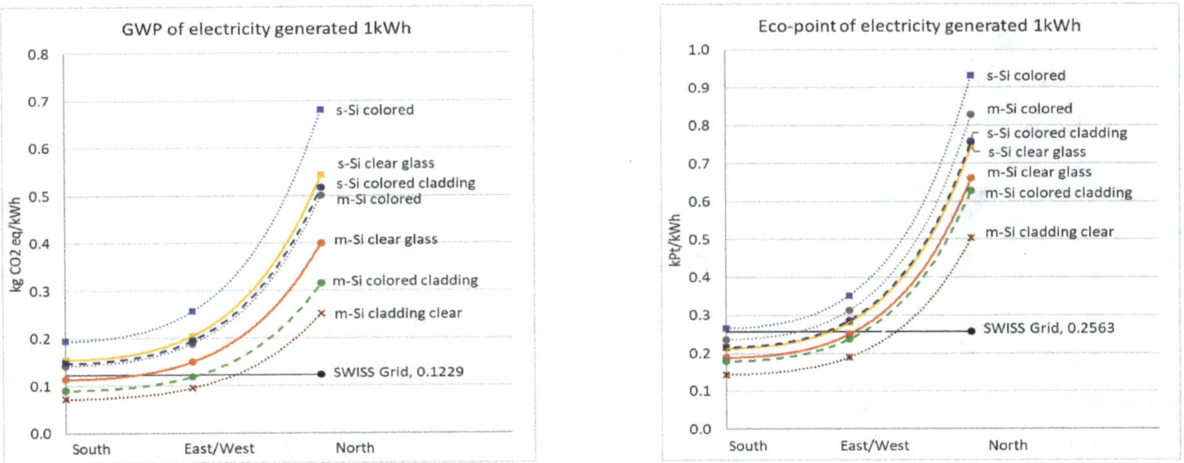

(a) (b)

Figure 9. Environmental impact of 1 kWh of electricity generated from each configuration and orientation of the PV façade in terms of global warming potential (GWP) (**a**) and eco-points (**b**).

- The north facing façade has the 2–5-fold GWP of the reference, and is unsuitable in any configuration. The worst-case GWP is infact comparable to that of coal power plants. The best-case GWP is, on the other hand, comparable to that of gas power plants.
- S-Si cells (with an efficiency of 17% in this study) are unsuitable under all conditions. They produce 13% more electricity than m-Si, but their GWP is also 54% higher. (see Table A1).
- Multi-coloured PV systems are superior to the Swiss low voltage electricity grid only when the laminates replace part of an existing rain cladding system. In this case, the GWP decreases by 36%: Solar glass (13%); the mounting system (19%); glass take-back and transportation (4%); and printing (0.4%).
- Clear glass m-Si photovoltaic on south facing façade is superior to the reference even when all the environmental impact is allocated to it.

Jayathissa [20] estimated GWPs of 0.144 to 0.331 kg CO_2 eq per kWh of electricity generated by a façade installation in Germany using different BIPV technologies (thin film and crystalline), disregarding the energy load of the building. In contrast, the GWP of all BIPV façade components in Table A1 of this study lies between 0.114 and 0.194 kg CO_2 eq per kWh, which reduces to 0.072–0.148 kg CO_2 eq per kWh when replacing the rain cladding. However, in relating these results, we note the following constraints, which only permit an indirect comparison:

- Different datasets (Ecoinvent 3.1 vs. 3.4)
- Lifespans of 20 years instead of 30 in this work.
- Annual yield of 855 kWh/m^2 (irradiation) $\times$ 0.11 (efficiency) $\approx$ 94 kWh/m^2 vs. annual yield of 700 kWh (facing south) $\times$ 24.6 kWp/252 $m^2 \approx$ 68 kWh/m^2.

Figure 10 shows the non-renewable CED per kWh of electricity generated by the façade. It decreases over time from its initial value of ca. 900 GJ prior to electricity production, which is the total CED of the m-Si façade. After a projected 30-year lifespan, this is reduced to 1.92 MJ/kWh. The EPBT of the south-facing multi-coloured PV system is 8 years, but when replacing a part of a rainscreen wall system, this drops to just 5.3 years. Other EPBT results are summarised below.

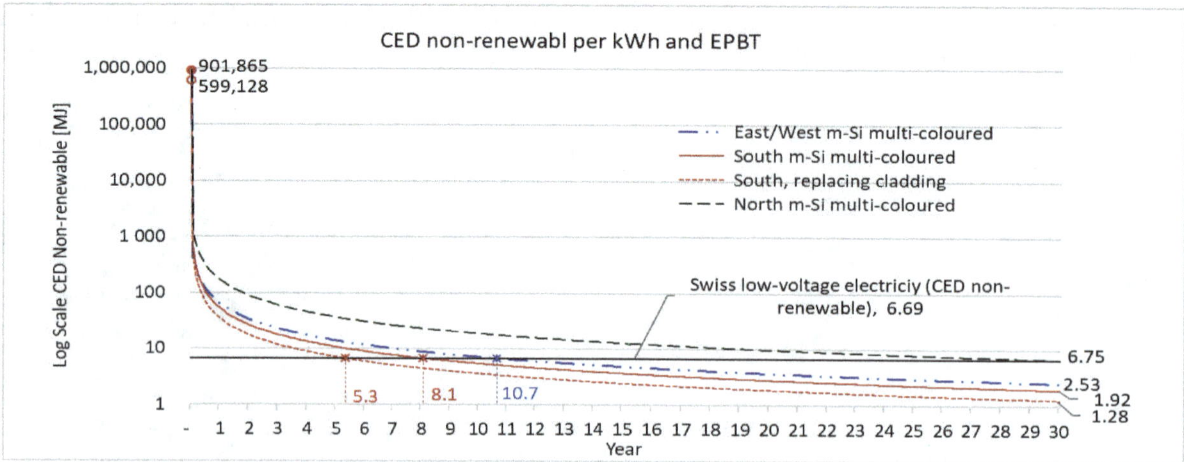

Figure 10. Cumulative energy demand (CED) non-renewable of an MCCG, m-Si PV façade over time, and energy pay-back time (EPBT) compared to Swiss low voltage electricity grid as reference.

- **EPBT m-Si:** The EBPT of a PV façade facing south with clear and coloured glass is 6 and 8 years, respectively. Facing east/west, this increases to 8 and 11 years, respectively, while facing north, the EBPT exceeds 20 years.
- **EPBT s-Si:** The EBPT of a PV façade facing south with clear and coloured glass is 8.4 and 10.6 years, respectively. Facing east/west, this increases to 11.2 and 14 years, respectively, while facing north, the invested energy does not pay off at all.

4. Conclusions

To achieve the goal of the Swiss energy strategy, namely to increase contribution of PVs to the national electricity supply by up to 20% by 2050, BIPVs need to be aesthetically appealing for visual integration while generating electricity with a lower environmental impact than that of the existing Swiss low-voltage grid. These results compare favourably with those of a similar study using the German grid electricity as reference [20].

The primary aim of this study was to support the BIPV industry by providing previously unavailable and updated LCI data and LCA results of a glass-glass PV laminate as a reference. This reference can be used in the future to assess the environmental impact of novel BIPV laminates. The secondary aim of this study was to provide a holistic view of the entire life cycle and supply chain of BIPV and relevant information to each involved party, by presenting the LCA results and LCI data in an organised manner suitable for upstream building industries.

The LCA was carried out based on the latest published data of crystalline silicon cell production (currently omitted in LCI databases), the actual material and energy flow data from the glass and PV laminate suppliers, and lastly a realistic façade use case suitable for practical installation planning to identify a viable application as electricity generation system.

The most obvious findings to emerge from the LCA is that printing on glass is negligible as it accounts for less than 0.2% of the total impact (in eco-points) of a PV façade and 1 kWh of electricity generated. By contrast, the impacts of the PV cells are significant, making up 41% for m-Si and 54% for s-Si. Single crystalline cells are not suitable in general cases as s-Si types contribute 68% higher eco-points than m-Si, while their efficiency is only 13% higher. Moreover, a north-facing PV façade generates far too little electricity to justify its environmental impact compared other façade orientations, as well as the Swiss low voltage electricity grid as reference.

The itemised LCA results revealed enormous optimisation potential in future glass production processes, for instance by reducing the idle time and losses, ideally due to increased demand and production volume. The dominant contributor to the impact of solar glass production in terms of eco-points is the float glass input with 79%. Lesser contributors are cutting and breakage losses with 25% of the input, while electricity consumption contributes 12% to the total eco-points.

Overall, the results support the notion that an itemised LCA can contribute to the building industry by presenting a holistic view to identify aggregated optimisation potential in the context of an individual BIPV product or a process in terms of its location and influence. However, to achieve progressive optimisation requires an iterative approach to LCA. The fragmented nature of the building industry hinders the gathering and arranging of data towards a collective effort. This therefore necessitates a systematic, inter-organisational approach to LCA.

The biggest limitation of the study may lie in the uncertainties inherent in LCA. Their quantitative effect is indicated by the exemplary options to reduce the environmental impact of the façade presented in Section 3.5: A relatively large deviation of 20% in solar glass production, for example, only affects 1.8% of the total. These interactions need to be further investigated in a systematic analysis. However, such an analysis is hampered by the lack of directly comparable data to confirm its veracity, which further emphasises the importance of a systematic LCA for the building industry.

Author Contributions: S.W. initiated this research, supported it with consultations, devised the use case, and wrote the abstract. D.H. gathered the LCI data, modelled and advised on the LCA, ran the software calculations and wrote the relevant sections. J.P. reviewed and compiled the inventory data, visualised and interpreted the results, and wrote the relevant sections comprising the bulk of the paper. All authors reviewed the publication.

Acknowledgments: Roland Schregle reviewed the paper. Peter Schaad (Glas Trösch AG, Bützberg, Switzerland) provided the LCI for the solar glass. Joachim Höhne (GES Gebäude- und Energiesysteme GmbH, Korbußen, Germany provided the LCI for the PV laminates.

Appendix A. BIPV Façade Results

Table A1. Lifetime electricity production and environmental impact by system configuration and orientation. If a configuration outperforms the reference (the Swiss grid electricity mix), it is marked in red.

	m-Si	m-Si	s-Si	s-Si
	Clear Glass	**Multi-Coloured**	**Clear Glass**	**Multi-Coloured**
Capacity [kWp]				
Laminate	0.22	0.176	0.25	0.2
System	30.8	24.64	35	28
Lifetime electricity generation [kWh]				
South (700 kWh/kWp)	582,054	465,643	661,425	529,140
East/West (530 kWh/kWp)	440,698	352,559	500,793	400,635
North (200 kWh/kWp)	166,301	133,041	188,979	151,138
Environmental Impact of all system components				
GWP [kg CO_2 eq]	66,503	66,620	102,758	102,876
UBP [kPt]	110,062	110,202	140,926	141,065
CED [MJ]	899,456	901,865	1,353,856	1,356,286

Table A1. *Cont.*

	m-Si	m-Si	s-Si	s-Si
	Clear Glass	**Multi-Coloured**	**Clear Glass**	**Multi-Coloured**
Environmental impact for rain cladding replacement (without glass, mounting system and transport)				
GWP [kg CO_2 eq]	41,965	41,965	78,221	78,221
UBP [kPt]	83,634	83,634	114,497	114,497
CED [MJ]	599,128	599,128	1,053,533	1,053,533
GWP [kg CO_2 eq/kWh (%)] for all components vs. reference (0.123 kg CO_2 eq)				
South	0.114 (−7.1%)	0.143 (16%)	0.155 (26%)	0.194 (58%)
East/West	0.151 (23%)	0.189 (54%)	0.205 (67%)	0.257 (109%)
North	0.399 (225%)	0.501 (307%)	0.544 (342%)	0.681 (453%)
Eco-points [kPt/kWh (%)] for all components vs. reference (0.256 kPt)				
South	0.189 (−26%)	0.237 (−7.7%)	0.213 (−17%)	0.267 (4.0%)
East/West	0.250 (−2.6%)	0.313 (22%)	0.281 (9.8%)	0.352 (37%)
North	0.662 (158%)	0.828 (223%)	0.746 (191%)	0.933 (264%)
GWP [kg CO_2 eq/kWh (%)] for rain cladding replacement vs. reference				
South	0.072 (−41%)	0.090 (−27%)	0.134 (9.3%)	0.148 (20%)
East/West	0.095 (−23%)	0.119 (−3.2%)	0.178 (44%)	0.195 (59%)
North	0.252 (105%)	0.315 (157%)	0.470 (283%)	0.517 (321%)
Eco-points [kPt/kWh (%)] for rain cladding replacement vs. reference				
South	0.144 (−44%)	0.180 (−30%)	0.197 (−23%)	0.216 (−16%)
East/West	0.190 (−26%)	0.237 (−7.4%)	0.260 (1.4%)	0.286 (12%)
North	0.503 (96%)	0.629 (145%)	0.689 (169%)	0.758 (196%)

References

1. Kaufmann, U. Schweizerische Statistik der Erneuerbaren Energien, Ausgabe 2016. Technical Report 2017.1002.01, Bundesamt für Energie BFE, 2017. Available online: http://www.bfe.admin.ch/php/modules/publikationen/stream.php?extlang=de&name=de_729286095.pdf (accessed on 30 August 2018).
2. Swiss Federal Office of Energy SFOE—Nuclear Energy. Available online: http://www.bfe.admin.ch/themen/00511/index.html?lang=en (accessed on 7 November 2018).
3. Wittkopf, S. Architektonische Veredelung von Photovoltaik für die Gebäudeintegration. In Proceedings of the 16th Nationale Photovoltaik-Tagung, Bern, Switzerland, 19–20 April 2018; Available online: https://www.swissolar.ch/fileadmin/user_upload/Tagungen/PV-Tagung_2018/Praesentationen/PVT18_5.2_Stephen_Wittkopf.pdf (accessed on 3 September 2018).
4. Ballif, C.; Perret-Aebi, L.E.; Lufkin, S.; Rey, E. Integrated Thinking for Photovoltaics in Buildings. *Nat. Energy* **2018**, *3*, 438–442, doi:10.1038/s41560-018-0176-2. [CrossRef]
5. Wittkopf, S. Coloured Cover Glass for Photovoltaic Module. Patent DE 102016001628 A1, 2017. Available online: https://patents.google.com/patent/DE102016001628A1/en (accessed on 19 October 2018).
6. Schregle, R.; Krehel, M.; Wittkopf, S. Computational Colour Matching of Laminated Photovoltaic Modules for Building Envelopes. *Buildings* **2017**, *7*, 72, doi:10.3390/buildings7030072. [CrossRef]
7. Schregle, R.; Wittkopf, S. An Image-Based Gamut Analysis of Translucent Digital Ceramic Prints for Coloured Photovoltaic Modules. *Buildings* **2018**, *8*, 30, doi:10.3390/buildings8020030. [CrossRef]
8. Masson, G.; Kaizuka, I. Trends 2017 in Photovoltaic Applications—Survey Report of Selected IEA Countries between 1992 and 2016. Technical Report IEA PVPS T1-32:2017, International Energy Agency Photovoltaic Power Systems Programme, 2017. Available online: http://www.iea-pvps.org/fileadmin/dam/public/report/statistics/IEA-PVPS_Trends_2017_in_Photovoltaic_Applications.pdf (accessed on 30 October 2018).
9. Green, M.A. Commercial progress and challenges for photovoltaics. *Nat. Energy* **2016**, *1*, 15015, doi:10.1038/nenergy.2015.15. [CrossRef]
10. Söderström, T.; Papet, P.; Ufheil, J. Smartwire Connection Technology. In Proceedings of the 28th European Photovoltaic Solar Energy Conference and Exhibition, Paris, France, 30 September–4 October 2013; pp. 495–499, doi:10.4229/28thEUPVSEC2013-1CV.2.17. [CrossRef]

11. Chown, A.; Brown, W.; Poirier, G. Evolution of Wall Design for Controlling Rain Penetration. Construction Technology Update 1997, No 9. Available online: https://pdfs.semanticscholar.org/7b84/6cc2d5d7d5fbcd0f7244a454b4f9b2ab5445.pdf (accessed on 10 September 2018).
12. CPD 10 2017: Glass Rainscreen Cladding Systems. Available online: https://www.building.co.uk/cpd/cpd-10-2017-glass-rainscreen-cladding-systems/5088631.article (accessed on 13 September 2018).
13. Ayres, R.U. Life Cycle Analysis: A Critique. *Resour. Conserv. Recycl.* **1995**, *14*, 199–223, doi:10.1016/0921-3449(95)00017-D. [CrossRef]
14. Passell, P. The Garbage Problem: It May Be Politics, Not Nature. *The New York Times*, 26 February 1991. Available online: https://www.nytimes.com/1991/02/26/science/the-garbage-problem-it-may-be-politics-not-nature.html (accessed on 5 November 2018).
15. Technical Committee ISO/TC 207/SC 5. Environmental Management—Life Cycle Assessment—Principles and Framework. Technical Report ISO 14040:2006(en), International Standards Organisation ISO, 2006. Available online: https://www.iso.org/obp/ui/#iso:std:iso:14040:ed-2:v1:en (accessed on 7 November 2018).
16. Technical Committee ISO/TC 207/SC 5. Environmental Management—Life Cycle Assessment—Requirements and Guidelines. Technical Report ISO 14044:2006(en), International Standards Organisation ISO, 2006. Available online: https://www.iso.org/obp/ui/#iso:std:iso:14044:ed-1:v1:en (accessed on 7 Novermber 2018).
17. Wang, Y.; Tian, W.; Ren, J.; Zhu, L.; Wang, Q. Influence of a building's integrated-photovoltaics on heating and cooling loads. *Appl. Energy* **2006**, *83*, 989–1003, doi:j.apenergy.2005.10.002. [CrossRef]
18. Baljit, S.; Chan, H.Y.; Sopian, K. Review of building integrated applications of photovoltaic and solar thermal systems. *J. Clean. Prod.* **2016**, *137*, 677–689, doi:10.1016/j.jclepro.2016.07.150. [CrossRef]
19. Zhang, T.; Wang, M.; Yang, H. A Review of the Energy Performance and Life-Cycle Assessment of Building-Integrated Photovoltaic (BIPV) Systems. *Energies* **2018**, *11*, 34, doi:110.3390/en11113157. [CrossRef]
20. Jayathissa, P.; Jansen, M.; Heeren, N.; Nagy, Z.; Schlueter, A. Life cycle assessment of dynamic building integrated photovoltaics. *Sol. Energy Mater. Sol. Cells* **2016**, *156*, 75–82, doi:10.1016/j.solmat.2011.12.016. [CrossRef]
21. Ng, P.K.; Mithraratne, N. Lifetime performance of semi-transparent building-integrated photovoltaic (BIPV) glazing systems in the tropics. *Renew. Sustain. Energy Rev.* **2014**, *31*, 736–745, doi:10.1016/j.rser.2013.12.044. [CrossRef]
22. Quintana, M.; King, D.; McMahon, T.; Osterwald, C. Commonly observed degradation in field-aged photovoltaic modules. In Proceedings of the Conference Record of the 29th IEEE Photovoltaic Specialists Conference, New Orleans, LA, USA, 19–24 May 2002; pp. 1436–1439, doi:10.1109/PVSC.2002.1190879. [CrossRef]
23. Frischknecht, R.; Heath, G.; Raugei, M.; Sinha, P.; de Wild-Scholten, M. Methodology Guidelines on Life Cycle Assessment of Photovoltaic Electricity, 3rd Edition. Technical Report IEA-PVPS T12-08:2016, International Energy Agency Photovoltaic Power Systems Programme, 2016. Available online: http://www.iea-pvps.org/fileadmin/dam/public/report/technical/Task_12_-_Methodology_Guidelines_on_Life_Cycle_Assessment_of_Photovoltaic_Electricity_3rd_Edition.pdf (accessed on 6 November 2018).
24. Frischknecht, R.; Itten, R.; Sinha, P.; de Wild-Scholten, M.; Zhang, J.; Fthenakis, V.; Kim, H.C.; Raugei, M.; Stucki, M. Life Cycle Inventories and Life Cycle Assessments of Photovoltaic Systems. Technical Report IEA-PVPS T12-04:2015, International Energy Agency Photovoltaic Power Systems Programme, 2015. Available online: http://www.iea-pvps.org/fileadmin/dam/public/report/technical/IEA-PVPS_Task_12_LCI_LCA.pdf (accessed on 30 September 2018).
25. Wernet, G.; Bauer, C.; Steubing, B.; Reinhard, J.; Moreno-Ruiz, E.; Weidema, B. The Ecoinvent Database Version 3, Part I: Overview and Methodology. *Int. J. Life Cycle Assess.* **2016**, *21*, 1218–1230, doi:10.1007/s11367-016-1087-8. [CrossRef]
26. Allocation at the Point of Substitution–Ecoinvent. Available online: https://www.ecoinvent.org/database/system-models-in-ecoinvent-3/apos-system-model/allocation-at-the-point-of-substitution.html (accessed on 6 November 2018).
27. De Wild-Scholten, M. Life Cycle Assessment of Photovoltaics Status 2011, Part 1: Data Collection. Technical Report; SmartGreenScans: Groet, The Netherlands, 2014. Available online:

http://smartgreenscans.nl/publications/SmartGreenScans-2014-Life-Cycle-Assessment-of-Photovoltaics-Status-2011-Part-1-Data-Collection--Sample-Pages.pdf (accessed on 7 November 2018).

28. Pilkington, L.A.B. Review Lecture: The Float Glass Process. *Proc. R. Soc. Lond. A* **1969**, *314*, 1–25, doi:10.1098/rspa.1969.0212. [CrossRef]
29. ift Rosenheim GmbH. Umweltdeklaration (EDP) für Flachglas, Einscheibensicherheitsglas, Verbundsicherheitsglas. Technical Report M-EPD-FEV-002005, Euroglas GmbH, 2017. Available online: https://www.glastroesch.ch/fileadmin/content/euroglas/Deutsch/Service/Zertifizierungen/EPD/2018-03-20_Euroglas_M-EPD_FG_ESG_VSG__002_.pdf (accessed on 16 September 2018).
30. Usbeck, V.C.; Pflieger, J.; Sun, T. Life Cycle Assessment of Float Glass. Technical Report, Glass for Europe, 2010. Available online: https://www.agc-yourglass.com/agc-glass-europe/au/de/pdf/lca/LCA.pdf (accessed on 7 November 2018).
31. Tschümperlin, L.; Stolz, P.; Wyss, F.; Frischknecht, R. Life Cycle Assessment of Low Power Solar Inverters (2.5 to 20 kW). Technical Report 174-Update Inverter_IEA PVPS_v1.1, Swiss Federal Office of Energy SFOE, 2016. Available online: http://www.treeze.ch/fileadmin/user_upload/downloads/Publications/Case_Studies/Energy/174-Update_Inverter_IEA_PVPS_v1.1.pdf (accessed on 7 November 2018).
32. Wambach, K. Life Cycle Inventory of Current Photovoltaic Module Recycling Processes in Europe. Technical Report IEA-PVPS T12-12:2017, International Energy Agency Photovoltaic Power Systems Programme, 2017. Available online: http://www.iea-pvps.org/index.php?id=369&eID=dam_frontend_push&docID=4239 (accessed on 26 October 2018).
33. Intergovernmental Panel on Climate Change. Anthropogenic and Natural Radiative Forcing. In *Climate Change 2013—The Physical Science Basis*; Cambridge University Press: Cambridge, UK, 2014; pp. 659–740, doi:10.1017/CBO9781107415324.018. [CrossRef]
34. Frischknecht, R.; Büsser Knöpfel, S. Ökofaktoren Schweiz 2013 gemäss der Methode der ökologischen Knappheit—Methodische Grundlagen und Anwendung auf die Schweiz. Technical Report UW-1330-D, Bundesamt für Umwelt BAFU, 2013. Available online: https://www.bafu.admin.ch/bafu/de/home/themen/wirtschaft-konsum/publikationen-studien/publikationen/oekofaktoren-2015-knappheit.html (accessed on 30 August 2018).

9

Life-Cycle Asset Management in Residential Developments Building on Transport System Critical Attributes via a Data-Mining Algorithm

Umair Hasan [1,*], Andrew Whyte [1] and Hamad Al Jassmi [2]

1 School of Civil and Mechanical Engineering, Curtin University, Perth, WA 6845, Australia; andrew.whyte@curtin.edu.au
2 Department of Civil and Environmental Engineering, United Arab Emirates University, Al Ain 17666, UAE; h.aljasmi@uaeu.ac.ae
* Correspondence: umair.hasan@postgrad.curtin.edu.au

Abstract: Public transport can discourage individual car usage as a life-cycle asset management strategy towards carbon neutrality. An effective public transport system contributes greatly to the wider goal of a sustainable built environment, provided the critical transit system attributes are measured and addressed to (continue to) improve commuter uptake of public systems by residents living and working in local communities. Travel data from intra-city travellers can advise discrete policy recommendations based on a residential area or development's public transport demand. Commuter segments related to travelling frequency, satisfaction from service level, and its value for money are evaluated to extract econometric models/association rules. A data mining algorithm with minimum confidence, support, interest, syntactic constraints and meaningfulness measure as inputs is designed to exploit a large set of 31 variables collected for 1,520 respondents, generating 72 models. This methodology presents an alternative to multivariate analyses to find correlations in bigger databases of categorical variables. Results here augment literature by highlighting traveller perceptions related to frequency of buses, journey time, and capacity, as a net positive effect of frequent buses operating on rapid transit routes. Policymakers can address public transport uptake through service frequency variation during peak-hours with resultant reduced car dependence apt to reduce induced life-cycle environmental burdens of buildings by altering residents' mode choices, and a potential design change of buildings towards a public transit-based, compact, and shared space urban built environment.

Keywords: sustainable-development; life-cycle social analysis; public-engagement; modal-variability; transit-policy; work-commute; travel-satisfaction

---

1. Introduction

Municipal residential areas and new developments are often marked by economic growth and high population density, where respective higher environmental emissions affect the air quality [1,2]. Indeed, the high private automobile traffic in the established residential areas affects the structural integrity of buildings in such developments [3] with knock-on negative impact on the residents [4]. Similarly, sustainable development and the creation and maintenance of a built environment necessarily requires measures of construction material transportation [5,6]. For example, studies have shown that for a building element such as roofing, a sizable proportion of a built asset's life-cycle assessment (LCA) relates explicitly to transportation which has been measured to contribute 10% to a residency's new-build/building-renovation's overall carbon footprint within a LCA system boundary covering the entire life-cycle of building-product usage [7,8]. Built environment development whole-costs are similarly affected by community transportation links; locations affect life-cycle costs with congestion apt to delay resources deliveries and often result in vehicle damage across respective congested transportation routes [9].

Regional constraints notwithstanding, mass-transit system plans are generally developed by municipal and transportation agencies to reduce ever-increasing traffic congestion on road networks by affecting mode choices of building residents [10]. These plans are simultaneously targeted as being environmentally conservative for the existing municipal residential areas or any upcoming residential developments in the region. The implication of adequate public transport accessibility and reduced reliance on private automobile usage towards sustainable residential areas; e.g., compact neighbourhoods, walking habits of residents, urbanisation and shared-space designs, construction of shops and other facilities in buildings and neighbourhoods is abundant in the literature [11,12]. Jabareen [13] further include sustainable transport as one of the key design concepts of sustainable building and urban design plans. Other researchers, such as Zimring, Joseph [14] and Cervero [15] propose that sustainable residential built environments, transport systems and urban forms should be designed so as to promote sustainable modes of transport, e.g., public transport, and devise policies to discourage individual car usage, among building residents.

On the other hand, transportation researchers such as de Luca [16] and Leyden, Slevin [17] argue that planning and provision of any (public) transport system is largely "flawed" as alternatives and policies are prioritised subjectively by decision-makers alone, with very little public engagement from local building occupants or consultation at the initial stages. Often public feedback generated is neglected or only marginally used to improve the existing system, which is argued by above literature to significantly increase the risk of implementing a public transport service and an overall transportation system incoherent with public expectations. This may lead to higher private automobile dependence among the building residents of any municipal area, causing heavy traffic congestions on road networks adjacent to these residential buildings [18]. In addition to disrupting the supply-chain by delaying construction material transport for new buildings or haulage of disposal material from demolished buildings in the area, traffic congestions can result in increased cost and environmental burdens depending upon the traffic load and transport system typology. For example, in case of high traffic volume roads, around 95% emissions are traffic-related [19], while Stephan, Crawford [20] showed that the residents' daily commute energy load for Belgian passive (i.e., energy-efficient) houses was approximately 27% of the entire life-cycle energy burdens of these passive houses. Stephan, Crawford [20] also proposed that the transport energy of these residents may be reduced by approximately 31%, provided a shift in mode use occurs from private vehicle transport in favour of public transport, corresponding to a reduction of 8.4% in the houses' life-cycle energy consumption. Since environmental impact assessment rating tools for buildings, such as the United States Green Building's Leadership in Energy and Environmental Design (LEED) system, also include the residents' daily commute energy and environmental load as an integral part of the total life-cycle burden of the buildings due to the building-induced impacts (more in Section 2.2), therefore, in order to reduce the overall life-cycle impact of buildings, the mode choice redistribution in favour of public transport should be researched [10,18].

In mode use redistribution or diversion research, Diana [21] and De Vos, Mokhtarian [22] suggest that traveller expectations influence commuter satisfaction, which ultimately influences the variation in their ridership preference over time (public vs. private transit), hereby referred to as modal variability. Establishing the satisfaction of an individual traveller may be difficult, aligning satisfaction with travel patterns of daily commuters is important for municipal policymakers aiming to influence mode choice and travel survey datasets are used for this purpose by building [23] and transport [24] researchers alike. However, the presence of large quantities of variables in the travel dataset complicates the pattern discovery process of deducing the potential for commuter mode choice diversion in favour of public transport [25] and optimising the *consumer mode choice – satisfaction* dynamics. Golob and

Hensher [26] and Diana and Pronello [27] maintain that although the patterns in large categorical variable sets may be discovered to some extent through multiple correspondence analysis, scatter plots and cross-tabulation techniques, such methods are fairly limited when many variables are to be jointly considered. Limitations of frameworks only applying these statistical analyses to study travel survey datasets can thus be avoided by data mining [28,29] through its potential to handle a large number of interrelated variables [30].

This study is part of a larger project that aims to deliver a multi-criteria decision-making framework towards the objective of sustainable cities and future green buildings served by a connected and sustainable road transportation system that meets commute demands of building residents. The proposed framework consists of: social aspect of government and user stakeholders' demands [31], life-cycle cost and environmental in/outflows [32]. In Hasan, Whyte [33], a multipartite model was developed by the authors to establish commuter satisfaction from the level of service (measured on network coverage and frequency) as the antecedent of building residents' mode choice. Moreover, market segmentation based on mode choice, satisfaction from service level and service perception as value for commuter money was also performed in Hasan, Whyte [33] to analyse the function of underlying exogenous factors in the sample residential area. The explicit objectives of the research presented in this paper are described below:

1. Examine the travel patterns in a representative intra-city (Abu Dhabi) dataset for existing residential areas or upcoming residential developments.
2. Present a systematic way of assessing critical factors eliciting mode choice in commuter market segments towards optimising the social, i.e., stakeholder demands, aspect of the overall LCA of transportation systems and in the process, reduce the user-transport life-cycle energy and environmental load of residential buildings.
3. Determine bus service desiderata for policymakers to develop an ameliorated bus service in future, which may divert more building residents to the improved bus service. This can potentially reduce the life-cycle costs and environmental (greenhouse gas emissions, smog, resource and energy use) burdens besides tacking the social parameters (stakeholders' perspectives) towards optimising overall life-cycle burdens of residential buildings as well improving future building designs, i.e., lesser parking area requirements, more shared walkable spaces, better accessibility etc.

In order to meet the study objectives, a custom travel data processing algorithm unifying association rules mining, and statistical analysis techniques is developed to identify two classes of variable combinations; statistically-significant and validated association-only. The two different groups aim to provide policymakers with the maximum information about the commute habits of local building residents in the city. Whilst statistical analysis is informative and interesting, it may fail to uncover the underlying modal variability patterns and commuter behaviour which may be visualised through association rules [34–36]. For this purpose, validation of the filtered-out association rules against an internal validation set was conducted to further generate a set of association rules [37,38] followed by data reduction (similar to [36] and [39]) to remove redundant association rules.

2. Theoretical Background

2.1. Multiscale Effect of Road Network Transport System on Residential Buildings

Researchers not only propose sustainable waste disposal from building constructions sites but also advocate use of recycled construction and demolition waste for new residential developments to reduce the overall environmental impact. The presence of an adequate transportation system at regional scale is critical for facilitating the disposal and reuse, for example, Building Research Establishment (BRE) assigns a weighted contribution of 8% to building construction material transportation in the overall life-cycle of buildings [40,41]. Other national and international guidelines on LCA of built assets that are used for *environmental impact assessment* of buildings such as EN 15978 [42], CEN/TC350 [43] and ISO/TS 12720 [44] also recommend estimating the material transport in both construction and disposal stages for accurate calculations. At the same time, a number of government regulations, subsidies and incentives exist for promoting recycled material usage in residential buildings and housing development sectors. Decree 205/2010 in Italy mandates at least 15% recycled material use in building construction [45] while Environmental Product Declaration on construction is now practiced in many European countries [19]. The United Kingdom Aggregates Levy for extracting quarry materials of £2 per tonne [46] has boosted recycled aggregates usage in building constructions by approximately 25% [47]. However, material supply-chain and costs due to far-off recycling sites, for e.g., accounting for ~70% costs for a 350 tonne/h facility in a Portugal-based study [48] and increased travel time of material transporting trucks due to on-route traffic congestion often discourage building contractors from recycled material usage. Similarly, any excessive time spent on road networks due to congestion may reverse some of the positive environmental benefits from using recycled materials due to fuel consumption. For example, using a standard 15-tonne lorry to move 1 tonne of aggregate material over 1km requires 0.014 kg of diesel, depending upon the local traffic conditions [40]. Thus, the question of traffic congestion mitigation should also be tackled by the policymakers aiming to promote recycled material usage in buildings (see Section 2.2).

Besides reducing traffic congestion on road networks and relieving residents' transport energy of buildings [18], public transit services also affect property prices of buildings and development projects depending upon the relative proximity to mass-transit services. Mulley, Ma [49] studied residential property market in Brisbane, Australia and found that bus rapid transit (BRT) affected housing prices at the rate of 0.14% per 100m closer to the transit service routes while train service inferred a negative effect of 0.15% reduction per 100m closer to the train service line. These effects were also strongly dependent upon urban form and spatial distributional densities of buildings. Their study also highlighted the importance of studying local mode use travel patterns, commuter perceptions and the key position of underlying service attributes of service frequency, on-board crowding and the journey time in determining the relative impact of BRT service on the residential property development sector.

2.2. Interrelated Urban Form, Transport System and Buildings LCA: Public Transport Accessibility and Residents' Mode Choices

The interrelation between urban form, transportation systems (private and public transit modes), and the overall life-cycle impacts of buildings is a complex issue which has been explored by many researchers focusing on the synergy between developing low-energy buildings and sustainable transport [10,50–52]. The lower density urban form undoubtedly puts extra load on private vehicle transport as the preferred mode choice of municipal residents, for example, accounting for ~70% in European passenger transport with predominantly semi-detached houses, whereas limited parking in buildings may promote public transit usage based upon proximity to service line and accessibility [51]. One of the earliest studies of the 21st century by Steemers [51] on developing a less polluted urban built environment noted the link between the overall building energy and transport mode choice of building residents. Similarly, the "Le Plan Bâtiment Durable" (French: "*The Sustainable Building Plan*") launched by the Government of France in 2009 aimed at reducing the existing buildings' energy consumption by 38%, also enlists the mode choice of building residents and developing strategies for promoting public transportation usage among the criteria for assessing the overall performance of buildings. Furthermore, other European Union states also highlighted the importance of understanding the interrelation between urban design, transportation system, building design, residents' mode choice pattern, "public transport connected buildings", compacted shared spaces and accessibility as environmental hotspots for future sustainable building designs [10]. Similar points on compact and shared urban form were also highlighted by Norman, MacLean [53] in their economic input-output LCA on buildings and by Cuéllar-Franca and Azapagic [54] who showed terraced house to carry 309 t CO_2 eq. compared to the 455 t CO_2 eq. of a semi-detached house in the UK.

Stephan, Crawford [20] observed that the majority of so-called low-energy passive houses in Belgium are designed as suburban detached single-family dwellings compared to the public transport-connected city buildings. They further commented that any energy savings by these low-energy houses are offset by increased mode use of private vehicles compared to public transport. Earlier research by the same authors [55,56] on life-cycle energy requirements of low-energy houses has shown that over the life-cycle (50–100 years) of these low-energy residential buildings, residents' transport and embodied energy are >50% of the overall life-cycle energy load. They proposed that in order to reduce the overall energy demands of residential buildings, the transport mode choices of building residents must be tackled towards public transport uptake in addition to using recycled building components and sustainable energy measures as the mode use alone represented 27% of the built assets' life-cycle energy [20].

Two studies on buildings and environmental analysis (i.e., greenhouse gas emissions and energy use) of buildings were conducted by Anderson, et al. [23,57]. They proposed transportation systems and residents' mobility patterns (i.e., mode choice distribution) as key parameters for interlinking residential buildings and the urban environment. They emphasised the strong tie between transportation and buildings and argued that the induced environmental impacts in terms of transport habits due to the design, locality and interaction of buildings with the urban form need to be captured. They used streamlined LCA to capture this effect on urban Munich residents in Germany using detailed mode use and travel behaviour surveys [23]. They found that the overall life-cycle impact of buildings and the induced transportation environmental impact due to mobility pattern of residents is dependent upon the daily commute distance, mode choice and the location of the buildings. For example, in their study the lowest impacts were found for Munich central business district (CBD) residents due to buildings typology and short commute distance of residents, where the largest share was claimed by overall transportation impacts of CBD residents as 1160 kg CO_2e/capita/annum while CBD residential building emissions were 1088 kg CO_2e/capita/annum. Both studies [23,57] concluded that although the design and material usage in buildings needs to be upgraded to be more sustainable, the real criticality still lies with improving the mode use behaviour of building residents. Similar findings were noticed by Dodd, Donatello [10] in their review on LCA of European buildings with the residents' transport energy demands claiming between a 19% share to even completely overwhelming the life-cycle energy load of some buildings.

The critical role of mobility and the mode choice of building residents induced by the interrelation between urban form, locality, transport service provision and the building design was also acknowledged by the LEED system weighting methodology by assigning a total of ~ 15% credit weightage to transport in terms of reduced parking, bicycling and better quality public transit accessibility near buildings. Promoting sustainable mode choice among residents, proximity or connection to public transport services and bicycling and shared vehicle facilities were also included in the property/real-estate based Investment Property Databank of the triple alliance between Sustainable Building, Bureau Veritas and Barclays (~ 16% weightage); the Invesco, Allianz, AXA and AEW founded Green Rating Alliance; and, the Global Real Estate Sustainability Benchmark [10]. The extensive focus on the mobility pattern of building residents in municipal residential zones/areas proposed by these works also recommended understanding the underlying factors behind mode choice to postulate policies towards uptake of more sustainable public transport usage.

2.3. Factors Affecting Travel Perception and Mode Choice

Researchers studying factors affecting mode choice behaviour, such as Dell'Olio, Ibeas [58] and de Oña, de Oña [59] proposed that generally any bias possessed by commuters, as *for or against* a particular travel mode (referred to as travel bias, henceforth) depending upon their unique characteristics, influences their perception of the quality attributes offered by a particular mode and may translate to a subsequent mode choice. The factors behind mode choices and variation of travel behaviour were also focused in a study conducted by Schmid, Schmutz [60]. They performed a hypothetical study of a low carbon post-private car world and found public transport to be the most popular mode amongst survey respondents, followed by bikes and car-sharing. Stradling [61] found that the modal variability of residents in a case-study Scottish housing area is temporally constrained, with 20% of private automobile users actually acting as multimodal, i.e., using several transport modes, when the analysis period is stretched over a week. Building upon this, the present study gathered commuter behavioural responses and travel pattern data over a month to investigate the sensitivity of modal shifts in long-term mobility patterns.

Built environment around the public transit service, accessibility near commuter accommodations and offices may also influence their mode choice. An empirical study on the association between travel behaviour of urban residents in New York City and the density of the urban neighbourhoods was conducted by Chen, Gong [62]. They noted that the density and built environment affected mode choice of the survey respondents. They also proposed that government authority planners may attract more users towards public transit modes by providing better accessibility near workplaces. Similarly, Cervero [15] proposed that residential developments with lesser shared-spaces for cyclists and pedestrians may affect the mode choice, particularly an inclination towards individual car usage, more so as the building designs may also carry abundant nearby parking facilities instead of using the space for mixed-use walkable areas.

Friman and Fellesson [63] and Diana [21] found modal variability and level of commuter satisfaction from public transport as highly dependent upon the local built environment and regional characteristics of the study area. They noted that the travellers near the city centres tended to sway towards higher public transport usage compared to their urban/suburban counterparts, making the results highly sensitive to the zonal distribution (urban vs suburban) as well as number of zones covered in the study area. The current study addresses this issue by surveying commuters on the intra-city travel routes across all zones of a metropolitan city (Abu Dhabi) to capture the maximum variability of responses. The City of Abu Dhabi has witnessed an increase in population accompanied by extensive residential developments and an increasing commuter dependence on private vehicles [64], resulting in traffic congestion and as such, is a suitable case study area for the purpose of this research. Furthermore, the excessive private vehicle ridership is responsible for high greenhouse gas emissions, for example in United Arab Emirates alone, it annually exceeds 11735.6 Gg CO_2 equivalent [65] from road transport. The local government agencies in UAE have aimed to reduce its environmental load [66]. The research presented in this paper and its accompanying works [19,32] may aid in reducing the environmental burden from road and building residents' daily transit dimensions towards an overall sustainable future city development.

2.4. Data Mining for Analysing Travel Datasets

Provided the competitive benefit of outlining commuter expectations, perceptions and rankings of existing service level to create pro-public transport policies, a large database of underlying variables (e.g., journey time, ride quality, transit fare, accessibility, nodes, socio-demographics of commuters etc.) are generally collected from representative population samples. A recent study by Liu, Xu [67] on improving the public mass-transit service quality in China acknowledges that an efficient data mining framework is capable of extracting useful information from transit data and representing it into clear and succinct policy-related recommendations. Liu, Xu [67] developed a data mining algorithm with data cleaning and filtration options. However, their proposed algorithm focused only on the time parameters and the pre-journey (accessibility, location, train/bus-station) and on-board (crowding, seating arrangements, fare and ride quality) parameters were not involved in the algorithm design.

Similarly, a Weka classifying algorithm-based data mining study on the occurrence of faults on tram lines, local atmospheric conditions and safety was performed by Gürbüz and Turna [36]. The study noted the ability of data mining algorithms to filter through large mobility datasets and association rules to visualise the interrelation between various variables in the collected dataset. Building upon these studies, the work presented in this current paper contends that many of the variables collected through travel surveys or response diaries are interdependent or associated to varying degrees. Studying them in isolation may be inadequate if environmental impact analysts aim to study the complex relationship of categorical and nonmetric variables to recommend apposite policies to promote public transport, specifically as emission-controlling measure for residential buildings, in addition to the Intergovernmental Panel on Climate Change (IPCC) [68] recommendations of using recycled materials as green construction practices [69–71]. Current literature has no such framework that can help policymakers to segregate the most desired transit service attributes from the several underlying variables affecting commuter psychometrics, thereby quantifying the stakeholder demands as part of the social aspect of LCA on any transportation system. This aspect is of particular importance for deducing marketing tactics aimed at attracting users that have somewhat of a neutral perception of either travel mode (bus or car) as well as retainment of the loyal public transport users.

Inspired by Ponte, Melo [72], who used the Gini coefficient primarily used in economics, for measuring the public bus-transit travel time heterogeneity in municipal areas and Diana [35] who used association rules for studying travel pattern, this study explores the applicability of data mining techniques used in economics and marketing research. Apriori association rules data mining is used here for achieving the objective of identifying the motivations behind commuter mode choices. Introduced by Agrawal, Imieliński [73] and Agrawal and Srikant [74] for mining frequent item-sets in transaction data without using any underlying relative and distributional assumptions, it has now developed into a robust market analysis technique. Association rules identify the frequency of different variables appearing together in an observation (data) set. An association analysis reads every single row of variables on the travel datasets to produce association rules, which are of the implication form given in Equation (1).

$$x_i, x_j, \ldots, x_n \Rightarrow y \quad (1)$$

where the left-hand side is the antecedent (a set of predictor variables) and the right-hand side is the consequent (and represents the response variable). Periodically used for frequent pattern mining in fields besides management sciences, e.g., health [75] and weather forecasting [76], its application in transportation research is relatively new. Nonetheless, researchers [35,77] have explored the efficacy of associations rules for handing large and complex travel and mobility datasets. A primary advantage of using Apriori over traditional multivariate statistical analyses is its ability to rapidly find an association between large sets of metric and non-metric variables. Once an association is established between the dataset variables, traffic flow patterns and the individual public transport service attributes behind commuters' observable modal variability can be extrapolated for policy recommendations in favour of low carbon and energy conserving transit among the local building residents and travellers.

Conversely, one of the main drawbacks of the Apriori algorithm is its tendency to produce a large number of association rules, rendering the technique to be an ineffective process if no other controls are provided. Pruning the rules is the next step [78] to filter out meaningless and inutile rules that may be misleading for policymakers. Association rules are constrained by the number of times a certain rule is supported by the dataset and the strength of rule (i.e., confidence) which is the fraction of dataset rows that contain both consequent and antecedent of the rule. However, confidence is an asymmetric measure which may provide erroneous results if the consequent has a large probability [79]. Moreover, if the antecedent and consequent variables are independent of each other, the generated rules may be unsuitable, irrespective of high confidence [80], for establishing inference relations for policy development. Wu, Zhang [81] have introduced an additional measure called "*interestingness*" to discard unsuitable association rules (Equation (2)).

$$Interest(X, Y) = |Support(X \cup Y) - Support(X) \bullet Support(Y)| \quad (2)$$

3. Materials and Method

3.1. Survey Design and Analysis

A questionnaire was designed to gather data on the factors that affect user perception of existing bus network and modal variability among the local building residents and travellers [33]. A total of 11 questions were developed for the survey, aimed at capturing commuter responses on the different variables according to Likert-type scales, dichotomous and multiple-choice options. Previous researchers [21,82] have noted the adequacy of weekly mode choice data in modelling the mode use of surveyed commuters, the sensitivity of commuter responses to the local built environment and zonal distributions as well as the importance of zonal characteristics in defining the observed transport pattern of building residents and the calculated transport parameter in buildings' overall environmental impacts [23,83]. This study targets the commuters in all zonal distributions (urban and suburban) of Abu Dhabi city. The survey questionnaire used for this study and the different zonal distributions are shown in Table A1 and Figure A1, respectively. The eight arterial and medium to high traffic routes serving the daily commute demands of the residents of various types of buildings in the city were selected for data collection (Table 1) to capture an optimally representative data of the city's usual commuters and weekly data over six weeks was collected to improve the number of observations.

Table 1. Bus routes selected for the purpose of study.

Outer-Urban or Suburban Routes	Route Number
Al Mina Souq ↔ Khalifa Park	056
Petroleum Institute ↔ Tourist Club Municipality	054
Abu Dhabi Courts ↔ Al Marina	034
Downtown City and Urban Routes	**Route Number**
Al Mina Fish Harbour ↔ Al Marina Mall	011
Al Mina Souq ↔ Al Marina Mall	009
Al Mina Road Tourist Club ↔ Al Marina	008
Tourist Club Municipality ↔ Al Marina	007
Al Mina ISC and Tourist Club ↔ Ras Al Akhdhar	006

The completed questionnaires collected by Department of Transport survey teams [84] were then analysed. Logic checking of data consistency was performed by the authors to address data sparseness, outliers and missing data based upon the guidelines by Osborne [85]. This resulted in a useful sample of 1520 completed questionnaires (inclusive of weekdays and weekends surveys) which identified (detailed under separate cover as Hasan, Whyte [33]) ten commuter segments, namely:

1. Modal variability: distribution of generated trips for each mode (i.e., bus and car travel). Five segments as *pro-sustainability* (PS) passengers (i.e., regular bus travellers and non-users of cars), *occasional multimodal* (OMD) travellers, *frequent car/taxi travellers* (FrCT) and *environmentally insensitive* (EI) commuters (i.e., non-users of public bus service) item.
2. Perception of bus service as value for money (VfM): trade-off between quality of ride and level of fare. Three segments of *good value for money, borderline value for money* and *bad value for money*.
3. Commuter satisfaction from level of service (LoS): ranked based on perception of the current level of network coverage and frequency of buses. Three segments of *good level of service, borderline level of service* and *bad level of service*.

The statistical distribution of the collected commuter responses is shown in Table A2. The study objective is finding service desiderata critical for policymakers to deter private automobile use and increase uptake of bus usage among local building residents and travellers to promote environmental conservation by developing an improved public transit service. A data mining algorithm is thus proposed below which is used to analyse the above case study commuter segments to generate variable associations for explaining the daily commute behaviour of the local building residents and travellers.

3.2. Conceptual Framework—Overall Proposed Modelling Approach

Current research suggests a market-based analysis that unifies "statistical measurement" and "associate data mining". The combination has been used to some extent in marketing literature (e.g., Shaharanee, Hadzic [38]) for filtering out redundant rules. Work here contributes by building a multi-tier modelling approach for travel survey dataset that filters "interesting rules" to split them in two distinct categories: associated rules; and, statistically-significant models. Initially, the generated rules are filtered using an "interestingness measure" from Eq. (2). The filtered rules are then used to build ordinal regression models, followed by (only) selecting meaningful rules at the filtration stage 2. "Meaningful rules" are defined as those that provide association between the variables that have not been already construed by the other rules passing the minimum thresholds, thereby rejecting the repetitive or redundant sets of rules. The problem is further explained in Figure 1.

Definition 1: Let $V = \{x_1, x_2, \ldots, x_m; y_1, y_2, \ldots, y_n\}$ be the set of variables, where x_m and y_n are respectively the antecedent and consequent variables. Apriori algorithm predicts sets of association rules R_1, R_2, R_3 as $R = \{R_1, R_2, R_3, \ldots, R_{xy}\}$ or $R = \{(x_1, x_2 \Rightarrow y_1); (x_1, x_2, x_3 \Rightarrow y_2); \ldots; (x_1, \ldots, x_m \Rightarrow y_n)\}$. Note: there may be some rules in set R whose variables imply the same relation or association. Towards a simplified form of the classic set cover problem to provide an approximation solution with M as a set of meaningful rules, defined by the following relation:

$$\mathrm{M} \subseteq \mathrm{R} \quad \exists\forall \text{ rule in R} \in \text{rules in M}$$

Figure 1. Definition of "meaningfulness measure" to filter repetitive set of rules.

This twofold technique proposed in this study filters out redundant rules and provides final sets of statistically-significant rules. The rejected rules are then analysed in the second phase by a validation technique using a validation set. It is argued here that this technique will remove any misleading rules while still retaining a significant number of rules for policymaking without compromising the analysis accuracy. The proposed conceptual framework is presented in Figure 2. It should be noted that there are several off-the-shelf packages available for performing the Apriori association mining [35]. However, as this paper proposes a unified framework, a tailored java script was used for data mining due to its compatibility with SPSS 23.0 regression functions, which can then be used to perform ordinal regression on the observed rules.

Initial travel dataset "ζ" was cleaned to ensure a high quality of dataset be fed into the analysis algorithm. Data cleaning is an important step in the knowledge discovery process from analysis of predictor and response variables as it removes missing or anomalous values [85]. Numerical values of household car-ownership and time to reach bus stations were also transformed into nominal variables using equally distributed classes to maintain uniformity in the data variable classes and ease the analysis procedure. The resulting dataset was then randomly split into a training dataset (ζ_{ts}, containing $\approx$ 70% data) and, a validation dataset (ζ_{vs}). The training dataset was then used for generating an initial set of association rules: frequency of bus travel (FBT); frequency of car/taxi travel (FCT); satisfaction with network coverage (NetCov); quality of ride (QoR); level of fare (LvlFare); and, frequency of buses (FreqBus) which were used interchangeably as response variables based upon the desired output.

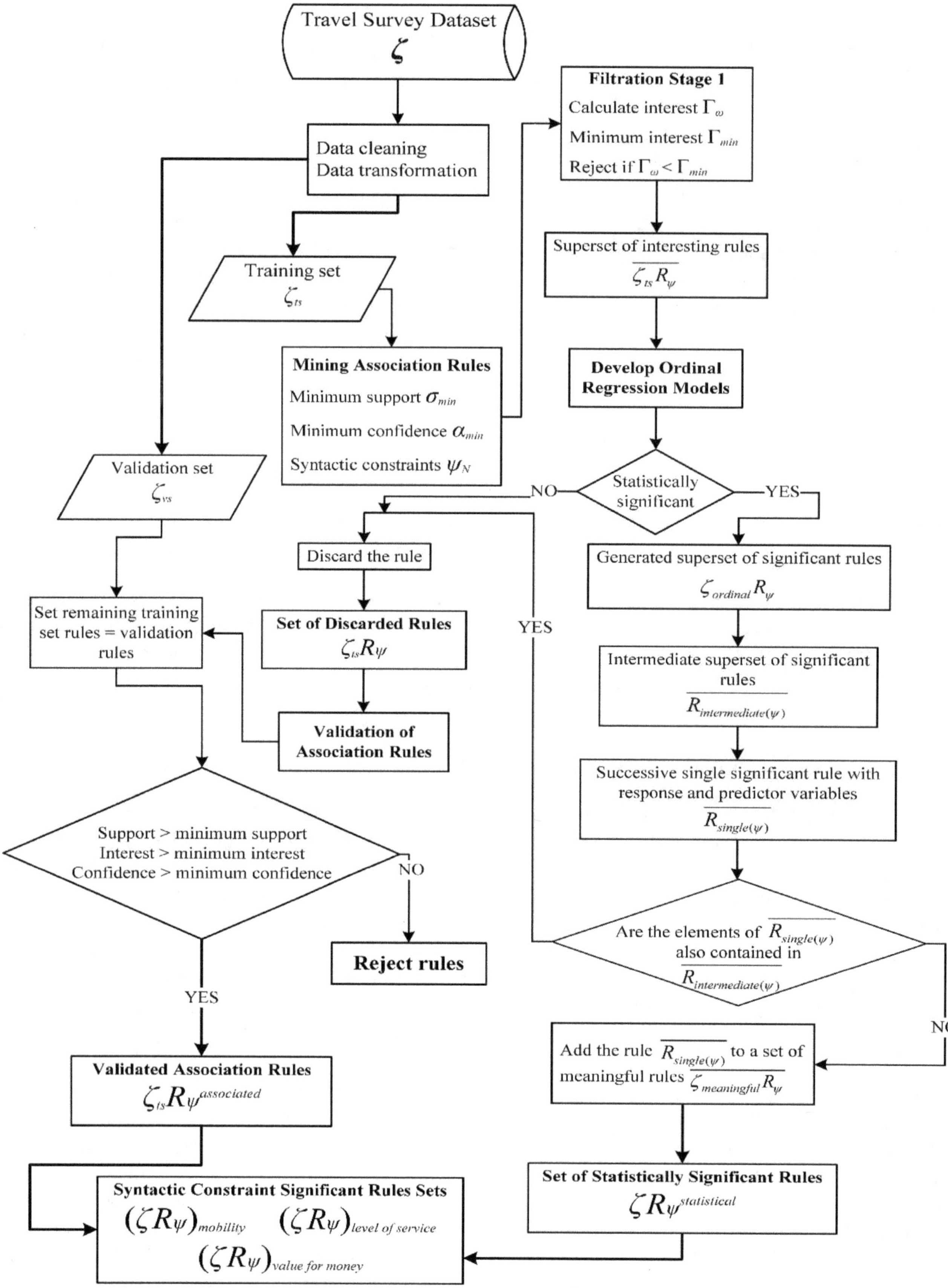

Figure 2. Conceptual framework of the proposed algorithm to mine a travel dataset.

In order for the algorithm to generate association rules, there should be at least one tuple in the dataset for each observation. Given our target of finding transit policy-related hotspots to promote sustainable transport usage among local building residents and travellers covered within the travel dataset, syntactic constraints were put towards the consequent-side based upon the market segments. This constrained analysis is interesting for policymakers [35] as it may help to define service attributes that may be improved upon in future services that meet the public needs regarding the local bus service. According to the principles of Apriori data mining, only one response variable can be set as a consequent. Therefore, response variables were interchangeably used as a consequent for obtaining the association rules for the respective category and obtain sets of rules specific to each segment, e.g., frequent car travellers, occasional multimode travellers and commuters ranking bus service as good value for money, and so on.

Past literature has suggested that higher support and confidence relate to a better category of rules but that lower values of support and confidence are still recommended. Therefore, to capture the most "interesting" rules, minimum support and confidence were respectively set here at 0.1 and 65% [34,35]. After appropriate data transformation and setting syntactic, support and confidence constraints, a number of frequent item rules were obtained for each category. The "interestingness measure" was used in the first filtration stage by setting a minimum interest value "Γ_{min}". The association rules mining through Apriori algorithm permits the analysts to set the minimum threshold for selecting useful models from a large set of created rules. Large numbers of association rules are initially generated which are then pruned by the analysts to reject falsely produced rules from the rule set that fail to qualify the established minimum threshold.

Evaluating the minimum threshold in the analysis may be guided through past experience or research literature. Therefore, in the present analysis, a minimum interest value of 0.05 was selected [16,86] with any future application of the proposed algorithm able to use a higher or lower value depending upon the study objectives. The retained rules were then used for ordinal regression analysis. The ordinal regression analysis calculated *log likelihood* and *p-values* for the models (the models were based on the association rules). Models which showed considerable change in *log likelihood* compared to null model and *p–values* < 0.05 were retained. The repetitive rules were then rejected using the theory defined in Figure 1 and the retained rules were used to represent the statistically-significant rules. The remaining rules were then used in phase II of the analysis to validate against the validation set while controlling for the minimum support, interest, confidence and non-repetitive criteria as discussed earlier. The algorithm proposed in this study is illustrated in Figure 3.

Input: Travel survey dataset ζ consisting of 31 travel attributes, minimum confidence, minimum support, minimum interest and number of rules across 1,520 questionnaires.

Syntactic Constraint Parameters: *FBT*, *FCT*, *NetCov*, *FreqBus*, *QoR*, *LvlFare* (**interchangeable**).

Output: $(\zeta R_{\psi}^{statistical})_t, (\zeta R_{\psi}^{associated})_t$;
as significant rules $| \; t \in$ {mode use, level of service, value for money}.

1. Data transformation: transform numeric attributes into nominal variables.
2. Dataset partition: split the dataset into training dataset ζ_{ts} and validation dataset ζ_{vs}.
3. **Phase I:** Generation of rules using ζ_{ts} based on syntactic constraints.

Using minimum support σ_{min}, minimum confidence α_{min} and syntactic constraints ψ_N for

$$N \in \{FBT, FCT, NetCov, FreqBus, QoR, LvlFare\}$$

a. ; **return** $R_{initial\ set(\psi)}$ as a superset of association rules.

b. **Filtration stage 1**, rejecting uninteresting rules based on minimum interest Γ_{min}

For any association rule $\overline{R_{\psi}}$ contained in superset $\overline{R_{initial\ set(\psi)}}$ as a subset, $\overline{R_{\psi}}$ describes association between predictor variable(s) "X" and response variable "Y", **calculate** interestingness as per Eq. (2).

If interest Γ_{ω} of rule $\overline{R_{\psi}}$ is less than Γ_{min}; **reject rule** $\overline{R_{\psi}}$ as uninteresting

Return remaining interesting rules as subsets in superset $\zeta_{ts}R_{\psi}$.

c. **Develop** ordinal regression models $\ln(\psi_k)$ for each response variable ψ in $\overline{\zeta_{ts}R_{\psi}}$.

Compute; log likelihoods *LL*,

coefficients β_i of predictor variables x_i and *p–values*.

d. **Filter** rules in superset $\overline{\zeta_{ts}R_{\psi}}$ based on computed $\ln(\psi_k)$ models.

For all $\ln(\psi_k) \in \overline{\zeta_{ts}R_{\psi}}$

If $\beta_i x_i$ insignificant; **reject** the model $\ln(\psi_k)$

Return retained rules passing the statistical significance test in the superset $\zeta_{ordinal}R_{\psi}$

e. **Filtration stage 2**, rejecting redundant rules postulating similar results

For the superset $\zeta_{ordinal}R_{\psi}$, **construct** an intermediate superset $\overline{R_{intermediate(\psi)}}$ containing all rules in $\zeta_{ordinal}R_{\psi}$, **while** $R_{single(\psi)}$ is any single association rule from $\zeta_{ts}R_{\psi}$ containing response and predictor variables

Select $R_{single(\psi)} \in \zeta_{ts}R_{\psi}$ by maximising $\left|R_{single(\psi)} \cap \overline{R_{intermediate(\psi)}}\right|$

At each stage $R_{intermediate(\psi)} \leftarrow R_{intermediate(\psi)} - R_{single(\psi)}$

And $\zeta_{meaningful}R_{\psi} \leftarrow R_{intermediate(\psi)}$, as sets of meaningful rules

f. **Denote** retained rules as $\zeta R_{\psi}^{statistical}$;

g. **Set** rejected rules from both **Stages d and e** as $\zeta_{ts}R_{\psi}$

4. **Phase II:** Validate rules $\zeta_{ts}R_{\psi}$ on ζ_{VS}.

Set $\zeta_{VS}R_{\psi} = \zeta_{ts}R_{\psi}$. **Compute** support σ and confidence α for each rule (X⟹Y) on ζ_{VS}

Reject rules with $\sigma < \sigma_{min}$, and $\alpha < \alpha_{min}$

Only repeat Step 3b (Filtration Stage 1) and **Step 3e (Filtration Stage 2)** to further cleanout uninteresting rules as retain only non-repetitive rules.

Label retained rules as $\zeta R_{\psi}^{associated}$ for which $\exists \zeta R_{\psi}^{associated} \in \overline{\zeta_{ts}R_{\psi}}$ and $\zeta R_{\psi}^{associated} \notin \zeta R_{\psi}^{statistical}$

5. **Phase III: Group** remaining rules into required syntactic constraint significant rules sets as following: $(\zeta R_{\psi})_{mode\ use}$, $(\zeta R_{\psi})_{level\ of\ service}$, $(\zeta R_{\psi})_{value\ for\ money} \subset (\zeta R_{\psi}^{statistical} \cup \zeta R_{\psi}^{associated})$

Figure 3. Proposed data mining algorithm for a travel dataset analysis of 1,520 questionnaires.

4. Results

4.1. Performance of the Proposed Algorithm

The initial application of the proposed algorithm produced a large set (~1559) of association rules between the collected variables in the travel dataset that passed the minimum support ($\sigma_{min} = 0.1$) and minimum confidence ($\alpha_{min} = 65\%$) thresholds set in Step 3a of the algorithm proposed in this study (Figure 3). Filtration stage 1 (Step 3b) was then applied to remove uninteresting rules based on Eq. (2) with the minimum interest threshold ($\Gamma_{min} = 0.05$). This way only 351 or 22.5% of the association rules were retained in the so-called intermediate superset as interesting rules. These 22.5% remaining association rules were then subjected to the algorithm stages in Step 3c and Step 3d and ordinal regression models were developed for each rule. All the association rules with insignificant p-values for coefficients were filtered out as a separate superset. Approximately 231 or ~65.8% of the remaining association rules passed the ordinal regression performance tests and were collected in a superset of ordinal rules. Although all of these 231 association rules may be useful for developing some pro-public transit policies for local building occupants and travellers, the large number of rules may be impractical for policy analysis and may also have some redundant or reoccurring relations as also suggested in the literature (see Gürbüz and Turna [36]). The algorithm steps of feature selection and Filtration Stage 2 were thus used for data-reduction to filter out redundant rules using the superset of the 231 ordinal regression rules, as explained below.

First, an intermediate superset was created as an exact copy of ordinal rules superset and using variable selection, any smaller association rule subset (e.g., {FBT 5 or more times a week, Weekday ⇒ Very satisfied with level of fare}) implying an association or interrelation already present in a larger subset association rule (e.g., {FBT 5 or more times a week, Employed full-time, Weekday ⇒ Very satisfied with level of fare}) already present in the intermediate superset was removed. The steps were automatically repeated for all the remaining association rules subsets as rules describing an existing or reoccurring association were redundant since they were already contained in the larger subset in the intermediate superset. This way only 25 or ~11% of the 231 ordinal regression rules were retained in the so-called intermediate superset as statistically-significant rules in Step 3f, while the filtered-out rules (total of 326 rules) from both Step 3d and Step 3e were collected in a separate superset. Using the validation travel dataset, 213 rules or approximately 65% of these discarded rules were found to be validated in Phase II, where again filtration approach based on interestingness measure (Step 3b) and meaningfulness (Step 3e) resulted in retainment of 22% or 47 rules in the superset of association-only rules. Finally, both supersets were grouped to give three association rules sets for (1) mode use, (2) level of service, and (3) value for money. Figure 4 presents the results of applying various filtration techniques in the developed algorithm where Phase III results present the grouping of phase I and II association rules. The overall the algorithm application managed to filter out a total of 95% (~1487 rules) of initially obtained association rules (total rules: ~1559) as redundant or insignificant to the total travel dataset, generating a total of ~5% or 72 rules for policymaking purposes.

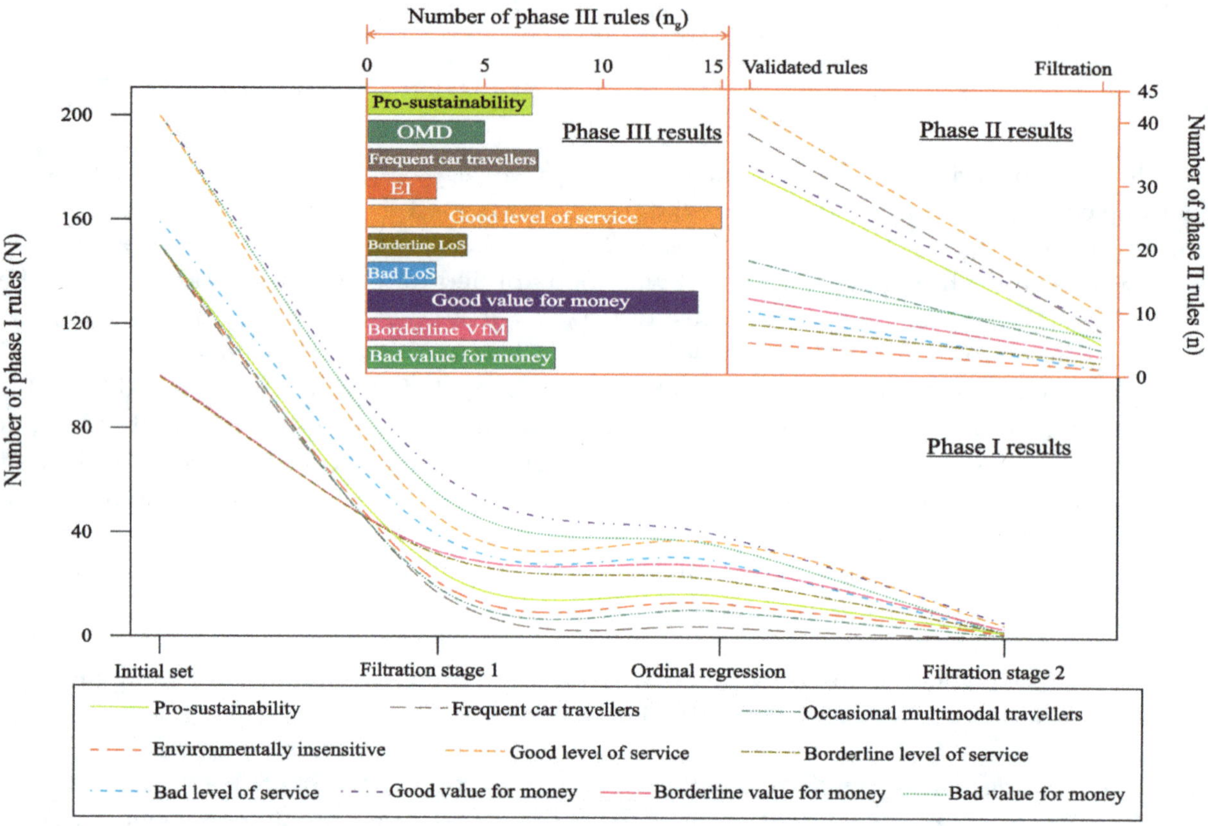

Figure 4. Filtration results of different phases in a collected travel dataset based on the developed algorithm.

The following sections present results of the proposed data mining algorithm on the selected case-study of intra-city commuters in Abu Dhabi. The results from each phase are presented separately to elucidate the respective findings. An importance index is also introduced to summarise the policy-relevant findings highlighted by public consultation regarding existing services. Policy recommendations are also given on how the transit service can be improved to promote an environmentally, resource and energy conservative transportation system by improved public (bus) transport ridership among the local residents and travellers, specifically the predominant (~55%) apartment building residents.

4.2. Phase I Results—Public Responses From Statistically-Significant Rules

The input and output variables were primarily categorical in nature with upper and lower bounds of each attribute explained in the previous sections. Ordinal regression showed that a majority of the commuters reported bus travel to be easy and that 36% were willing to pay more for seat. Full-time workers were more flexible towards bus use than the younger respondents. Furthermore, as the collected data was skewed towards male commuters, some of the association rules were found to be affected by the gender of the respondents. On the other hand, only a few of the quality attributes of the bus service were found to be statistically-significant. In addition, the statistically-significant models also failed to produce inference between the purpose of the commuters' journey and their mode choice, thereby requiring further analysis in phase II. The ordinal regression performance results of the meaningful association rules, retained after the second filtration stage, are presented in Table 2. Interestingly, the residential building typology (e.g., apartment building, villa-style house, etc.) had no statistically-significant influence on the mode choices of the residents, requiring further investigation to understand the association between these two factors as proposed by literature (e.g., [23,57]).

4.2.1. Affordability and Constrained Users

In the first block of mode use results, the odds of a full-time employed commuter to choose public transport was 6-7 times that of their unemployed counterparts (*PS rule 1*: $\chi^2 = 72.68, p < 0.001$) confirming market segmentation observations made previously by this project and identified under separate cover [33]. The commuters admitting affordability of private automobiles were 1.3 ($p < 0.05$, $\chi^2 = 140.07$: *OMD rule 1*) times more likely to be multimodal and 0.85 ($p < 0.05$, $\chi^2 = 39.96$: *EI rule 1*) as likely to be first time travelling by bus compared to others. While as expected, the commuters with zero household car-ownership were 1.86 times ($p < 0.05$, $\chi^2 = 67.18$: *PS rule 2*) as likely to use a bus service more than twice a week. Further exploring the affordability of the bus service exhibited that the commuters with a "very satisfied" perception of level of fare, also had 6.72 times likelihood of travelling 5 or more times/week by bus transport (*Good VfM rule 2*). The commuter satisfaction of public transport service level exhibited a slight decline relative to passenger travel time to the nearest bus-stop (see *rule 2* of *borderline level of service* and *rule 1* of *bad value for money*). Similarly, modal variability was affected when analysed against the same attribute as both *rule 2* of *pro-sustainability* and *rule 2* of *environmentally insensitive* market segments. Moreover, the commuters reporting buses to be crowded, were also twice as likely to belong to the environmentally insensitive group, indicating its significance in defining this market segment.

Table 2. Predictors of modal variability and service perception from association rules passing ordinal regression performance test.

Rules	Predictor Variables	Odds Ratio	Response Variable Category	Model χ2 ***	– 2LL †	Group
1	Employed full-time Willingness to pay more for seat	6.71*** 0.47***	Bus travel > 2 times a week	72.68	72.68	PS
2	5–15 min to bus stop No cars	2.22** 1.86*	Bus travel > 2 times a week	67.18	67.18	PS
1	Car/taxi is inexpensive Weekday	1.30* 0.27***	Once a week by car	140.07	1.54	OMD
1	Living near friends & family is important Car/taxi is inexpensive	0.47*** 0.85*	First time by bus	39.96	46.46	EI
2	Buses are crowded 5–15 min to bus stop	2.04*** 2.21**	First time by bus	57.41	41.62	EI
1	Male Bus travel is easy	14.1*** 1.27*	Very satisfied with frequency of bus	42.48	57.41	Good LoS
2	Very satisfied with network coverage Very Satisfied with quality of ride	2.88* 2.67**	Very satisfied with frequency of bus	352.27	167.16	Good LoS
3	Very satisfied with frequency of buses	7.49***	Very satisfied with network coverage	354.67	140.07	Good LoS
4	Male Employed full-time Bus travel is easy	6.06** 5.61** 1.30*	Satisfied with network coverage	39.23	139.62	Good LoS
5	Satisfied with journey time	13.22***	Satisfied with network coverage	214.14	42.48	Good LoS
1	Neutral on network coverage	7.08***	Neutral on frequency of buses	348.09	140.74	Borderline LoS
2	Willingness to pay more for seat 5–15 min to bus stop	1.87*** 0.56*	Neutral on network coverage	47.70	352.27	Borderline LoS

Table 2. *Cont.*

Rules	Predictor Variables	Odds Ratio	Response Variable Category	Model χ2 ***	− 2LL †	Group
1	Car/taxi is inexpensive Buses are crowded	0.53*** 1.62***	Dissatisfied with frequency of bus	38.145	155.53	Bad LoS
2	Dissatisfied with frequency of buses	10.45***	Dissatisfied with network coverage	354.67	385.31	Bad LoS
1	Satisfied with frequency of buses	9.65***	Satisfied with quality of ride	164.79	38.15	Good VfM
2	FBT 5 or more times a week Employed full-time Weekday	6.72*** 3.66*** 3.69***	Very satisfied with level of fare	671.05	214.14	Good VfM
3	Very Satisfied with quality of ride	2.86***	Very satisfied with level of fare	97.02	47.71	Good VfM
4	Employed full-time Car/taxi is inexpensive	3.48*** 1.24*	Very satisfied with level of fare	54.11	354.67	Good VfM
5	Satisfied with journey time Employed full-time	3.68** 6.36***	Satisfied with level of fare	66.09	26.66	Good VfM
1	Neutral on network coverage	3.89***	Neutral on quality of ride	210.79	354.67	Borderline VfM
2	Neutral on frequency of buses	8.48***	Neutral on quality of ride	164.79	39.23	Borderline VfM
3	Employed full-time Willingness to pay more for seat	4.95** 1.16*	Neutral on level of fare	62.22	7.68	Borderline VfM
1	Dissatisfied with bus-station 5–15 min to bus stop	14.3*** 2.68***	Dissatisfied with quality of ride	288.16	21.18	Bad VfM
2	Willingness to pay more for seat Delayed by traffic congestion Weekend	0.71** 0.75* 0.04***	Very dissatisfied with level of fare	632.49	164.80	Bad VfM

PS = pro-sustainability, OMD = occasional multimodal travellers, FrCT = frequent car travellers, EI = environmentally insensitive, LoS = level of service, VfM = value for money. $*p < 0.05$, $**0.001 \leq p < 0.01$, $***p < 0.001$. † difference in -2 log likelihood of final model and null model.

4.2.2. Crowded Buses

Rules pertaining to crowded buses were found in both less frequent bus users (*EI rule 2*) and bad service level (*Bad LoS rule 1*) groups. The results are consistent with the previous findings by Batarce, Muñoz [24] where the disutility of public transportation in Chilean commuters was observed to be correlated with in-vehicle crowding. Comparison results of value for money groups further suggested a number of priority areas in this regard. The commuter market segment willing to pay more for a seat in the bus, were also 1.16 times more likely to hold a neutral perception of the level of fare (*Borderline VfM rule 3*: $p < 0.05$, $\chi^2 = 62.22$) and 0.71 times more likely to be very dissatisfied ($p < 0.01$, $\chi^2 = 632.49$: *Bad VfM rule* 2). This implies that merely increasing the number of seats may at first persuade borderline users, but may not affect dissatisfied travellers to the same extent. Interestingly, travellers labelling buses as crowded were 1.62 times as likely to be "dissatisfied" by the frequency of buses (*Bad LoS rule* 2). Even though a direct relationship between the variables may have been unperceived, a deeper correlation may be present as less frequent buses on the travel routes serving the commute demands of these local building residents may impose a higher load on the existing buses. This finding indicates the presence of a potential market (of borderline and dissatisfied commuters) for policymakers by improving the public transit service based upon the public need of frequent service. The improved public transit ridership due to a corresponding reduction in private vehicle usage may then reduce the overall life-cycle environmental and energy load of the residential buildings in these studied Abu Dhabi City zones due to lesser contribution from the daily transport energy demands of their residents.

Proposing investment in size or number of units of supply (buses) to meet the local building occupants' daily commute needs can then be countered by slightly higher fares and it may optimise the financial elasticity of the publicly-owned transport agencies. A follow-up life-cycle study on the cost of running an improved public bus transport service, i.e., more frequent (shorter headways) shall also be conducted. It is reiterated by the observance of travellers "satisfied" with journey time, as 13.22 times likely to be also satisfied with the network coverage (*Good LoS rule 5*: $p < 0.001$, $\chi^2 = 214.14$). It should also be noted that the two indicators of level of service, namely *network coverage* and *frequency of buses* were found to be highly correlated as the travellers who were "very satisfied" with the coverage were 2.88 times more likely to have the same perception of the frequency (*Good LoS rule 2*: $p < 0.05$, $\chi^2 = 352.27$). Commuters were also around 7 times more likely (*Borderline LoS rule 1*: $p < 0.001$, $\chi^2 = 348.09$) to hold a neutral perception of network coverage, causing a neutral opinion of the bus frequency, and 10.45 times more likely to be dissatisfied with both; thereby exhibiting a strong correlation between the two attributes.

4.2.3. Dynamics of Bus Fare, Quality and Frequency

Exploring the reciprocity between fare and quality added to this recommendation of a more frequent bus service supported by higher fares. The commuters unequivocally associated "levels of fare" with "quality of ride" (*rule 3* of *good VfM; OR* =2.86, $p < 0.001$, $\chi^2 = 97.02$), so the transit by bus was regarded as a value worthy of riders' money. Statistically, quality of ride was also found to be constrained by the frequency of buses as commuters satisfied with frequency of buses were 9.65 times more likely to be also satisfied by the quality of ride (*rule 1 of good VfM*; $p < 0.001$, $\chi^2 = 164.79$) and similar trends of interdependency were observed for the neutral perception of these three variables. This current study implies that unlike findings of previous studies (see Bachman and Katzev [87], Savage [88]), and fortifying the observations by Tirachini [89] and Tirachini, Hensher [90], decreasing levels of fare may not be solely responsible for influencing the perceptions of commuters regarding existing public transport services.

A slight increase in fares may be justified by more frequent buses or a higher ride quality; even though more frequent bus travellers exhibited an unwillingness to pay more for seats, a sizeable increase in frequent bus commuters may be achieved by optimising the "fare-frequency-quality" dynamic due to modal diversion, i.e., occasional multimode (both bus and car) users and frequent car travellers shifting towards higher (public) bus service ridership. Transportation economists and strategists traditionally propose a balance between frequency of buses, journey time and the level of fare [88], but fall somewhat short of indicating how the balance affects commuter behaviour and perception. As the fares were recently increased by the Department of Transport—Abu Dhabi, the observations of the current study may be of significance to the policymakers to understand the straightforward implications of such decisions in the studied traveller market. A benefit of the proposed algorithm is that the endogenous variables determining the perception of these variables among the commuter market segments are analytically filtered across a broad range of travel attributes.

4.3. Phase II Results—Policy Insights from Validated Association Rules

Mode choice dependency trends and bus ridership characteristics were further studied by analysing the association rules filtered out from phase I. Table 3 presents the validated association rules, grouped analogous to market segmentation analysis [33]. Some interesting econometric models about the journey purpose and backgrounds of the commuters were found; contrary to past research, this study found that statistical analyses alone may be deficient for capturing detailed characteristics of the traveller market. Inner correspondence of association rules was also partially detected by *rule 1* of PS where first time car travellers at the time of the survey were found to be more frequent bus users.

4.3.1. Targeting Work Commutes and Full-Time Workers

Studies on building residents working full-time have found that a majority of them use private automobiles for work commutes [18,23,91], yet research targeting work-related commutes to encourage public transport use among these building residents is usually neglected. Policymakers may need to further investigate the quality improvements for attracting these consumers as this study found full-time workers commuting to/from work as predominant in both occasional multimodal (*OMD rule 3*) and frequent car travellers (*FrCT rule 1*). These findings are also consistent with the previous findings by Horner and Mefford [92] where the potential of public transit for work-related commutes was recognised. Furthermore, the need to live close to place of work and/or friends and family was one of the primary elicited reasons across all segments (e.g., *PS rule 4*, *Good LOS rule 7*, *Borderline LOS rule1* and *Good VfM rule 5*).

4.3.2. Impact of Residential Building Typology, Locality and Nodal (Bus-Stop) Characteristics

The type of residential building, local urban form and location of the buildings were found to affect the daily mode choice of the residents [23,93,94]. However, no association was found between origin-destination areas and mode choice for the commuter segments surveyed here. Yet some influence of the commuters' residence building type (e.g., apartment, villa house, labour camps etc.) were observed (*rule 3* of *PS*). These observations may indicate the generalisability of the conclusions drawn from the results of the current study, specifically the dominant (> 80%, as per Abu Dhabi Government [95]) expatriate population of Abu Dhabi also being the dominant group among the survey respondents as apartment building dwellers [96]. The residual effect of commuter satisfaction with the node, i.e., bus-station waiting area, on the overall quality of ride was also exposed by the results; namely that the commuters who were dissatisfied with the waiting area were also likely to be dissatisfied with the quality of ride (*rule 2* of *bad VfM*). This represents a probable tendency of commuters to carry forward a negative perception that was developed before embarking on the public transit journey.

4.3.3. Budgetary Constraints, Bus Service Frequency and Network Coverage

This work found that the environmentally insensitive commuters, i.e., those who primarily use private automobiles for daily travel (*rule 1* of *EI*), were largely satisfied with the public transport fare level. It can be inferred that the budgetary constraints and pricing of the service may be of little or no concern to such users (high social or financial background, employment status etc.) and there can be several underlying reasons that may motivate their mode choice shift to public buses, if at all. Frequent car travellers indicated a willingness to pay further after observing a need for seat as a major concern (*rule 5* of *FrCT*) and attributed no delays due to traffic congestion during their daily commute. Phase II results also showed that some commuters may have been compelled to use public transport due to necessity, *rules 1* and *4* of *PS* (no driving license or car-ownership). Investigation of the rules on neutral and dissatisfied consumers of both *LoS* and *VfM* also revealed that financially (or otherwise) obliged travellers might have settled for public transport against a more idealised mode.

Consumer satisfaction from the level of fare was also found to be dependent upon the quality of ride, thereby confirming initial perceptions that commuters perceive fare as a function of ride quality, as presented in *rule 2* of *VfM*. This is also largely consistent with the findings of Hensher, Stopher [97] and fare resilience addressed in the literature review by Redman, Friman [98] where fare was largely associated with service quality. It is likely that the solution to motivating transit mode choice shift from private vehicles and/or improving the service perception may not be a reduction in fares but delivering consumer-expected quality for the charged price.

One interesting finding of phase I and II was the tendency of a majority of commuters to purchase cash tickets instead of monthly passes regardless of the transit service perception. Marketing policies for monthly fare collection may prove to be significant in this regard. Additionally, follow up analysis attributed to *rules 1* and *2* (*good LoS*) and *rules 1, 4, 5, 6* and *7* (*good VfM*) exemplifies that contrary to past research (see Redman, Friman [98]), economic restraints only partly diminish the commuters' positive cognitive assessment of the public transport. Moreover, perception of the service quality attributes, such as journey time was also found to influence the satisfaction from level of bus service and should also be carefully considered by the policymakers.

4.3.4. Journey Time and Ride Quality

Rule 3 of *bad VfM* suggests that the commuters mentioning a neutral opinion of the journey time were very dissatisfied about the quality of the ride. It implies the role of long commute time in developing stress among travellers, as past studies [99] also found commuter sensitivity to journey time. Longer journey time is also significant due to: high vehicle operating costs to local building residents [100,101]; higher fuel consumption [102,103] and high environmental burden to the buildings' overall embodied life-cycle [10].

The local government policymakers should, therefore, be mindful of the journey time detriments caused by any variations in the service route or frequency. In addition to confirming previously noticed significance of journey time on commuter satisfaction [97] and mode choice [104], phase II of the proposed algorithm was also able to pinpoint the contingent service level factors (namely frequency of buses and crowded buses). Policymakers can then predict the relative influence of changing any of the antecedent variables on commuter satisfaction from remaining variables and the subsequent mode choice. Another implication was the relative nature of quality of ride and level of service attributes of network coverage (Rule 4 of *good LoS*), also noticed in ordinal regression performance.

Table 3. Validated association rules for interdependency of studied transit-related variables ††.

Rules	Antecedent		Consequent	Group	Support	Confidence	Interest
1	>5 times/week by bus, Weekday, No cars	⇒	First time by car	PS	0.526	0.837	0.493
2	Work-related commute, Employed full-time	⇒	>5 times/week by bus	PS	0.389	0.702	0.215
3	Residential apartment, 5–15 minto bus stop	⇒	>5 times/week by bus	PS	0.380	0.705	0.115
4	No driving license, Living near friends & family is important	⇒	2–4 times/week by bus	PS	0.206	0.724	0.076
1	Cash ticket, No driving license, Pay seat, Buses are crowded	⇒	1–3 times/month by bus	OMD	0.319	0.681	0.295
2	Male, No driving license, No delays by traffic congestion	⇒	1–3 times/month by bus	OMD	0.291	0.662	0.264
3	Work commute, Employed full-time, Weekday	⇒	1–3 times/month by car	OMD	0.198	0.850	0.160
4	Male, Employed full-time, 5–15 minto bus stop, Weekday	⇒	Less often by car	OMD	0.239	0.633	0.113
1	Employed full-time, Buses are crowded, Weekday	⇒	Once a week by car	FrCT	0.441	0.857	0.422
2	25–34 years old, Weekday, Employed full-time	⇒	Once a week by car	FrCT	0.310	0.725	0.291
3	Work commute, Employed full-time, Pay for seat	⇒	Once a week by car	FrCT	0.296	0.791	0.282
4	No delays by traffic congestion, Weekday	⇒	Once a week by car	FrCT	0.235	0.884	0.213
5	Male, Willing to pay more for seat	⇒	Less often by bus	FrCT	0.251	0.826	0.201
1	Cash ticket, Satisfied with level of fare, 5–15 minto bus stop	⇒	>5 times/week by car	EI	0.221	0.830	0.214
1	Very satisfied with journey time, Employed full-time	⇒	Very satisfied with frequency of buses	Good LoS	0.467	0.756	0.406
2	Very satisfied with network coverage, No cars	⇒	Very satisfied with frequency of buses	Good LoS	0.432	0.835	0.376
3	Very satisfied with journey time, Male	⇒	Very satisfied with network coverage	Good LoS	0.417	0.883	0.357
4	Very satisfied with quality of ride, Male	⇒	Very satisfied with network coverage	Good LoS	0.394	0.854	0.329
5	Satisfied with quality of ride, Male	⇒	Satisfied with frequency of buses	Good LoS	0.411	0.829	0.217
6	Male, 5–15 minto bus stop	⇒	Very satisfied with network coverage	Good LoS	0.342	0.752	0.185
7	Living near friends & family important, Bus travel is easy	⇒	Satisfied with network coverage	Good LoS	0.352	0.773	0.158

Table 3. *Cont.*

Rules	Antecedent		Consequent	Group	Support	Confidence	Interest
8	Male, Cash ticket, No delays by traffic congestion	⇒	Very satisfied with network coverage	Good LoS	0.247	0.730	0.111
9	Rent under 10,000 AED per annum	⇒	Very satisfied with frequency of buses	Good LoS	0.206	0.628	0.094
1	Employed full-time, Living near work important, No cars	⇒	Neutral on frequency of buses	Borderline LoS	0.362	0.660	0.260
2	No driving license, Employed full-time, No cars	⇒	Neutral on frequency of buses	Borderline LoS	0.258	0.607	0.148
1	Work commute, Male, Living near work important, Weekday	⇒	Very dissatisfied with frequency of buses	Bad LoS	0.169	0.715	0.164
1	>5 times/week by bus, Cash ticket, Employed full-time	⇒	Satisfied with level of fare	Good VfM	0.518	0.937	0.354
2	Very satisfied with quality of ride	⇒	Very satisfied with level of fare	Good VfM	0.400	0.770	0.335
3	Cash ticket, Employed full-time, Weekday	⇒	Very satisfied with level of fare	Good VfM	0.249	0.799	0.165
4	Satisfied with journey time, Male	⇒	Satisfied with quality of ride	Good VfM	0.351	0.881	0.163
5	Living near friends & family is important, No cars	⇒	Satisfied with quality of ride	Good VfM	0.369	0.826	0.105
6	Bus travel is easy, No cars	⇒	Very satisfied with quality of ride	Good VfM	0.270	0.832	0.103
7	South Asian, Weekend, Cash ticket	⇒	Satisfied with level of fare	Good VfM	0.307	0.819	0.091
1	No driving license, 5–15 minto bus stop, No cars	⇒	Neutral on quality of ride	Borderline VfM	0.268	0.736	0.188
2	Buses are crowded, Cash ticket	⇒	Neutral on level of fare	Borderline VfM	0.278	0.742	0.159
1	Cash ticket, Employed full-time, Weekend	⇒	Dissatisfied with level of fare	Bad VfM	0.427	0.877	0.403
2	Cash ticket, Dissatisfied with bus-station	⇒	Dissatisfied with quality of ride	Bad VfM	0.266	0.922	0.262
3	Cash ticket, Neutral on journey time, Weekend	⇒	Very dissatisfied with quality of ride	Bad VfM	0.188	0.739	0.186

†† The reported confidence, support and lift are based upon the original observations in the training dataset. Some values are rounded-off.

4.4. *Public-Accorded Desiderata for Value-Added Bus Service for Local Building Residents*

Public demand-based policy implications for the bus service to improve bus ridership among local building residents and travellers towards reduction in the residents' transport energy and environmental component of the local residential buildings are distributed across the generated "meaningful rules". In order to encapsulate the findings of the algorithm proposed in this study on the case study dataset, this work also sought to solve an importance index (I_{in}) to estimate the policy recommendations consistent with the public demands. For a response variable ψ in the phase III results, with categorical values ranging from *1* to *N*, the importance index can be calculated for a predictor variable x_i as shown in Equation (3). It should be worth noting, as the predictor variables are also nominal in nature, only one value η is used at a time to determine the index (I_x) for the specific predictor variable. The results of applying Equations (3) and (4) on the grouped association rules of phase III are illustrated in Figure 5 and have been summarised to some degree to eliminate similar findings from any particular commuter market segment. For example, the variable category *Weekday* was kept for only frequency of car travel despite its occurrence in the results for frequency of bus travel models:

$$I_x^\eta = \left[\sum_{\psi=1}^{N} (x_i)_\psi \right]_\eta \tag{3}$$

$$I_{in} = \frac{I_x^\eta}{max(I_x^\eta)} \times 100 \qquad \forall\ I_x \in phaseIII\ rules \tag{4}$$

Findings here suggest that full-time workers in the 25–34 age group formed the major portion of the target market, and frequently commuted on the case study routes during weekdays. These study observations were distinct from some of the past studies that analysed only segregated student and young traveller markets [105], whereas a majority of the young users studied in this paper travelled by private automobiles. Further findings are detailed in the text below.

4.4.1. Residential Apartment Building Designs May Need to Be Upgraded

Residential apartments scored higher than any other dwelling types in the existing residential zones of the surveyed Abu Dhabi city (I_{in} = *16.6%*) as majority of the commuters responded as living in such buildings. This may reflect a pattern of denser land-use with an abundance of multi-storeyed buildings either on streets traversing or facing main arterial roads serviced by adequate public bus transport routes, especially since most commuters travelled short distance to reach the bus-stop (further details in Section 4.4.2), similar to the observations made by Thøgersen [106] and Hrelja [107]. Improving public transport service in these residential areas may impact the parking requirements in the residential apartment buildings [108]. Further work on building design implications may be developed from this study for upcoming residential development projects with adequate by provided local or building-connected public transport, as fewer parking floors may be needed in such buildings in the studied Abu Dhabi City area.

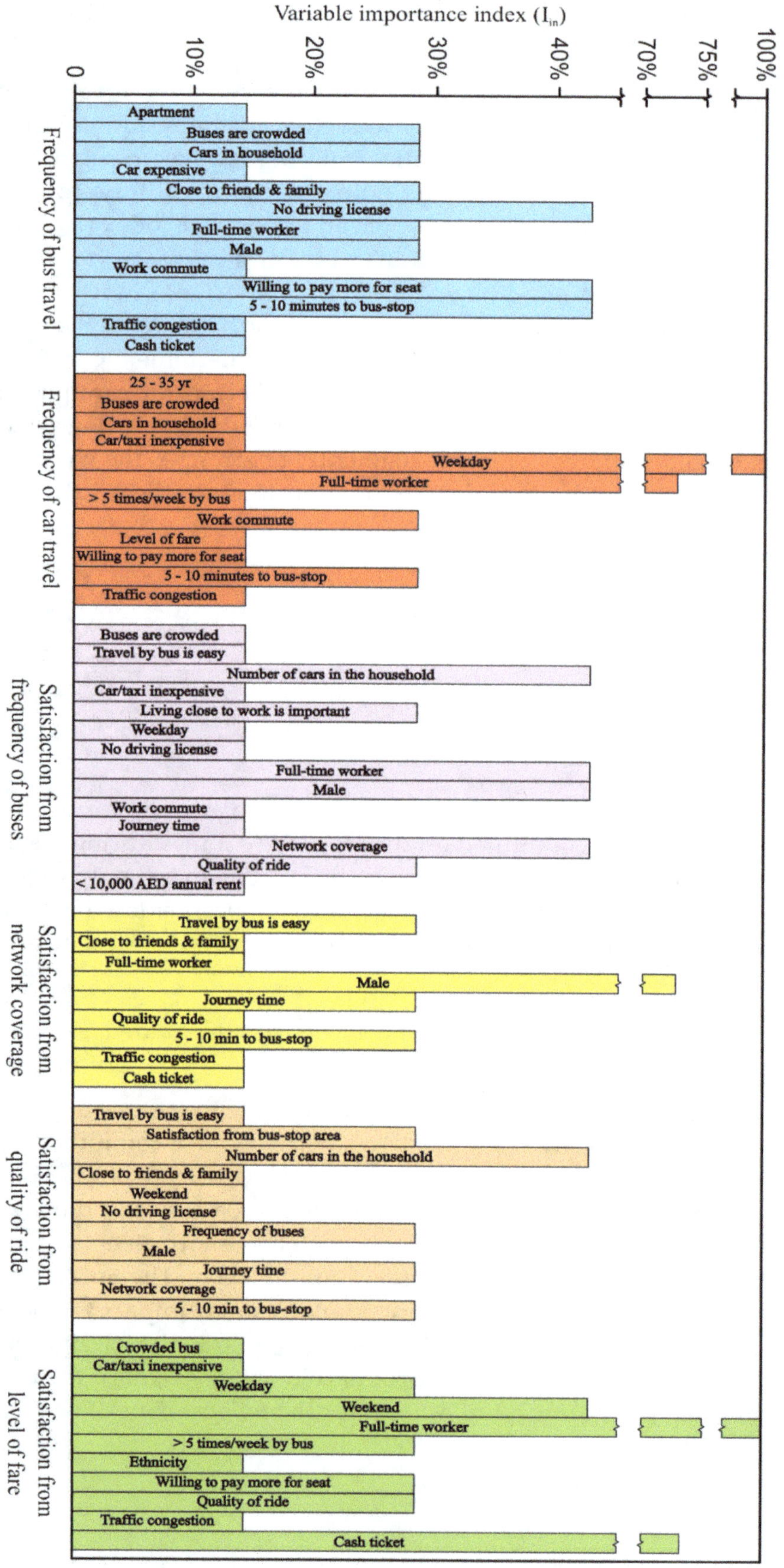

Figure 5. Importance index values for the phase III association rules of satisfaction and frequency criteria.

4.4.2. Existing Nodal (Bus-Stop) Distribution is Adequate

The survey respondents generally stated travelling a short distance to reach the nodes (bus-stop) from their residential and office buildings across the studied travel routes and thus the time to reach bus-station from these buildings only had nominal effect on the modal variability of their occupants. For example, I_{in} = *45.57%* for the frequency of bus travel and I_{in} = *28.57%* for satisfaction from network coverage and ride quality. This study, therefore proposes that establishing more bus-stations or further updating the connection between these buildings and a public transit service to serve the local building residents on the studied Abu Dhabi City routes may not necessarily encourage them to shift towards bus transport. This is a policy-sensitive observation and different from some of the past studies (e.g., Chien and Qin [109]). However, as Heinen and Chatterjee [110] found in their review on the status of research in transportation literature, few of the studies extensively explored modal variability over long-term and a majority only utilised a limited set of categorical variables to study university students or employees living in near-work communities, which may have also restricted their findings in terms of application to real-world situations.

4.4.3. Optimise Journey Time to Impact Commuter Satisfaction

Intrinsically, importance indices from this study for network coverage and quality of ride show that the commuters assume upper and lower bounds of journey time regardless of mode choice, in line with past works [111]. Any deviation in either direction causes a positive/negative ripple in commuter predilection of the service depending upon the journey time, confirming findings from previous research (see Hensher, Stopher [97]). In general, the commuters satisfied with the journey time on public bus transit were also satisfied from its network coverage (I_{in} = *28.57%*) and ride quality (I_{in} = *42.86%*). This supports the findings of previous studies [89,112] where the travellers were found to associate public transport as a slow and inflexible travel mode and the importance of journey time in promoting modal diversion among local residents towards a more sustainable mode (e.g. buses) was identified. On the other hand, unlike some of the past works [113,114], results gathered by this work found that the commuters admitting affordability of private automobiles were more likely to be multimodal in their transit behavior.

4.4.4. Bus Service Frequency, Crowded Buses and Level of Fare

Results here argue that the public ranking of the ride quality was influenced more by the frequency of bus service (I_{in} = *28.57%*) and less by the level of fare. The commuters interviewed were also found to be mainly concerned with the low capacity (crowded) buses, as both frequent bus and car travellers labelled the existing buses to be crowded. Other similar interesting correlations were also found, as the commuters who commented on over-crowded buses were also dissatisfied with the frequency of buses (I_{in} = *14.29%*) and level of fare. It builds upon the deductions from the econometric ordinal models that even though the commuters satisfied with the fare level travelled the most by buses (OR = 6.72); neutral (OR = 1.16) and dissatisfied respondents were more willing to pay extra for seats. Validated association rules further explained these observations by indicating that the level of fare was of little to no concern to non-users of bus services (i.e., the environmentally insensitive commuter segment), unlike past works [114,115]. Commuters ranking level of fare as satisfactory also indicated willingness to pay more for a seat (I_{in} = *28.57%*).

Frequent car travellers were also willing to pay more for the bus service if they get a seat (I_{in} = *14.29%*) and face no delays due to traffic congestion (I_{in} = *14.29%*). In addition, the level of fare was mostly perceived by surveyed commuters in terms of ride quality and it is argued here that strong correlations exist between the two variables. These findings support a change in the current perspective of policymakers where agencies generally advocate a reduction in fare level [116] to encourage local building residents in the cities to choose public buses instead of private automobiles for the daily commute. This study proposes that instead of implementing a blanket reduction in the fare level, policymakers may be able to achieve better results in terms of user predilection towards bus service and modal use by increasing the frequency of buses, economically supported by a commensurate reasonable increase in the current level of fare. Furthermore, as journey time was of significant concern to the commuting public, travel should also be optimised by route improvements such as establishing a bus rapid transit (BRT) service to facilitate the daily commutes of the current and potential service users in Abu Dhabi. The cost and environmental implication of any improved transit system alternatives for future modifications based on public demand and the resulting journey time savings can then be compared through a detailed LCA study.

5. Discussion

There is a lack of research in the behavioural econometric analysis of the social aspect of commuter decision-making regarding transport mode choice, factors and attributes affecting these decisions such as biases, service quality, and frequency etc., which, according to IPCC (Intergovernmental Panel on Climate Change) [68], considerably hinders the selection of a suitable climate-conserving transport alternative for a sustainable municipal built environment. Apart from influencing daily commute pattern of local building residents in cities, inadequate transportation systems also impact the construction of sustainable buildings due to their direct impact on the material supply-chain and disposal options due to time delays caused by congestions on local road (Section 2.1). Therefore, by extension transportation systems are also argued to hinder sustainability performance of residential buildings. Stakeholders then seek to limit multiple-handling and transportation of feed-stock from building construction and demolition activities, as well as properly factor the impact of haulage and disposal distances and traffic congestion on the material choice of building developers. Furthermore, as explained in Section 2.1, the provision of an effective transportation system with less congestion and a better public transportation system may also influence the residency choices, sale and property development aspects of buildings.

In order to address these issues, the study presented here analyses a survey questionnaire of 1,520 responses gathered to capture the commuter perceptions on mode choice and predilection of an existing intra-city setting bus service towards improving the overall transportation system. Passenger responses can be used to influence their mode choice and reduce overall traffic congestion of the transport system after implementing the prospective changes in the transport service, based upon the public demand. Literature review covered in Section 2.1 explained that this reduction in traffic congestion via more uptake of public-transport contributes to an efficient distribution of building materials. While Section 2.2 explained that the reduction of embodied operational and life-cycle energy and environmental load of residential buildings is significantly influenced by the daily commute mode choices of its residents and may be reduced by relying on public transport instead of private vehicle transport.

To identify the motivations behind the public and private transport as preferred mode of building residents, this study proposed a data-mining algorithm unifying a modified version of the Apriori algorithm and ordinal regression to highlight policymaking proposals pertinent to the studied Abu Dhabi and Middle Eastern residents. Training portion of the travel dataset was subjected to the modified Apriori algorithm in order to generate the initial set of association rules that were partitioned according to the individual market segment candidates in the consequent of the rules. After initial *interestingness*-based filtration of rules, the results were exported to SPSS for externally executing ordinal regression analyses. Ordinal regression performance results of the obtained association rules, after passing a *meaningfulness* measure, were used to analyse the market segments. Throughout the ordinal regression experiments, a majority of the respondents stated travelling a short distance to reach the bus-stop from their residential and/or office buildings, and time to bus-station had nominal effect on the modal variability of commuters. This may indicate that the current connection between these buildings and the supplied public transit service is adequate.

Mode use pattern was found to be only partly related to the socio-demographic characteristics of the surveyed local Abu Dhabi City building residents. The gender of these respondents had no statistically-significant influence on modal variability but was found to influence their perception of the service as the males generally had a more positive outlook. As such, a majority of similar variables (e.g., building typology, location, age etc.) failed to pass the ordinal regression tests and the next stage (validation of association rules) was used to capture the influence of these variables. Towards travel constraints limiting mode choice of commuters, this study found that a significant share of frequent bus travellers may have been compelled to uptake of public transport due to socio-demographics (lack of access to cars and lack of driving license). Nonetheless, the economic restraints only partly prejudiced their negative perception of the bus service and a majority of the so-called "compelled" travellers viewed public buses as good value for money and deemed the current service level to be satisfactory.

It is noted here that *occasionally multimodal* and *frequent car travellers* exhibit a substantial willingness to travel by public buses provided the capacity is increased or the buses become less crowded, even at the cost of higher fare. The results show that commuter satisfaction from the quality of ride and level of fare were highly interrelated. Frequency of the public bus service, as a measure of the passenger satisfaction from the level of service, is found to influence the mode choices of local building occupants. Given the competitive nature of private automobiles and the tendency of commuters in the studied dataset to use public transport for work-related commute dissimilar to past findings, policymakers may need to optimise quality-frequency-fare dynamics especially targeting peak work hours to reduce residents' transport-related environmental load of local buildings. One of the common strategies that is often implemented in many countries around the world for this purpose is investing in mass-transit systems. Technical and regulatory aspects of implementing the mass-transit system may include smoother traffic flow due to lesser private vehicles on the road network, reduction in cost and environmental burdens, fewer energy and fuel requirements on the power and fuel supply grids in the municipal regions, reduced overall operational and user-transport energy of residential buildings and availability of more recycled material haulage options (from far-off locations) in some areas for building construction projects. The mode use changes as a result of any improved public transit service may also affect the urban form and building designs: accessibility and shared spaces, lesser parking needs routine maintenance due to lower traffic and noise vibration damages.

6. Conclusions

The life-cycle energy and environmental load inventory of residential buildings includes the impact of material usage choices due to haulage requirements from virgin material suppliers vs. recycling plants, traffic congestion impact on per tonne kilometre (tkm) energy load during the actual material transport, disposal feedback loops and the embodied building residents' daily transportation requirements. Public transport services are generally proposed for alleviating excessive private commuter traffic load on road networks and reduce the overall transport system energy and environmental burdens in the local municipal regions. Yet, the interrelations of mode choice to underlying social (i.e., stakeholder-related) attributes of journey time, location of residential and office buildings comparative to the public transit service node (bus-stop), on-board crowding and ride quality are somewhat unexplored.

To that end, the methodology proposed in this study aids the policymakers in finding the critical attributes for improving performance of any built asset by engaging the stakeholders in the life-cycle asset management process through a social component. The findings of the current study demonstrate its application to a real-world case study of Abu Dhabi City building residents and travellers. Based on results, it recommends that a bus rapid transit (BRT) service, more frequent at peak office hours, may entice significant car users towards the public bus transport service. Whilst the overall whole-cost and explicit environmental impact of meeting public demands by constructing and managing a proposed practicable BRT system can be explored through subsequent detailed LCA, the work presented here does find that addressing public transportation systems preferences can contribute to congestion reduction and impact positively to sustainable development.

The proposed framework used data-mining to filter through the large set of public response survey data for a transport pattern behavioural case study among local building occupants. However, future works may apply the same procedure for data-mining the stakeholder perspectives of other aspects in the residential building development: need for a recreational building facility, upgrading/modifying an existing building, façade renovations, and interior works for buildings, etc.

Author Contributions: Conceptualization, methodology, formal analysis, validation and writing—original draft, U.H.; Supervision, writing—review and editing, and validation, A.W.; Resources and data curation, H.A.J.

Acknowledgments: The authors thank the Integrated Planning Department at the Department of Transportation (DoT) Abu Dhabi for their support of data collection and assistance.

Appendix A

The appendix that contains details and data supplemental to the main text is provided below.

Table A1. Questionnaire sample (English version) used for this study †.

No.	Questions	Responses (*please circle as appropriate*)						
MU	*Modal variability (mode use) variables*							
MU1	How often do you travel by bus?	**1.** First time	**2.** Less often	**3.** 1–3 times/month	**4.** Once a week	**5.** 2–4 times/week	**6.** Over 5 times/week	**7.** Never
MU2	How often do you travel by private car or taxi?	**1.** First time	**2.** Less often	**3.** 1–3 times/month	**4.** Once a week	**5.** 2–4 times/week	**6.** Over 5 times/week	**7.** Never
LoS	*Level of service variables*							
LoS1	How satisfied are you with current frequency of buses on this route?	**1.** Very dissatisfied	**2.** Dissatisfied	**3.** Neutral	**4.** Satisfied	**5.** Very satisfied		
LoS2	How satisfied are you with current level of network coverage on this route?	**1.** Very dissatisfied	**2.** Dissatisfied	**3.** Neutral	**4.** Satisfied	**5.** Very satisfied		
VfM	*Value for money variables*							
VfM1	How satisfied are you with current quality of ride on buses on this route?	**1.** Very dissatisfied	**2.** Dissatisfied	**3.** Neutral	**4.** Satisfied	**5.** Very satisfied		
VfM2	How satisfied are you with current level of fare of buses on this route?	**1.** Very dissatisfied	**2.** Dissatisfied	**3.** Neutral	**4.** Satisfied	**5.** Very satisfied		
ST	*Travel Bias (Structural-type Constraints Questions)*							
ST1	Your accommodation type?	**1.** Villa	**2.** Apartment	**3.** Hotel	**4.** Labour camp	**5.** Other		
ST2	What is your employment status?	**1.** Retired/Other	**2.** Visitor	**3.** Student	**4.** Unemployed	**5.** Work part-time	**6.** Work full-time	
ST3	What is your annual rent? (AED)	**1.** Under 10,000	**2.** 10,000–20,000	**3.** 20,001–40,000	**4.** 40,001–60,000	**5.** 60,001–100,000	**6.** Over 100,000	
SP	*Travel Bias (Spatial-type Constraints Questions)*							
SP1	Where do you live?	**1.** Al-Bateen	**2.** Downtown	**3.** CBD	**4.** Al-Mina	**5.** Al-Wahdah	**6.** Shakhbout St to city edge	**7.** Out of city
SP2	Where did you start this journey?	**1.** Al-Bateen	**2.** Downtown	**3.** CBD	**4.** Al-Mina	**5.** Al-Wahdah	**6.** Shakhbout St to city edge	**7.** Out of city
SP3	Where are you travelling to?	**1.** Al-Bateen	**2.** Downtown	**3.** CBD	**4.** Al-Mina	**5.** Al-Wahdah	**6.** Shakhbout St to city edge	**7.** Out of city
SP4	Purpose of your journey today?	**1.** Work	**2.** Study	**3.** Business	**4.** Personal reason	**5.** Shopping	**6.** Leisure	
SP5	Type of ticket you purchased today?							
SD	*Travel Bias (Socio-demographic Constraints Questions)*							
SD1	Age (years)	**1.** Under 16	**2.** 16 – 24	**3.** 25 – 34	**4.** 35 – 44	**5.** 45 – 64	**6.** Over 65	
SD2	Number of cars in household	**1.** No cars	**2.** 1 to 2 cars	**3.** 3 to 5 cars	**4.** Over 5 cars			
SD3	Do you hold a UAE driving license?	**1.** Yes	**2.** No					
SD4	Ethnicity/Nationality	**1.** UAE	**2.** Caucasian	**3.** Middle Eastern	**4.** African	**5.** South Asian	**6.** Southeast Asia	**7.** Other
SD5	Gender	**1.** Male	**2.** Female					
SD6	Your gross monthly income in AED	**1.** Under 10,000	**2.** 1,000–3,000	**3.** 3,001–5,000	**4.** 5,001–10,000	**5.** 10,001–20,000		

Table A1. *Cont.*

No.	Questions	Responses (*please circle as appropriate*)				
MU	*Modal variability (mode use) variables*					
SQ	*Perceived Service Quality Questions*					
SQ1	I am satisfied with journey time	1. Strongly disagree	2. Disagree	3. Neutral	4. Agree	5. Strongly agree
SQ2	The buses are too crowded	1. Strongly disagree	2. Disagree	3. Neutral	4. Agree	5. Strongly agree
SQ3	Bus travel is the easiest way for me	1. Strongly disagree	2. Disagree	3. Neutral	4. Agree	5. Strongly agree
SQ4	I am satisfied with the bus-stops	1. Strongly disagree	2. Disagree	3. Neutral	4. Agree	5. Strongly agree
SQ5	Travel by car or taxi is expensive	1. Strongly disagree	2. Disagree	3. Neutral	4. Agree	5. Strongly agree
SQ6	Traffic congestion delays my journey	1. Strongly disagree	2. Disagree	3. Neutral	4. Agree	5. Strongly agree
SQ7	I chose to live further from work (i.e. near family and friends) and longer commute time is insignificant to me	1. Strongly disagree	2. Disagree	3. Neutral	4. Agree	5. Strongly agree
SQ8	I chose to live closer to work as shorter commute time is significant to me	1. Strongly disagree	2. Disagree	3. Neutral	4. Agree	5. Strongly agree
SQ9	Willing to pay more for bus travel if I always had a seat	1. Strongly disagree	2. Disagree	3. Neutral	4. Agree	5. Strongly agree
SQ10	I am satisfied with the existing distribution of bus-stops on the current travel route (Today it took me longer/many minutes to get to bus-stop)	1. Strongly disagree	2. Disagree	3. Neutral	4. Agree	5. Strongly agree

† Both English and Arabic versions were used along with multilingual teams to capture data across all demographics.

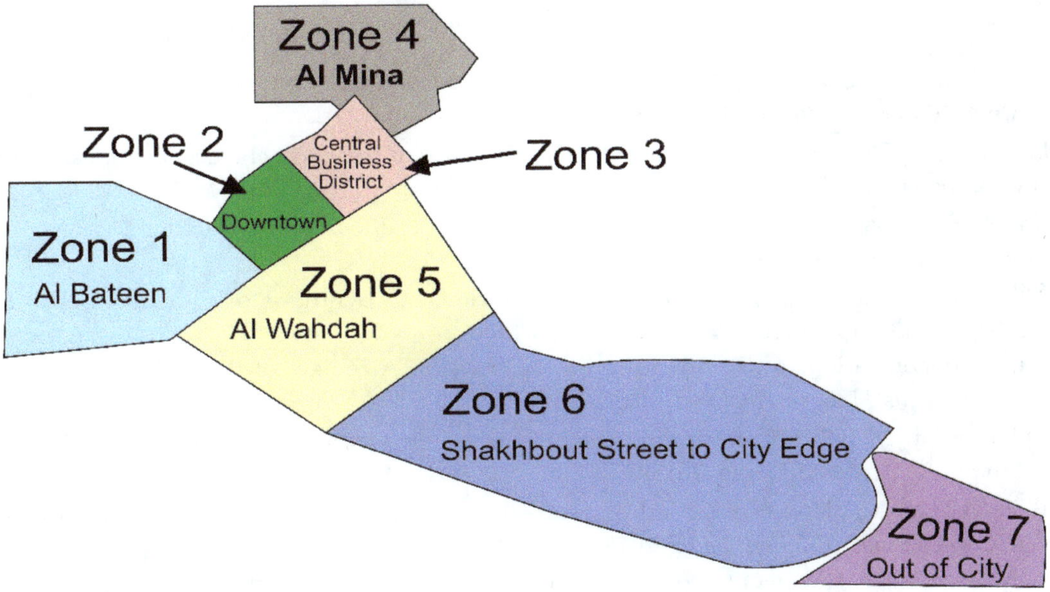

Figure A1. Zonal distribution of the Abu Dhabi City (based on AECOM (2015) [84].

Table A2. Statistical descriptive distribution of the collected variables.

Variables	N (Valid)	Mean	Standard Deviation	Variance
Frequency of bus travel	1517	5.20	1.149	1.321
Frequency of car travel	1305	2.94	1.414	2.000
Satisfaction with frequency of buses	1512	3.70	0.899	0.809
Satisfaction with network coverage	1494	3.74	0.890	0.793
Satisfaction with quality of ride	1501	3.98	0.976	0.953
Satisfaction with level of fare	1505	3.37	1.351	1.824
Your accommodation type?	1509	2.52	1.390	1.933
What is your employment status?	1505	5.55	1.103	1.216
What is your annual rent? (AED)	1384	2.09	1.252	1.566
Where do you live?	1519	3.76	1.814	3.291
Where did you start this journey?	1518	3.35	1.845	3.405
Where are you travelling to?	1515	3.38	1.823	3.323
Purpose of your journey today?	1514	3.25	2.130	4.539
Type of ticket you purchased today?	1516	1.32	0.732	0.536
Age (years)	1507	3.21	0.923	0.851
Number of cars in household	1440	0.17	0.392	0.153
Do you hold a UAE driving license?	1503	1.79	0.411	0.169
Ethnicity/Nationality	1507	5.02	1.070	1.145
Gender	1509	1.13	0.333	0.111
Your gross monthly income in AED	1385	2.47	1.048	1.099
I am satisfied with journey time	1508	3.95	0.826	0.682
The buses are too crowded	1519	0.60	0.489	0.240
Bus travel is the easiest way for me	1519	0.66	0.475	0.226
I am satisfied with the bus-stops	1496	3.38	1.125	1.265
Travel by car or taxi is expensive	1519	0.45	0.497	0.247
Traffic congestion delays my journey	1519	0.35	0.478	0.228
I chose to live further from work	1519	0.66	0.472	0.223
I chose to live closer to work	1319	4.70	2.70	7.301
Willing to pay more for bus seat	1519	0.36	0.479	0.229
Satisfied with existing bus-stop distribution	1519	2.10	0.676	0.457

References

1. Borrego, C.; Tchepel, O.; Barros, N.; Miranda, A.I. Impact of road traffic emissions on air quality of the Lisbon region. *Atmos. Environ.* **2000**, *34*, 4683–4690. [CrossRef]
2. Gray, D.; Laing, R.; Docherty, I. Delivering lower carbon urban transport choices: European ambition meets the reality of institutional (mis)alignment. *Environ. Plan. A Econ. Space* **2016**, *49*, 226–242. [CrossRef]
3. Crispino, M.; D'Apuzzo, M. Measurement and prediction of traffic-induced vibrations in a heritage building. *J. Sound Vib.* **2001**, *246*, 319–335. [CrossRef]
4. Rychtáriková, M.; Jedovnický, M.; Vargová, A.; Glorieux, C. Synthesis of a Virtual Urban Soundscape. *Buildings* **2014**, *4*, 139–154. [CrossRef]
5. Hasan, U.; Chegenizadeh, A.; Budihardjo, M.; Nikraz, H. A review of the stabilisation techniques on expansive soils. *Aust. J. Basic Appl. Sci.* **2015**, *9*, 541–548.
6. Hasan, U. Experimental Study on bentonite Stabilisation Using Construction Waste and Slag. Ph.D. Thesis, Curtin University, Perth, Australia, 2015.
7. Le, A.B.D.; Whyte, A.; Biswas, W.K. Carbon footprint and embodied energy assessment of roof-covering materials. *Clean Technol. Environ. Policy* **2018**. [CrossRef]
8. Mawed, M.; Al-Hajj, A.; Alshemery, A.A. The impacts of sustainable practices on UAE mosques' life cycle cost. In Proceedings of the Smart, Sustainable and Healthy Cities: 1st International Conference of the CIB Middle East and North Africa Research Network, Abu Dhabi, UAE, 14–16 December 2014; pp. 307–324.
9. Alqahtani, A.; Whyte, A. Evaluation of non-cost factors affecting the life cycle cost: An exploratory study. *J. Eng. Des. Technol.* **2016**, *14*, 818–834. [CrossRef]
10. Dodd, N.; Donatello, S.; Garbarino, E.; Gama-Caldas, M. *Identifying Macro-Objectives for the Life Cycle Environmental Performance and resource Efficiency of EU Buildings*; JRC EU Commission: Luxembourg, 2015; p. 117.
11. Badland, H.; Schofield, G. Transport, urban design, and physical activity: An evidence-based update. *Transp. Res. Part D Transp. Environ.* **2005**, *10*, 177–196. [CrossRef]
12. Yigitcanlar, T.; Kamruzzaman, M.; Teriman, S. Neighborhood Sustainability Assessment: Evaluating Residential Development Sustainability in a Developing Country Context. *Sustainability* **2015**, *7*, 2570–2602. [CrossRef]
13. Jabareen, Y.R. Sustainable Urban Forms: Their Typologies, Models, and Concepts. *J. Plan. Educ. Res.* **2006**, *26*, 38–52. [CrossRef]
14. Zimring, C.; Joseph, A.; Nicoll, G.L.; Tsepas, S. Influences of building design and site design on physical activity: Research and intervention opportunities. *Am. J. Prev. Med.* **2005**, *28*, 186–193. [CrossRef] [PubMed]
15. Cervero, R. Public transport and sustainable urbanism: Global lessons. In *Transit Oriented Development*; Routledge: London, UK, 2009; pp. 23–35.
16. De Luca, S. Public engagement in strategic transportation planning: An analytic hierarchy process based approach. *Transp. Policy* **2014**, *33*, 110–124. [CrossRef]
17. Leyden, K.M.; Slevin, A.; Grey, T.; Hynes, M.; Frisbaek, F.; Silke, R. Public and Stakeholder Engagement and the Built Environment: A Review. *Curr. Environ. Health Rep.* **2017**, *4*, 267–277. [CrossRef] [PubMed]
18. Stephan, A.; Stephan, L. Life cycle energy and cost analysis of embodied, operational and user-transport energy reduction measures for residential buildings. *Appl. Energy* **2016**, *161*, 445–464. [CrossRef]
19. Hasan, U.; Whyte, A.; Al Jassmi, H. Critical review and methodological issues in integrated life-cycle analysis on road networks. *J. Clean. Prod.* **2019**, *206*, 541–558. [CrossRef]
20. Stephan, A.; Crawford, R.H.; de Myttenaere, K. A comprehensive assessment of the life cycle energy demand of passive houses. *Appl. Energy* **2013**, *112*, 23–34. [CrossRef]
21. Diana, M. Measuring the satisfaction of multimodal travelers for local transit services in different urban contexts. *Transp. Res. Part A Policy Pract.* **2012**, *46*, 1–11. [CrossRef]
22. De Vos, J.; Mokhtarian, P.L.; Schwanen, T.; Van Acker, V.; Witlox, F. Travel mode choice and travel satisfaction: Bridging the gap between decision utility and experienced utility. *Transportation* **2016**, *43*, 771–796. [CrossRef]
23. Anderson, J.E.; Wulfhorst, G.; Lang, W. Expanding the use of life-cycle assessment to capture induced impacts in the built environment. *Build. Environ.* **2015**, *94*, 403–416. [CrossRef]

24. Batarce, M.; Muñoz, J.C.; de Dios Ortúzar, J.; Raveau, S.; Mojica, C.; Ríos, R.A. Use of Mixed Stated and Revealed Preference Data for Crowding Valuation on Public Transport in Santiago, Chile. *Transp. Res. Rec. J. Transp. Res. Board* **2015**, *2535*, 73–78. [CrossRef]
25. Stathopoulos, A.; Cirillo, C.; Cherchi, E.; Ben-Elia, E.; Li, Y.-T.; Schmöcker, J.-D. Innovation adoption modeling in transportation: New models and data. *J. Choice Model.* **2017**, *25*, 61–68. [CrossRef]
26. Golob, T.F.; Hensher, D.A. The trip chaining activity of Sydney residents: A cross-section assessment by age group with a focus on seniors. *J. Transp. Geogr.* **2007**, *15*, 298–312. [CrossRef]
27. Diana, M.; Pronello, C. Traveler segmentation strategy with nominal variables through correspondence analysis. *Transp. Policy* **2010**, *17*, 183–190. [CrossRef]
28. Tan, P.-N.; Steinbach, M.; Kumar, V. *Introduction to Data Mining*, 1st ed.; Addison-Wesley Longman Publishing Co., Inc.: Boston, MA, USA, 2005.
29. Elder, J. *Handbook of Statistical Analysis and Data Mining Applications*, 1st ed.; Nisbet, R., Miner, G., Eds.; Academic Press: Boston, MA, USA, 2009; p. 864.
30. Al-Hussaeni, K.; Fung, B.C.M.; Iqbal, F.; Dagher, G.G.; Park, E.G. SafePath: Differentially-private publishing of passenger trajectories in transportation systems. *Comput. Netw.* **2018**, *143*, 126–139. [CrossRef]
31. Laing, R.; Tait, E.; Gray, D. Public engagement and participation in sustainable transport issues. In Proceedings of the Construction, Building and Real Estate Research Conference of the Royal Institution of Chartered Surveyors, Las Vegas, NV, USA, 11–13 September 2012.
32. Hasan, U.; Whyte, A.; Al Jassmi, H. Framework for Delivering an AV-based Mass Mobility Solution: Integrating Government-Consumer Actors and Life-cycle Analysis of Transportation Systems. In Proceedings of the 46th European Transport Conference, Dublin, Ireland, 10–12 October 2018; p. 18.
33. Hasan, U.; Whyte, A.; Al Jassmi, H. Public-Transport System Management: Improving Service Satisfaction and Sustainable Uptake. **2018**. submitted.
34. Chu, K.; Chapleau, R. Augmenting transit trip characterization and travel behavior comprehension. *Transp. Res. Rec. J. Transp. Res. Board* **2010**, *2183*, 29–40. [CrossRef]
35. Diana, M. Studying patterns of use of transport modes through data mining: Application to us national household travel survey data set. *Transp. Res. Rec. J. Transp. Res. Board* **2012**, *2308*, 1–9. [CrossRef]
36. Gürbüz, F.; Turna, F. Rule extraction for tram faults via data mining for safe transportation. *Transp. Res. Part A Policy Pract.* **2018**, *116*, 568–579. [CrossRef]
37. Ordonez, C. Association rule discovery with the train and test approach for heart disease prediction. *IEEE Trans. Inf. Technol. Biomed.* **2006**, *10*, 334–343. [CrossRef]
38. Shaharanee, I.N.M.; Hadzic, F.; Dillon, T.S. Interestingness measures for association rules based on statistical validity. *Knowl. Based Syst.* **2011**, *24*, 386–392. [CrossRef]
39. Lazcorreta, E.; Botella, F.; Fernández-Caballero, A. Towards personalized recommendation by two-step modified Apriori data mining algorithm. *Expert Syst. Appl.* **2008**, *35*, 1422–1429. [CrossRef]
40. Whyte, A.; Laing, R. Deconstruction and reuse of building material, with specific reference to historic structures. In Proceedings of the 1st Australasia and South East Asia Conference in Structural Engineering and Construction (ASEA-SEC-1), Perth, Australia, 28 November–2 December 2012; pp. 171–176.
41. Whyte, A. *Life-Cycle Assessment of Built-Asset Waste Materials: Sustainable Disposal Options*; Lambert Aademic Publishing: Saarbrucken, Germany, 2012.
42. NSAI. *EN 15978:2011: Sustainability of Construction Works—Assessment of Environmental Performance of Buildings—Calculation Method*; British Standard Institute: London, UK, 2011.
43. AFNOR Normalisation. CEN/TC 350 Sustainability of Construction Works. Available online: http://portailgroupe.afnor.fr/public_espacenormalisation/CENTC350/index.html (accessed on 10 December 2018).
44. International Standards Organisation (ISO). *ISO/TS 12720: Sustainability in Buildings and Civil Engineering Works—Guidelines on the Application of the General Principles in ISO 15392*; ISO: Geneva, Switzerland, 2014.
45. Giorgi, S.; Lavagna, M.; Campioli, A. Guidelines for Effective and Sustainable Recycling of Construction and Demolition Waste. In *Designing Sustainable Technologies, Products and Policies: From Science to Innovation*; Benetto, E., Gericke, K., Guiton, M., Eds.; Springer International Publishing: Cham, Switzerland, 2018; pp. 211–221.
46. HM Government. Environmental Taxes, Reliefs and Schemes for Businesses. Available online: https://www.gov.uk/green-taxes-and-reliefs/aggregates-levy (accessed on 12 December 2018).

47. Bleischwitz, R. Towards a resource policy—Unleashing productivity dynamics and balancing international distortions. *Miner. Econ.* **2012**, *24*, 135–144. [CrossRef]
48. Coelho, A.; de Brito, J. Economic viability analysis of a construction and demolition waste recycling plant in Portugal—Part I: Location, materials, technology and economic analysis. *J. Clean. Prod.* **2013**, *39*, 338–352. [CrossRef]
49. Mulley, C.; Ma, L.; Clifton, G.; Yen, B.; Burke, M. Residential property value impacts of proximity to transport infrastructure: An investigation of bus rapid transit and heavy rail networks in Brisbane, Australia. *J. Transp. Geogr.* **2016**, *54*, 41–52. [CrossRef]
50. Strømann-Andersen, J.; Sattrup, P.A. The urban canyon and building energy use: Urban density versus daylight and passive solar gains. *Energy Build.* **2011**, *43*, 2011–2020. [CrossRef]
51. Steemers, K. Energy and the city: Density, buildings and transport. *Energy Build.* **2003**, *35*, 3–14. [CrossRef]
52. Tronchin, L.; Manfren, M.; Nastasi, B. Energy efficiency, demand side management and energy storage technologies—A critical analysis of possible paths of integration in the built environment. *Renew. Sustain. Energy Rev.* **2018**, *95*, 341–353. [CrossRef]
53. Norman, J.; MacLean, H.L.; Kennedy, C.A. Comparing High and Low Residential Density: Life-Cycle Analysis of Energy Use and Greenhouse Gas Emissions. *J. Urban Plan. Dev.* **2006**, *132*, 10–21. [CrossRef]
54. Cuéllar-Franca, R.M.; Azapagic, A. Environmental impacts of the UK residential sector: Life cycle assessment of houses. *Build. Environ.* **2012**, *54*, 86–99. [CrossRef]
55. Stephan, A.; Crawford, R.H.; de Myttenaere, K. Towards a comprehensive life cycle energy analysis framework for residential buildings. *Energy Build.* **2012**, *55*, 592–600. [CrossRef]
56. Stephan, A.; Crawford, R.H.; de Myttenaere, K. Towards a more holistic approach to reducing the energy demand of dwellings. *Procedia Eng.* **2011**, *21*, 1033–1041. [CrossRef]
57. Anderson, J.E.; Wulfhorst, G.; Lang, W. Energy analysis of the built environment—A review and outlook. *Renew. Sustain. Energy Rev.* **2015**, *44*, 149–158. [CrossRef]
58. Dell'Olio, L.; Ibeas, A.; Cecín, P. Modelling user perception of bus transit quality. *Transp. Policy* **2010**, *17*, 388–397. [CrossRef]
59. De Oña, J.; de Oña, R.; Calvo, F.J. A classification tree approach to identify key factors of transit service quality. *Expert Syst. Appl.* **2012**, *39*, 11164–11171. [CrossRef]
60. Schmid, B.; Schmutz, S.; Axhausen, K.W. Explaining mode choice, taste heterogeneity, and cost sensitivity in a post-car world. In Proceedings of the 95th Annual Meeting of the Transportation Research Board, Washington, DC, USA, 10–14 January 2016.
61. Stradling, S. Determinants of Car Dependence. In *Threats from Car Traffic to the Quality of Urban Life*; Garling, T., Steg, L., Eds.; Emerald Group Publishing Limited: New York, NY, USA, 2007; pp. 187–204.
62. Chen, C.; Gong, H.; Paaswell, R. Role of the built environment on mode choice decisions: Additional evidence on the impact of density. *Transportation* **2008**, *35*, 285–299. [CrossRef]
63. Friman, M.; Fellesson, M. Service supply and customer satisfaction in public transportation: The quality paradox. *J. Public Transp.* **2009**, *12*, 57–69. [CrossRef]
64. Menichetti, D.; Vuren, T.V. Modelling a low-carbon city. *Proc. Inst. Civ. Eng. Transp.* **2011**, *164*, 141–151. [CrossRef]
65. Environment Agency (Abu Dhabi). *Greenhouse Gas Inventory for Abu Dhabi Emirate—Inventory Results Executive Summary*; Environment Agency (Abu Dhabi): Abu Dhabi, UAE, 2012; p. 25.
66. Al Tayer, S.M. UAE is committed to reducing carbon footprint. In *Khaleej Times*; Galadari Printing and Publishing LLC: Dubai, UAE, 2018.
67. Liu, F.; Xu, R.; Fan, W.; Jiang, Z. Data analytics approach for train timetable performance measures using automatic train supervision data. *IET Intell. Transp. Syst.* **2018**, *12*, 568–577. [CrossRef]
68. IPCC. *Climate Change 2007: The Physical Science Basis. Contribution of Working Group I, II and III to the Fourth Assessment Report of the IPCC*; Pachauri, R.K., Reisinger, A., Core Writing Team, Eds.; Intergovernmental Panel on Climate Change: Geneva, Switzerland, 2008; p. 104.
69. Hasan, U.; Chegenizadeh, A.; Budihardjo, M.A.; Nikraz, H. Experimental Evaluation of Construction Waste and Ground Granulated Blast Furnace Slag as Alternative Soil Stabilisers. *Geotech. Geol. Eng.* **2016**, *34*, 1707–1722. [CrossRef]
70. Hasan, U.; Chegenizadeh, A.; Budihardjo, M.A.; Nikraz, H. Shear Strength Evaluation Of Bentonite Stabilised With Recycled Materials. *J. GeoEng.* **2016**, *11*, 59–73.

71. Hasan, U.; Chegenizadeh, A.; Nikraz, H. Nanotechnology Future and Present in Construction Industry: Applications in Geotechnical Engineering. In *Advanced Research on Nanotechnology for Civil Engineering Applications*; IGI Global: Hershey, PA, USA, 2016; pp. 141–179.
72. Ponte, C.; Melo, H.P.M.; Caminha, C.; Andrade, J.S., Jr.; Furtado, V. Traveling heterogeneity in public transportation. *EPJ Data Sci.* **2018**, *7*, 42. [CrossRef]
73. Agrawal, R.; Imieliński, T.; Swami, A. Mining association rules between sets of items in large databases. In Proceedings of the ACM SIGMOD International Conference on Management of Data, Washington, DC, USA, 25–28 March 1993; pp. 207–216.
74. Agrawal, R.; Srikant, R. Fast algorithms for mining association rules. In Proceedings of the 20th Very Large Data Bases (VLDB) Conference, Santiago de Chile, Chile, 12–15 September 1994; Bocca, J.B., Jarke, M., Zaniolo, C., Eds.; Morgan Kaufmann Publishers Inc.: San Francisco, CA, USA, 1994; pp. 487–499.
75. Nahar, J.; Imam, T.; Tickle, K.S.; Chen, Y.-P.P. Association rule mining to detect factors which contribute to heart disease in males and females. *Expert Syst. Appl.* **2013**, *40*, 1086–1093. [CrossRef]
76. Hu, L.; Zhuo, G.; Qiu, Y. Application of Apriori Algorithm to the Data Mining of the Wildfire. In Proceedings of the 2009 Sixth International Conference on Fuzzy Systems and Knowledge Discovery, Tianjin, China, 14–16 August 2009; pp. 426–429.
77. Pande, A.; Abdel-Aty, M. Discovering indirect associations in crash data through probe attributes. *Transp. Res. Rec. J. Transp. Res. Board* **2008**, *2083*, 170–179. [CrossRef]
78. Tan, P.-N.; Kumar, V.; Srivastava, J. Selecting the right objective measure for association analysis. *Inf. Syst.* **2004**, *29*, 293–313. [CrossRef]
79. Nosratabadi, H.E.; Pourdarab, S.; Nadali, A.; Khalilinezhad, M. Evaluating discovered rules from association rules mining based on interestingness measures using fuzzy expert system. In Proceedings of the Fourth International Conference on the Applications of Digital Information and Web Technologies (ICADIWT 2011), Stevens Point, WI, USA, 4–6 August 2011; pp. 112–117.
80. Ma, W.-M.; Wang, K.; Liu, Z.-P. Mining potentially more interesting association rules with fuzzy interest measure. *Soft Comput.* **2011**, *15*, 1173–1182. [CrossRef]
81. Wu, X.; Zhang, C.; Zhang, S. Efficient mining of both positive and negative association rules. *ACM Trans. Inf. Syst.* **2004**, *22*, 381–405. [CrossRef]
82. Cherchi, E.; Cirillo, C.; de Dios Ortúzar, J. Modelling correlation patterns in mode choice models estimated on multiday travel data. *Transp. Res. Part A Policy Pract.* **2017**, *96*, 146–153. [CrossRef]
83. Arbabi, H.; Mayfield, M. Urban and Rural—Population and Energy Consumption Dynamics in Local Authorities within England and Wales. *Buildings* **2016**, *6*, 34. [CrossRef]
84. AECOM. *Working Paper No. 1: Task 3 Review and Recommendations on the Proposed Travel Surveys*; Abu Dhabi Department of Transport: Abu Dhabi, UAE, 2015; p. 94.
85. Osborne, J.W. Conceptual and Practical Introduction to Testing Assumptions and Cleaning Data for Logistic Regression. In *Best Practices in Logistic Regression*; SAGE Publications: Thousand Oaks, CA, USA, 2014; p. 488.
86. Shaheen, M.; Shahbaz, M. An Algorithm of Association Rule Mining for Microbial Energy Prospection. *Sci. Rep.* **2017**, *7*, 46108. [CrossRef]
87. Bachman, W.; Katzev, R. The effects of non-contingent free bus tickets and personal commitment on urban bus ridership. *Transp. Res. Part A Gen.* **1982**, *16*, 103–108. [CrossRef]
88. Savage, I. The dynamics of fare and frequency choice in urban transit. *Transp. Res. Part A Policy Pract.* **2010**, *44*, 815–829. [CrossRef]
89. Tirachini, A. Estimation of travel time and the benefits of upgrading the fare payment technology in urban bus services. *Transp. Res. Part C Emerg. Technol.* **2013**, *30*, 239–256. [CrossRef]
90. Tirachini, A.; Hensher, D.A.; Rose, J.M. Multimodal pricing and optimal design of urban public transport: The interplay between traffic congestion and bus crowding. *Transp. Res. Part B Methodol.* **2014**, *61*, 33–54. [CrossRef]
91. Wang, T.; Chen, C. Attitudes, mode switching behavior, and the built environment: A longitudinal study in the Puget Sound Region. *Transp. Res. Part A Policy Pract.* **2012**, *46*, 1594–1607. [CrossRef]
92. Horner, M.W.; Mefford, J.N. Examining the Spatial and Social Variation in Employment Accessibility: A Case Study of Bus Transit in Austin, Texas. In *Access to Destinations*; Emerald Group Publishing Limited: New York, NY, USA, 2005; pp. 193–214.

93. Currie, G.; De Gruyter, C. Exploring links between the sustainability performance of urban public transport and land use in international cities. *J. Transp. Land Use* **2018**, *11*, 325–342. [CrossRef]
94. Kennedy, C.; Miller, E.; Shalaby, A.; Maclean, H.; Coleman, J. The Four Pillars of Sustainable Urban Transportation. *Transp. Rev.* **2005**, *25*, 393–414. [CrossRef]
95. Abu Dhabi Government. Abu Dhabi Emirate: Facts and Figures. Available online: https://www.abudhabi.ae/portal/public/en/abu-dhabi-emirate/abu-dhabi-emirate-facts-and-figures;jsessionid=zKnROF_msRJQoP3Lu22DJzCLFXxFVXf5ED6y_2KpOCR0wIG8-4k7!496354345!-733067616!1541108424678 (accessed on 2 November 2018).
96. Flint, S. Living in: Abu Dhabi. Available online: http://www.bbc.com/travel/story/20131203-living-in-abu-dhabi (accessed on 2 November 2018).
97. Hensher, D.A.; Stopher, P.; Bullock, P. Service quality—-Developing a service quality index in the provision of commercial bus contracts. *Transp. Res. Part A Policy Pract.* **2003**, *37*, 499–517. [CrossRef]
98. Redman, L.; Friman, M.; Gärling, T.; Hartig, T. Quality attributes of public transport that attract car users: A research review. *Transp. Policy* **2013**, *25*, 119–127. [CrossRef]
99. Sottile, E.; Cherchi, E.; Meloni, I. Measuring Soft Measures Within a Stated Preference Survey: The Effect of Pollution and Traffic Stress on Mode Choice. *Transp. Res. Procedia* **2015**, *11*, 434–451. [CrossRef]
100. Santos, G.; Bhakar, J. The impact of the London congestion charging scheme on the generalised cost of car commuters to the city of London from a value of travel time savings perspective. *Transp. Policy* **2006**, *13*, 22–33. [CrossRef]
101. Trigaux, D.; Wijnants, L.; De Troyer, F.; Allacker, K. Life cycle assessment and life cycle costing of road infrastructure in residential neighbourhoods. *Int. J. Life Cycle Assess.* **2017**, *22*, 938–951. [CrossRef]
102. Kuo, Y. Using simulated annealing to minimize fuel consumption for the time-dependent vehicle routing problem. *Comput. Ind. Eng.* **2010**, *59*, 157–165. [CrossRef]
103. Biswas, W.K.; Thompson, B.C.; Islam, M.N. Environmental life cycle feasibility assessment of hydrogen as an automotive fuel in Western Australia. *Int. J. Hydrogen Energy* **2013**, *38*, 246–254. [CrossRef]
104. Nguyen-Phuoc, D.Q.; Currie, G.; De Gruyter, C.; Young, W. How do public transport users adjust their travel behaviour if public transport ceases? A qualitative study. *Transp. Res. Part F Traffic Psychol. Behav.* **2018**, *54*, 1–14. [CrossRef]
105. Nobis, C. Multimodality: Facets and Causes of Sustainable Mobility Behavior. *Transp. Res. Rec. J. Transp. Res. Board* **2007**, *2010*, 35–44. [CrossRef]
106. Thøgersen, J. Promoting public transport as a subscription service: Effects of a free month travel card. *Transp. Policy* **2009**, *16*, 335–343. [CrossRef]
107. Hrelja, R. Integrating transport and land-use planning? How steering cultures in local authorities affect implementation of integrated public transport and land-use planning. *Transp. Res. Part A Policy Pract.* **2015**, *74*, 1–13. [CrossRef]
108. Rowe, D.H.; Christine Bae, C.H.; Shen, Q. Evaluating the Impact of Transit Service on Parking Demand and Requirements. *Transp. Res. Rec.* **2011**, *2245*, 56–62. [CrossRef]
109. Chien, S.I.; Qin, Z. Optimization of bus stop locations for improving transit accessibility. *Transp. Plan. Technol.* **2004**, *27*, 211–227. [CrossRef]
110. Heinen, E.; Chatterjee, K. The same mode again? An exploration of mode choice variability in Great Britain using the National Travel Survey. *Transp. Res. Part A Policy Pract.* **2015**, *78*, 266–282. [CrossRef]
111. Milakis, D.; Cervero, R.; Van Wee, B. Stay local or go regional? Urban form effects on vehicle use at different spatial scales: A theoretical concept and its application to the San Francisco Bay Area. *J. Transp. Land Use* **2015**, *8*, 59–86. [CrossRef]
112. Yao, B.; Hu, P.; Lu, X.; Gao, J.; Zhang, M. Transit network design based on travel time reliability. *Transp. Res. Part C Emerg. Technol.* **2014**, *43*, 233–248. [CrossRef]
113. Kingham, S.; Dickinson, J.; Copsey, S. Travelling to work: Will people move out of their cars. *Transp. Policy* **2001**, *8*, 151–160. [CrossRef]
114. Eriksson, L.; Friman, M.; Gärling, T. Stated reasons for reducing work-commute by car. *Transp. Res. Part F Traffic Psychol. Behav.* **2008**, *11*, 427–433. [CrossRef]

115. Thøgersen, J.; Møller, B. Breaking car use habits: The effectiveness of a free one-month travelcard. *Transportation* **2008**, *35*, 329–345. [CrossRef]
116. Trépanier, M.; Habib, K.M.N.; Morency, C. Are transit users loyal? Revelations from a hazard model based on smart card data. *Can. J. Civ. Eng.* **2012**, *39*, 610–618. [CrossRef]

10

Heat Stress Pattern in Conditioned Office Buildings with Shallow Plan Forms in Metropolitan Colombo

Upendra Rajapaksha

Department Architecture, Faculty of Architecture, University of Moratuwa, Moratuwa 10400, Sri Lanka; upendra@uom.lk

Abstract: This paper critically evaluates indoor overheating of multilevel office buildings in Colombo—a tropical warm humid city. The work questions the building morphological characteristics on thermal performance and indoor climate, thus the levels of Building Energy Indices (BEI) of air conditioned buildings. Pattern of heat stress on buildings due to building characteristics and its relationship to the BEI were identified. A study of 87 multilevel office buildings contributed to identify two critical cases in shallow plan form with similar morphological characteristics such as wall-to-window ratio, aspect ratio, orientation, occupant and equipment density, and façade architecture. A comprehensive thermal performance investigation on these two critical cases quantified the heat stress patterns on their facades and thus indoor thermal environments. Indoor air temperature during office hours in 3 m $\times$ 3 m multizones across the depths and lengths in these two buildings showed deviations up to 10.5 °C above the set point temperature level (24 °C). Findings highlight the severity of heat stress on air conditioned indoor environments and the need to address this issue for energy sustainability of urban office buildings in the tropics.

Keywords: Multilevel buildings; indoor overheating; operational energy; shallow plan forms

---

1. Introduction

The IPCC's Fifth Assessment Report notes that continued GHG emissions due to human-induced contribution at or above current rates would cause further warming and induce many adverse changes in the global climate system [1]. Studies that have quantified the effects of global warming and projections of ambient temperature increases [1] indicate that more warm days are expected in most parts of subtropics and tropics [2]. Warming climates increase internal temperatures of buildings and studies have shown that this relationship is linear [3]. Increase of extreme air temperatures may considerably impact the electricity demand for cooling [4] and thus emissions [5]. With the warming of cities [6], both free running and air conditioned office buildings designed for existing climatic conditions are increasingly prone to indoor overheating in future. With the projected increase in energy use and the demand for more comfortable indoor environments in office buildings, there is a growing concern for high energy consumption and its likely adverse impacts on the environment [7]. Further, overheating cannot be avoided completely under expected warming climate scenarios [8]. These behaviors would obstruct the efforts of emission reduction strategies and targets for the building sector [9] and require the understanding of thermal behavior patterns of buildings in warming climates if strategies need to be identified to control unnecessary indoor overheating [10].

A linear correlation between the increase of average external air temperature and the increase of building cooling load, and thus the total energy use, is established for air conditioned buildings [11,12]. The average energy consumption in most countries varies between 100 and 500 kWh/m^2/a due to many reasons including climate type, climate change factor, building characteristics, HVAC and lighting systems, office equipment, and operational schedules [13]. Further, urban environments are faced with the challenge of population inflation resulting in urban heat island (UHI) effects [14]. Hence, buildings, infrastructure and open spaces must be adequately "climate change proofed" to counter long term effects of urbanization and climate change effects [15]. Colombo, the commercial capital of Sri Lanka is no exception to high temperatures resulting from global warming and surface urban heat island (SUHI) effects due to the increase of building density, population and anthropogenic heat from traffic congestion, etc. [16].

Electricity use in the commercial building sector of Sri Lanka contributes 24% to the total consumption [17]. The annual energy consumption in office buildings in Colombo, which is known as a typical warm humid city in Asia, is ~250 kWh/m^2/a [17]. Studies that do exist for Sri Lanka highlights that demand side management of operational energy use by buildings are primarily focused on highly sophisticated end use equipment which may be commercially available but not financially viable due to negative cost options. Further, it is also highlighted that apart from limited hydropower capacity, other available technologies such as wind and dendro power impose a cost penalty on GHG mitigation [18].

1.1. Climate Response for Energy Sustainability: An Adaptation and Mitigation Option

Adaptation [19] and mitigation are complementary approaches for reducing global warming impacts on buildings and thereby emissions [1,20]. Improving energy sustainability of building operations through climate responsive building design is definitely the most significant and cost effective action the commercial office-building sector can take in its direction to reduce dependency on non-renewable energy resources and GHG emissions. Energy in buildings is consumed mainly for cooling, heating, and lighting [21] while a portion is used for equipment. Integration of energy conservation interventions involves both technological and nontechnological aspects of building design and operation.

Many believe that a well implemented mix of regulatory instruments such as mandatory codes, carbon energy tax policies and tradable permits, and voluntary instruments such as unilateral and negotiated agreements and voluntary programs are effective nontechnological aspects for achieving energy efficiency which could incur lower implementation costs as well [8].

Bioclimatic design [22,23] is seen as an appropriate basis for technological aspects of climate responsive design which involves the way buildings filter the climate for occupants' comforts involving four equally important interlocking variables—climate, biology, technology, and architecture. Bioclimatic design involves the conscious decision to operate buildings in the "selective mode", where the external environment impinges on the building and indoor environments are achieved from building–climate interplay. The "exclusive mode", which seeks to barricade the indoor environment from climate influence, effectively excludes benefits such as passive cooling and passive heating [24].

The building envelope, cross-section, and form are main bioclimatic drivers that foster interventions to reduce negative impacts of outdoor air temperature, summer heat gain, winter heat loss, optimization of daylight, and appropriate solar control. Morphological characteristics of buildings in terms of plan form, sectional form, envelope, orientation, and fenestration details and their effects on energy consumption are known [25]. Of them, plan form is a major contributor that controls the level of "building–climate interplay" [26] and thus the indoor air temperature levels in buildings.

Consideration of a building's geometry as an application of "air fluid composition" based on "plan form" consists of several design drivers such as structure, space planning, number of floor plates, disposition of open plan cellular spaces, the number of stacks, and style of atriums and light wells; this seemingly offers benefits for operational energy saving, indoor air quality improvement and, likely, tolerance to climate change effects in buildings [27].

In multilevel buildings, height and lesser roof area in relation to external facade area means that the heat transfer between indoor environment and outdoor climate is significant through the façade and fenestration. Three ways of heat gain due to environmental and internal load transfer are known—conduction through opaque surfaces, conduction through glazed areas, and radiation through glazed areas [28]. The bioclimatic approach in heat gain control offers an excellent opportunity to enhance a modified indoor environment as well by improving quality of daylight without heat gain, reducing conductive environmental heat gain through envelope, and better control of indoor temperature.

Heat gain due to internal loads from equipment and higher occupancy rates can be greater in office buildings in certain climates with daytime ambient temperatures ranging within the comfort zone. Gaun's [6,24] simulation study predicts more indoor overheating potentials primarily due to internal loads in commercial office buildings in moderate climates, where summer air temperatures move closer to the comfort zone. In these climates, a reduction in summertime air conditioning use could be dominated by the use of internal thermal mass [29]. On the other hand, in tropical zones with severe warm humid conditions, controlling sun penetration offers better opportunities for delivering improved energy efficiency in conditioned buildings. The bottom line is that we should be in a position to make a deliberate and positive impact on improving energy efficiency through building facades [30–32] and plan depths [33]. Acquiring the most effective plan depth and envelope has value for climate responsiveness and thus advanced energy performance, increased human thermal comfort and climate sensitive role of buildings [34].

1.2. Heat Stress through Building Façades in Tropics

Heat stress [35] on building occupants is usually measured in terms of Wet Bulb Globe Temperature (WBGT) that combines air temperature, humidity, radiation, and air flow into a single value. Studies have quantified that in warm humid climates, levels of heat stress for occupants in free running buildings are mostly higher than the preferred comfort level of WBGT in the whole year [36]. This suggests the criticality of the climate in tropics and the need for careful integration of appropriate design interventions to avoid environmental heat stress through building facades.

Patterns of environmental heat gain through building facades due to direct radiation on different types of plan forms are known [37,38]. Shallow and linear plan forms with narrower facades along east–west axis promote least direct exposure to direct solar heat gain in equatorial climates. However, with high levels of cloud cover in tropics, presence of diffused radiation is a major challenge in making a building resistant to heat stress.

Percentage of glazing area of a wall know as window-to-wall ratio (WWR) of a façade can have a significant impact on heat gain potential and thus energy consumption levels in terms of cooling and heating in buildings [39]. Aspect ratio of a plan form is another key geometric parameter that defines the building surface area by which heat is transferred between the interior and exterior environment [40] and the amount of façade area that is subject to solar gain. The extent to which this can be beneficial or detrimental depends on the climate type. Heat transfer through opaque materials in the façade is another concern to be addressed.

In Sri Lanka, building heat stress remains a largely less investigated area. The few studies that do exist are focussed on energy efficiency measures of orientation, building envelope and lighting on free running indoors. Particularly, heat stress through facades has been least investigated for air conditioned buildings of any form. Reducing the risk of heat gain into air conditioned buildings is an essential phenomenon to study. When a building is air conditioned, the excessive heat gains through facades may not be perceptive to the occupants due to the conditioned environment inside, but nevertheless contribute to the increase of air conditioning energy load, thus exacerbating the emissions and global warming problem [41–43]. An in-depth understanding of performance improvement for building design for air conditioned indoors in tropics is yet to be achieved.

2. Climate in Colombo

Warm humid tropical climates are found in the region extending 15° North and South of the equator (Figure 1). Colombo (latitude 5°55′ to 9°49′ N and longitude 79°51′ and 81°51′ E) is an example of this climate and is characterised by lack of seasonal variations in temperature. The mean monthly temperatures range from 27 °C in November to 30 °C in April and relative humidity varies from 70% to 80% during a typical year. The daily maximum temperatures are high as 25 °C to 38 °C and the daily pattern in the dry season (September to November and March to May) has diurnal temperature range of 7 to 8 °C. Air temperature in these climates does not vary substantially over the daily typical cycle, thus limiting the opportunity for thermal mass for passive cooling. While solar effects are consistent, seasonal rain patterns affect temperature level (perceptions), but rainfall is increasingly unpredictable.

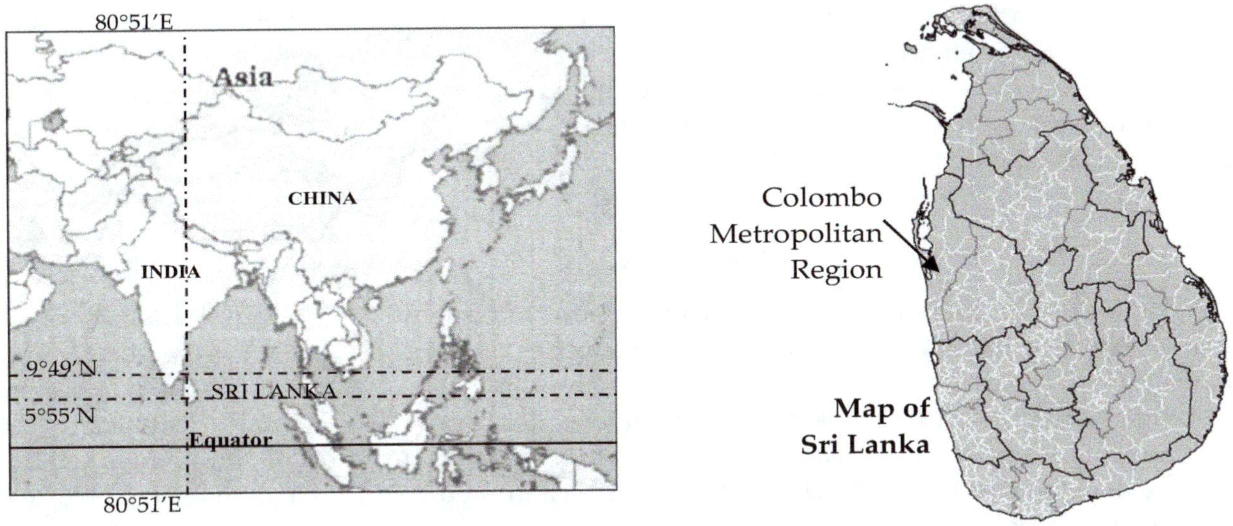

(a) (b)

Figure 1. (**a**) Sri Lanka in the world map. It is situated between 9°49′N and 5°55′N of Equator. (**b**) Sri Lanka.

KŐppen Geiger climate classification demonstrates that nearly 60% of the geographical area of Sri Lanka represents type "Af"—an equatorial fully humid climate. Of the 8760 hourly data points of dry bulb temperature in the psychrometric chart developed for this study using Climate Consultant 5.5 software, nearly 60% and 40 % of data points represent air temperatures in the range of 27–38 °C and 21–27 °C respectively (Figure 2). Annual monthly mean, maximum temperature, and relative humidity in Colombo vary within the ranges of 26.2 to 28.8 °C, 29.2 to 31.4 °C, and 73.8 to 83.6, and 90.2 to 97.9%, respectively.

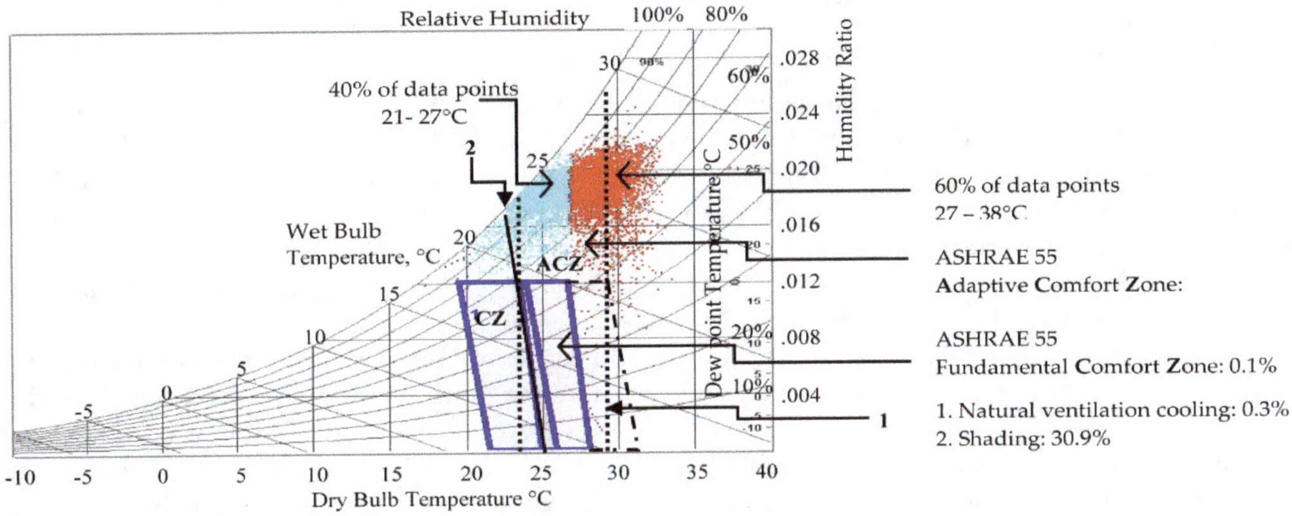

Figure 2. Psychrometric chart for Colombo developed from Climate Consultant 5.5 during the study.

The main characteristic of the warm humid climate in Colombo, from the human comfort and building design viewpoint, is a combination of high temperature and high humidity which in turn reduces the dissipation of body's surplus heat. Irradiation in Colombo ranges from 400 to 6000 W/m^2 throughout the year. Olgyays [22] criterion of ''optimum shape''—taking into account solar radiation, temperature, and other climatic conditions—calls for buildings freely elongated in the east–west direction. This provides shelter from the Sun for north–south facades and, in addition, facilitates the air movement entering from the south–west of buildings. Moreover, the directional effects of airflow inlets and building orientation can help to ameliorate the indoor airflow situation [38]. The presence of solar radiation relates to the temperature increase in the ambient air. Thus, prior to incoming air travelling through the indoor spaces, its sol-air temperature should always be reduced at the source by shading. The most efficient way of protecting a building is to shade the windows and other openings to improve immediate microclimate [44,45]. Shading against solar radiation reduces the effective temperature experienced by an occupant by up to 8 °C and could reduce the environmental heat transfer as well.

2.1. The Urban Climate and Building Stock in Colombo

World's urban population is expected to increase to 68% by 2050 and projections show that 70% of global urbanization will concentrate in countries of Asia and Africa by the year 2050 [46]. Urbanization in tropics accounts for a significant proportion in Global GHG emissions [47]. Colombo Metropolitan Region (CMR) is no exception to this, and indicates Surface Urban Heat Island (SUHI) effects intensifying from 2007 to 2017 due to rapid urbanization [16]. In 2005 the extent of UHI was 42% of the total land area of CMR. This UHI is growing and predicted for an annual increase of 1.75%. Moreover, an increasing trend is apparent in the extent of built-up area, which increased from 74% in the year 2005 to 97.3% in the year 2013 [48]. Scientific disclosure of the influence of UHI on increasing energy demand is well established.

Building stocks in many developed countries have been evaluated for the relationship between energy consumption and built characteristics [26,49,50]. They have been further investigated to identify the effect of building morphology such as shape, composition, orientation, and fenestration details for assessment of building performance and the end user energy demand [51–53]. However, similar studies on building stocks in Sri Lanka are less prioritized. The available few studies on energy consumption of buildings have focused on energy efficiency measures of lighting and simulations to assess indoor environmental quality of residential buildings [54,55]. Furthermore these studies are based on secondary data, which evidences the lack of a comprehensive national database that integrates the characteristics of the building stock and the energy utility intensities.

CMR, which has the highest office building density in Sri Lanka, was selected to explore the national urban building stock in respect to thermal performance and end-user energy demand.

3. Method of Study

The study was conducted in 3 phases involving 87 commercial multilevel office buildings covering an approximate floor area of 73,000 m^2. Phase I consisted of a walk through survey of all buildings and an onsite thermal performance investigation of selected buildings to evaluate the impact of building morphology on Building Energy Index (BEI), for which office buildings of each division is presented as the mean BEI. It was defined as the ratio of total annual energy used in kWh per Meter Square per annum. A walkthrough field investigation was performed in 35 Grama Niladhari (GN) divisions in the CMR (GN division is the smallest administrative division in Sri Lanka). Data collection protocol was structured with the use of geographical information system (GIS) data and activity zone maps of the CMR. The identified office building stock excluded mixed administrative buildings but included buildings exclusively used for administrative activities of private and government organizations including banks and other financial offices. Building morphology data was focused on building design parameters such as orientation, plan shape, construction materials, and fenestration characteristics such as WWR and aspect ratio (façade length/depth). Technical and operational characteristics were recorded as working hours, space conditioning systems, and usage of office equipment. Occupied hours, occupancy profiles, air conditioning systems, and type and number of equipment used were observed as nearly similar in these buildings. Thermal variables of indoor air and surface temperatures were recorded using HOBO UX100-003 data loggers, and monitored continuously for a week in randomly selected samples from the 87 buildings during one of the hottest months of March 2017. Climatic data of Meteorological Department of Sri Lanka shows that March and April as the hottest months in a typical year. A statistical analysis of the entire building stock in respect to BEI and morphological characteristics was conducted at the end of Phase I, developing a classification of the building stock in terms of BEI in an integrated framework of orientation, plan shape, WWR and Aspect Ratio.

Phase II of the research project involved a comprehensive thermal performance investigation of two critical case buildings identified from the statistical analysis of Phase I. The two buildings were identified as having two extreme levels of BEIs despite having similar morphological characteristics, occupant and equipment densities with shallow plan forms with perfect orientations along the east–west axis. Indoor air temperature variations across the depths of their shallow plan forms from perimeter to perimeter were quantified in air conditioned mode using 22 HOBO UX 100-003 meters from 6 am to 6 pm over a week in April 2018. A typical floor plate was considered as a collection of 3 m × 3 m multizones and each zone had a HOBO meter. Measuring 3 m × 3 m zones contributed to give one average value for the entire floor plate and was considered more accurate than considering just one or few points in the respective floor plates. The two critical cases are orientated with the longer facades facing north and south, due to which it can be assumed that heat stress on facades due to direct solar exposure is minimal. The objective of Phase II was to investigate the levels of indoor air temperature inside these two buildings and then to judge any correspondence between BEI and level of indoor air temperature in constant conditioned mode.

Phase III was an extension to Phase II and continued for further two weeks, day and night from 26th July to 16th August 2018 and measurements were taken in multizones of 3 m × 3 m at 10-min intervals. Measurements included both inside and outside perimeter wall surface temperatures, indoor air temperatures and RH in all 3 m × 3 m multizones, and air temperature just outside the exterior walls on all four orientations. Wall surface temperatures were taken using ONSET thermo couples UX120-014M.

4. Phase I of the investigation

4.1. Dispersion of the Office Building Stock

The office building stock is concentrated towards Northwest and West locations of the CMR. Figure 3 shows the distribution of the stock and BEIs of the critical GN divisions in CMR. Thirteen of 35 GN divisions have a substantial number of office buildings as shown in Figure 3a. The percentage of office buildings to total buildings of these GN divisions varies from 5 to 19%. The highest building stock of offices is in Keselwatte GN Division and the lowest is visible in Milagriya GN Division with percentages of 19 and 5, respectively. Descending order of GN divisions from the 2nd highest percentage of 16 up to 10% of the office stock are Suduwella, Colpetty, Bambalapitiya, Hunupitiya, Slave Island, Fort, and Wekanda. Positioning of office buildings along high-density traffic arteries is common in these zones, and the corresponding roads are Galle road, Dharmapala Mawatha, Sangharaja Mawatha, Sir James Peiris Mawatha, and D.R Wijewardana Mawatha. Thus the "road side sealed offices" represent the overarching character of national urban office building stock in CMR.

Phase I included three main Steps of investigations. The objective of the 3 steps is to sample the identified building population in the following manner:

- Identifying dispersion pattern of office building stock in CMR—Step 1.

- Characterizing 87 selected buildings in concentrated, dense areas in CMR in respect to BEI with plan forms, façade configurations/orientations, and compositions with neighboring buildings—Step 2.
- Statistical analysis to sample the building population mentioned in Step 2 in respect to BEI, orientation, and morphological characteristics of WWR and Aspect Ratio—Step 3.

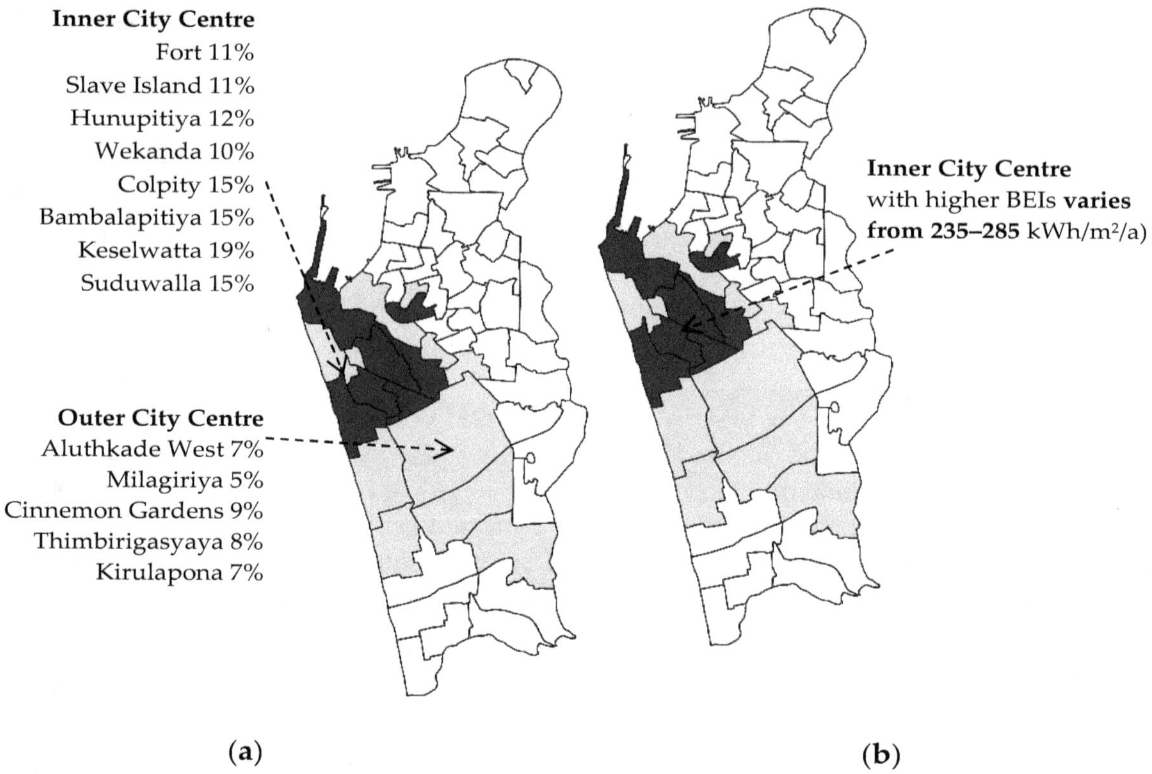

Figure 3. Map of Colombo Metropolitan Region showing (**a**) dispersion of the office building stock and (**b**) an inner city center with buildings with higher Building Energy Indices (BEIs).

4.2. Percentage Distribution of the Building Stock in Terms of BEI—Step 1

BEIs of the office building stock were assessed for 06 critical GN divisions. These six GN divisions represent a considerable number of office buildings and correspond to a range of 19 to 10 percent of office buildings in each division. BEI for office buildings of each division is presented as the mean value. Figure 3b presents the BEIs of six critical GN divisions. The results indicate that the annual BEI varies from 235 to 285 $kWh/m^2/a$. The highest demand is apparent in Kollupitiya GN Division followed by Hunupitiya GN Division with an office building stock of 15% and 12%, respectively. Results prove that there is no relationship between the percentage of the building stock and BEI (Table 1), demanding the importance of investigating morphology of office buildings in critical GN divisions.

Table 1. Percentage distribution of three ranges of BEIs of the building stock.

Levels of BEI Ranges	Values of BEI Ranges	Percentage Distribution
Lower Range (LR)	100–150 kWh/m^2 per annum	35.72%
Average Range (Av.R)	150–250 kWh/m^2 per annum	54.08%
Critical Range (CR)	Above 250 kWh/m^2 per annum	10.20%

Three levels of BEI ranges were identified in the pilot investigation. They are Lower Range (LR), Average Range (AvR) and Critical Range (CR), see Table 1.

The lower range was decided to be 110 to 120 KWh/m^2/a—a commonly known acceptable practice [56]. Nearly 36% of buildings are found to be in this region. The average range claims BEIs between 150 and 250 kWh/m^2/a, which is the national average of office buildings in Sri Lanka [17] but still considered as high. The critical range includes more than 10% of the stock with BEIs over 250 kWh/m^2/a.

a. Dispersion of BEIs in the Office Building Stock

The annual BEI in these office buildings varies within a range from 91 to 412 kWh/m^2/a and 95% of the buildings are above 110 kWh/m^2/a, which represents the accepted standard for energy efficient building codes [56]. Within this building stock, 56% of the buildings show an annual BEI above 200 kWh/m^2/a, of which almost half of the buildings are above 250 kWh/m^2/a. The mean BEI for critical office building stock in CMR is 212 kWh/m^2/a and thus solidifies that the building stock is energy obsolete beyond the standards (Figure 4). The outcome advises further investigation of the thermal performance of the building stock and related parameters, which has an effect in the end use energy demand of this stock.

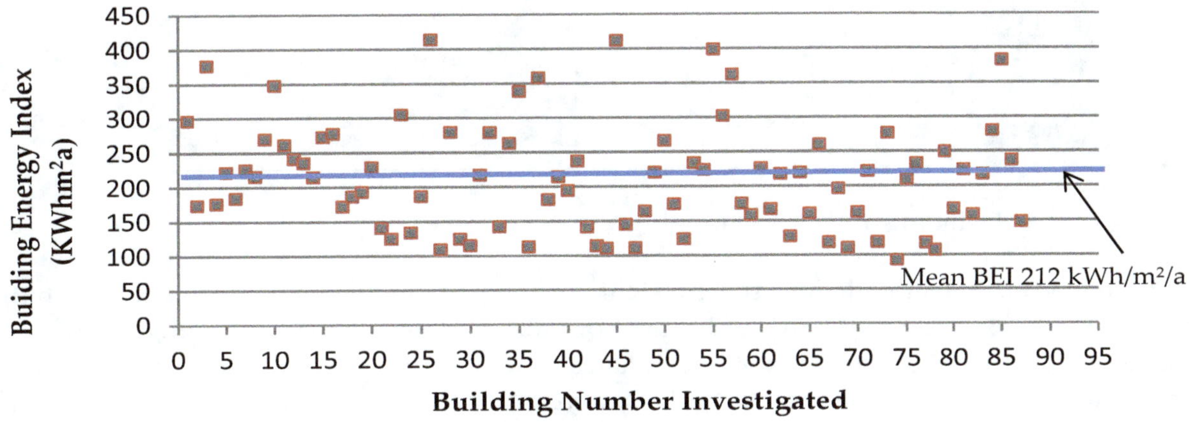

Figure 4. Building energy indices of 87 buildings.

4.3. Characteristics of the Building Morphology and Sampling—Step 2

Characterizing building morphology with specific concerns to plan form, fenestration material details, orientation and composition with neighboring buildings was assessed.

a. Plan Forms

This study proposes to categorize the office building stock into the predominant six types. These can be generalized as basic plan forms and composite plan forms. The basic plan forms consist of square, linear, and circular forms, while composite plan forms are derived from a combination of these and demonstrate the shapes L, U, and trapezium. Table 2 depicts the identified classification and its percentage distribution within the stock. Basic plan forms are commonly used, and of these, 85.05% demonstrate linear and square forms of 62.07% and 22.98%, respectively.

Among the composite plan forms, L shape is prominent with a percentage distribution of 8.1%. Although there are office buildings with circular, trapezium and U-shape plan forms they have not been used widely. Circular and trapezium plan forms are equal percentage of 2.30% in the building stock. Composite built form of L shape is a combination of two linear plan forms or a linear and a square plan form. Moreover the amalgamation of three linear plan forms represents the U-shape. These composite plan forms in CMC region consists of 2.3% of U and 8.0% of L shape plan forms, respectively.

Table 2. Common plan form classification of the building stock in Colombo—linear and shallow plan forms are more common in the stock.

Basic Plan Forms		Composite Plan Forms	
	Linear plan form 62% Shallow 		L Shape plan form 8.1% 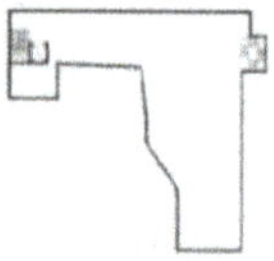
	Square plan form 23% Deep 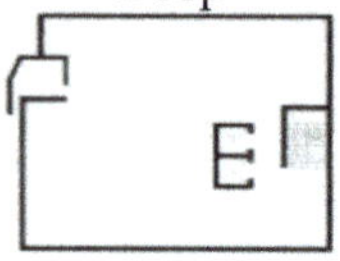		U-shape plan form 2.3% 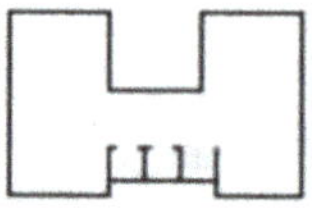
	Semi-circular plan form 2.3% 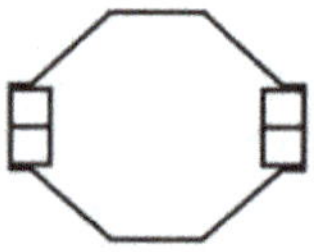		Trapezium plan form 2.3% 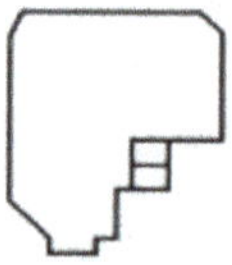

b. Building Facade Detailing and Orientation

Building facade detailing demonstrates a significant impact on the thermal behavior of a building determined by "glazed to wall" ratio and facade orientation [56]. Increased glazing proportions of the facades facing critical orientations of hot climates promotes higher gain in solar radiation and the overheated interiors increase the cooling load of air conditioned offices. On the other hand, perimeter areas benefit from the higher percentage of glazing in terms of daylight. However, the large amount of daylight, which enters a space through highly glazed areas, often reduces the quality of visual comfort due to glare problems. Therefore, this study identified a characterization of the building stock in this respect. Table 3 presents material compositions of front facades and percentage distribution of each facade type within the office building stock.

Table 3. Fenestration details of road facing façades and percentage distribution within the stock.

Glazed with Solid Facades		Aluminum Cladding Facades		Composite and other Facades	
	Nearly 17% within the stock		Nearly 26% within the stock		Nearly 57% within the stock Majority from this group
	17%		26%		57%

Facades are predominantly composed of glass and aluminum cladding. The percentage distribution of these materials in front facades is 17%, 26%, and 57%, respectively. Glazed facades are primarily curtain walls and composed of fixed and few operable panels with blinds to control direct solar radiation. These facades are orientated in different directions within the building stock. Table 4 represents the details of the facade orientations, their percentage distributions, and the relationship between the orientation and mean BEI. Office buildings in this stock represents four main orientations such as East–west (EW), North–south (NS), Northeast–Southwest (NE–SW), and Northwest–Southeast (NW–SE). Building forms have similar envelope properties of high mass concrete and brick. Floor to floor height varies between 3 and 3.5 m in all buildings.

Table 4. Prevalent orientations of front façade facing the road and mean building energy index (BEI).

Façade Orientation	Percentage Distribution	Mean BEI kWh/m^2/a
East–west	41.4%	211.6
North–south	13.8%	200.2
Northeast–Southwest	25.6%	200.1
Northwest–Southeast	19.2%	205.7

Table 4 presents the percentage distribution of office buildings for EW, NS, NE–SW, and NE–SW orientations which are 41%, 14%, 26%, and 19%, respectively. The major characteristic is the orientation of the front façade towards East or West direction along the major roads running from north to south. This has become a design challenge for architects but no efforts are seen in making these facades more defensive from direct solar gain from east and west in the inevitable situation of having to use this specific orientation. Thus the building stock is evident for less attention on climate responsive design strategies which has affected the end use energy demand. Buildings with EW-oriented front facades demonstrate the highest mean BEI of 212 kWh/m^2/a. These findings confirm the criticality of the building morphology in addition to inappropriate orientation and material usage irrespective to climatic forces of the locality.

c. Composition with Other Buildings

Buildings within an urban block can be either attached to other buildings or exist independently, detached. This office building stock consists of a mix of both compositions with 63% detached buildings and 37% attached buildings. Attached buildings have variations, such as attached from single and both facades. Varied compositions show marginal differences in annual BEI with 211 and 210 kWh/m^2/a for detached and attached buildings, respectively. Buildings with both sides attached to other buildings display better control of solar heat gain to office interiors. However, this means that the use of artificial lighting is increased due to daylight restrictions. Lack of difference between BEIs of the two types highlights the energy demand for lighting, and thus informs the importance of a comprehensive study to calculate the tenancy energy demand of this stock. However, lack of regulatory mechanisms to initiate sub metering of the national building stock is a limitation.

4.4. Statistical Analysis with BEI, WWR, Aspect Ratio, and Orientation

Multiple linear regression models were applied to identify any nexus between façade composition (window-to-wall ratio—WWR), plan form (aspect ratio), building orientation, and physical configuration to BEIs of the building sample population. Aspect ratio is the footprint in a ratio of length and width relative to the east–west or north–south. A change in the aspect ratio can vary the façade area subject to solar radiation [40]. WWR and aspect ratio define the glazing area of a façade and, therefore, can have a clear impact on cooling or heating [56]. The work attempted to model the relationship between two or more explanatory variables and a response variable by fitting a linear equation to observed data. The multiple linear regression equation is as follows.

$$y = B_0 + b_1X_1 + b_2X_2 + \ldots b_nX_n \quad (1)$$

where y is the predicted or expected value for the linear model with regression coefficients of dependent variable b_1 to b_n and y intercept b_0 when the values for the predictor variables are X_1 to X_n, which is the building energy index. X_1 through X_n are n distinct independent variables, which are the building morphological parameters. B_0 is the value of y when all of the independent variables (X_1 through X_n) are equal to zero, and b_1 through b_n are the estimated regression coefficients. Each regression coefficient represents the change in y relative to a unit change in the respective independent variable.

The regression analysis was carried out in three stages for 87 buildings. First stage was to analyze whether independent variables are correlating with each other to prevent multi colinearity in the regression models. The evaluation of independent variables was based on analyzing correlation coefficient and *p*-values. Second stage was to perform subset regression analysis to ascertain the relationship between dependent variable of BEI with independent variables of building morphological characteristics such as WWR, aspect ratio, and orientation. Subset regression analysis helped to evaluate all possible regression models. These models were revaluated based on R^2 value, R^2 adjusted value, and Mallow's Cp value. The highest R^2 value and R^2 adjusted value with lesser Mallow's Cp value was taken into account when determining the best fit models. The minimum R^2 value was taken as 50%, where 50% of the predicted value of BEI was explained by the model. Stepwise regression was conducted to evaluate the predictors in the selected models. R^2 value and *p*-value were considered when reviewing the best fit models. Polynomial regression analysis was preceded during the occasions when R^2 value was marginal.

p-values of the independent variables, namely WWR of all four facades, the aspect ratio of the plan, and building orientation with the BEI, remained less than 0.05 in 95% of confidence intervals in regression models with four major prevailing orientations of the sample building stock. The four orientations were North–South axis, East–West axis, Northwest–Southeast axis, and Northeast–Southwest axis. The analysis was carried out for all three groups based on the BEIs of the stock namely Lower Range, Average Range and Critical Range. Table 5 shows summative results of the analysis showing a nexus between BEIs, WWRs and orientations.

Plan forms with longer facades facing north and south directions maintain relatively lower BEIs compared to other orientations in all three groups of BEI ranges. Results show that façade orientations have greater impacts on the BEIs and thus plan forms on the east–west axis performs better. However, it is interesting to note that few buildings in this group claim BEIs as high as 250.31 kWh/m^2/a, falling in to the critical range.

With two orientations of plan forms on NS (north–south) and EW (east–west) axes, BEIs claim a diversity showing a significant change of impact from facades facing NS and EW orientations. This was a common behavior in all three ranges, i.e., lower, average, and critical. The change of impact due to other orientations was less significant. Multiple regression models developed to configure a nexus between aspect ratio and WWR of NS- and EW-oriented buildings in all three BEI ranges highlight a criticality of plan form with higher aspect ratios. This was evident with high BEI averages and they can be identified as shallow plan forms orientated on EW orientations. Despite these forms are ideal for tropics, an increase of the aspect ratio is seen as a strong independent variable contributing to higher BEIs, the critical range (Table 6). Outcome demands a field investigation of these buildings.

Table 5. Summative results of the analysis showing a nexus between BEIs, window-to-wall ratios (WWRs), and orientations.

BEI Range	**Orientation**															
	N-S axis orientation				E-W axis oriented				NW-SE axis oriented				NE-SW axis oriented			
Three main categories in respect to BEI	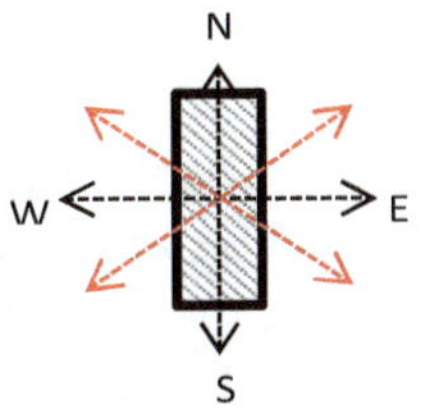				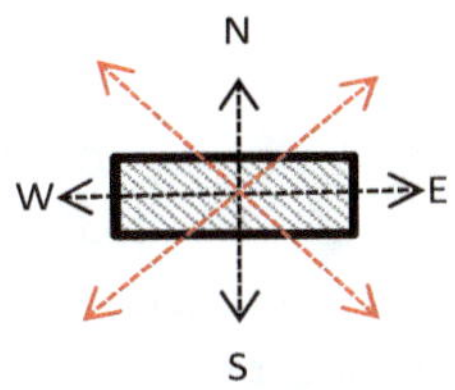				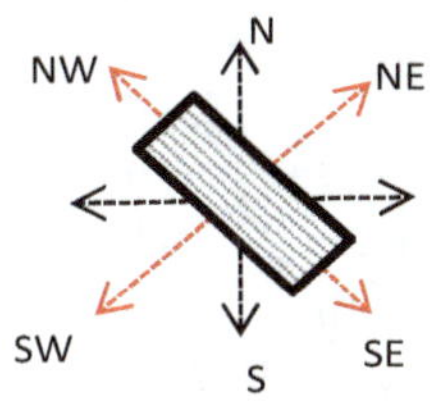				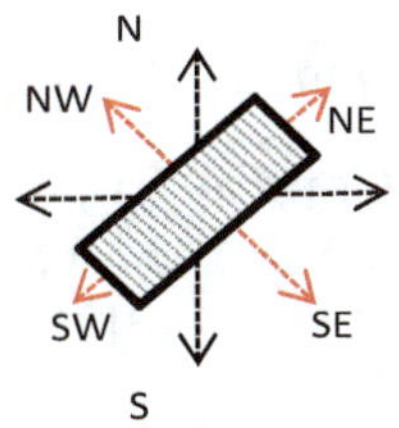			
Lower Range	**Average BEI 133.45 kWh/m²/a**				**Average BEI 122.71 kWh/m²/a**				**Average BEI 130.66 kWh/m²/a**				**Average BEI 125.79 kWh/m²/a**			
	WWR				**WWR**				**WWR**				**WWR**			
	E	W	N	S	E	W	N	S	E	W	N	S	E	W	N	S
	0.36	0.49	0.11	0.08	0.39	0.18	0.32	0.53	0.07	0.13	0.68	0.53	0.33	0.17	0.41	0.20
Total = 30	**Number of units 09**				**Number of units 05**				**Number of units10**				**Number of units 06**			
Average Range	**Average BEI 204.29 kWh/m²/a**				**Average BEI 170.94 kWh/m²/a**				**Average BEI 186.91kWh/m²/a**				**Average BEI 187.64 kWh/m²/a**			
	WWR				**WWR**				**WWR**				**WR**			
	E	W	N	S	E	W	N	S	E	W	N	S	E	W	N	S
	0.76	0.63	0.13	0.02	0.42	0.47	0.20	0.34	0.54	0.54	0.27	0.31	0.38	0.39	0.48	0.40
Total = 46	**Number of units 17**				**Number of units 09**				**Number of units 12**				**Number of units 08**			
Critical Range	**Average BEI 272.34 kWh/m²/a**				**Average BEI 202.31 kWh/m²/a**				**Average BEI 262.14 kWh/m²/a**				**Average BEI 262.33 kWh/m²/a**			
	WWR				**WWR**				**WWR**				**WWR**			
	E	W	N	S	E	W	N	S	E	W	N	S	E	W	N	S
	0.81	0.63	0.14	0.06	0.66	0.68	0.91	0.64	0.65	0.65	0.70	0.55	0.70	0.45	0.40	0.50
Total = 11	**Number of units 05**				**Number of units 03**				**Number of units 01**				**Number of units 02**			

Table 6. Summative results of the nexus between aspect ratios, BEIs, WWRs, and orientations of buildings on the North–South and East–West axes.

Orientation and Plan Form	**WWR and Aspect Ratio in Three BEI ranges in East and West Orientations**											
East–West axis 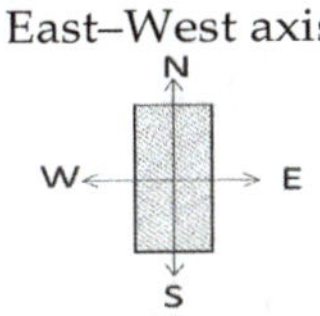	Lower Range 100–150 kWh/m²/a				Average Range 150–250 kWh/m²/a				Critical Range Above 250 kWh/m²/a			
	WWR Glazing exposure				WWR Glazing exposure				WWR Glazing exposure			
	E	W	N	S	E	W	N	S	E	W	N	S
	0.36	0.49	0.11	0.08	0.76	0.63	0.13	0.02	0.81	0.63	0.14	0.06
Street layout 	Aspect Ratio along NS axis (Length/width) 0.70				Aspect Ratio along NS (Length/width) 0.30				Aspect Ratio along NS (Length/width) 0.26			
	Average no of floors 5				Average no of floors 6				Average no of floors 6			
	Average BEI 133.45 kWh/m²/a				Average BEI 204.29 kWh/m²/a				Average BEI 272.34 kWh/m²/a			
North–South axis	**WWR and Aspect Ratio in three BEI ranges in North and South orientations**											
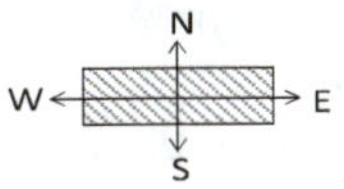	WWR Glazing exposure				WWR Glazing exposure				WWR Glazing exposure			
	E	W	N	S	E	W	N	S	E	W	N	S
	0.03	0.18	0.32	0.53	0.42	0.47	0.20	0.34	0.76	0.68	0.91	0.64
Street layout 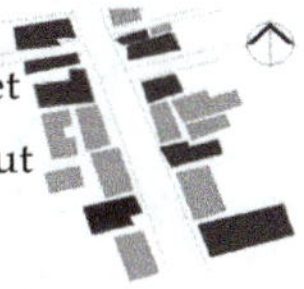	Aspect Ratio along EW axis (Length/width) 1.49				Aspect Ratio along EW axis (Length/width) 3.22				Aspect Ratio along EW axis (Length/width) 4.4			
	Average no of floors 5				Average no of floors 6				Average no of floors 6			
	Average BEI 122.71 kWh/m²/a				Average BEI 170.94 kWh/m²/a				Average BEI 202.31 kWh/m²/a			

5. Phase II and Phase III of the Investigation

Phase II was more focused on the shallow plan forms. As discussed in the Phase I, buildings with higher aspect ratios on the East–West direction have a shallow plan form with two extremes of BEI values, i.e., lower and critical ranges.

5.1. Sampling of the Building Stock for Phase II

Conducting a thermal performance investigation on two sample buildings with shallow plan forms was aimed at in Phase II. The classification of building morphological characteristics showed that 63% of building stock has shallow plan forms, compared to 26% of deep and 12% of composite forms. Of the shallow forms, 17 buildings are seen orientated on the EW axis in different locations. Two buildings were then sampled from these 17 buildings that showed different levels of BEIs ranging from 110 to 320 kWh/m^2/a in the outcome of statistical analysis in Phase I.

Buildings attached or close to other structures were eliminated from the beginning of the study. Only free standing buildings were considered so that impact of urban climate on all facades of any particular building could be assessed.

The two sample office buildings—"A" and "B"—are composed of similar morphological character (WWR and aspect ratio), orientation, operational profile, and occupant and equipment density but extensively different from the BEI point of view (Table 7). Both buildings are occupied from 8 am to 6 pm on weekdays and are located in similar urban contexts just a kilometre apart.

Table 7. Plan form and physical characteristics of two investigated buildings in Phase II and I.

Building "B"—Shallow Plan Form	**Building "A"—Shallow Plan Form**
 Left- side (north) and right-front elevations	 Left- Side (North)and right-south elevations
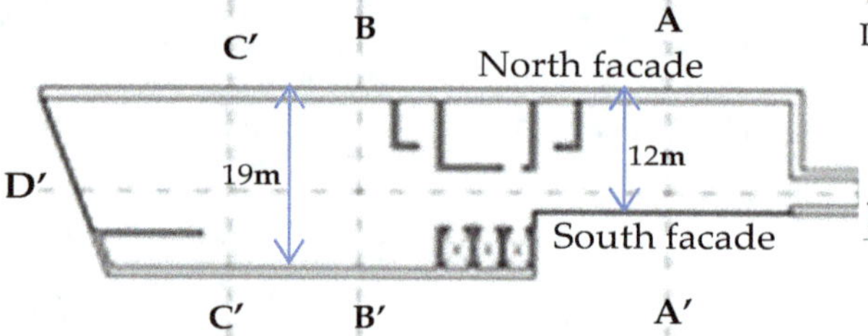 **Shallow Plan form** – size 12 m × 50 m (19 m on west)	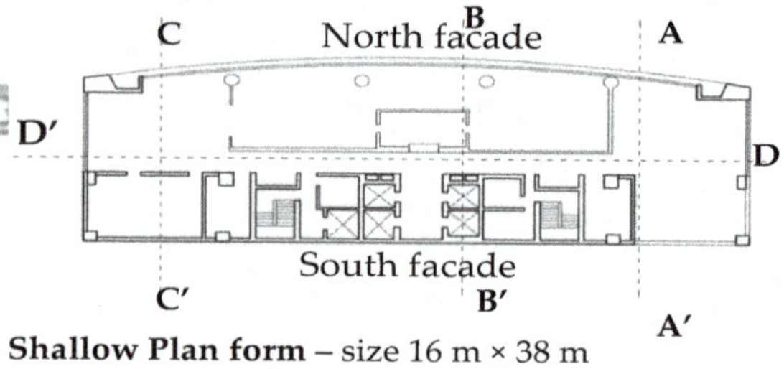 **Shallow Plan form** – size 16 m × 38 m
Common thermal performance characteristics of both buildings are listed below Orientation: longer axis along east–west facing north and south with nearly 60% glass on north Set point temperature: 24 °C degrees during office hours from 08.00 am to 18.00 pm Occupant population 6 persons/20 m^2 approximately—occupancy increases with visitors in the morning Wall and slab construction—cement and lime plastered brick and concrete Average U-value of external envelope: 0.22 W/m^2 K Front façade is facing east with glazing properties of 5.0 w/m^2/k Solid to void ratio is nearly 40:60—fully packed floor plates with no voids between floors Occupied hours: 8 am to 6 pm on weekdays (weekends nonfunctional and A/C is off) Lifts and service spaces are on the south orientation on floors	

The objective of the focused field investigation was to assess the distribution pattern of heat stress on indoor environment depending on the plan depth. Building "A" has BEI as high as 340 kWh/m^2/a whereas the Building "B" has a moderate level of 120 kW/hm^2/a, which is close to acceptable level. Front narrower façade with the main entrance is positioned facing east orientation in both cases. Table 7 explains similarity of the operational and physical characteristics of the two buildings. The only difference between the two building forms is the floor area where a typical floor plate of Building "B" is relatively larger than the same of Building "A". However, occupant and equipment density are comparatively similar in both cases.

Multizone indoor air temperature reading in 3 m × 3 m zones across the plan depth (and along the building length as explained in the method) were taken using HOBO data loggers during three working days in the month of April 2018 (23rd–27th of April 2018) at 10 min intervals and then averaged to hourly values. The measurement rationale was to ascertain a number of comparisons that are as follows.

- Dynamics of air temperature distribution in peripheral and central zones on a typical floor plate to assess the collective heat stress effect of façades and plan depth.
- Dynamics of air temperature deviation from the set point temperature to assess the thermal load on the air conditioning system.

The study acknowledges that any dynamics in the indoor air temperature demonstrates an effect of environmental loads on air conditioning and the indoor thermal behaviour. Microclimatic data just outside the building were recorded onsite. The central air conditioning system of both cases is similar in specifications (15 tons per 1 AHU). However, Building "A" has a higher number of diffusers (1 for each 13 m^2) than in Building "B" (1 for each 36 m^2) per floor. Readings were taken in three typical floors above 6th level in both buildings. Temperature fluctuations on the day that the study was conducted when the building was in non-functional mode (weekends) demonstrates the effect of environmental load on indoor thermal behaviour.

5.2. Phase III

Phase III was aimed at repeating the investigation performed in Phase II but continuously over 20 days from 26th July to 16th August 2018. Measurements of each 3 m × 3 m multi-zones recorded at 10-mi intervals were later averaged to hourly values, which in turn averaged to a single value for the entire floor. Objective of Phase III was to ascertain the following.

- Hourly pattern of dynamics of average indoor air as a single value in a typical floor against the coordinating set point temperatures.
- Quantify the extremes of heat stress on air conditioned environments and identify contributing factors.

6. Results and Discussion

Set-point temperature of all office and useable floors of both buildings is 24 °C degrees. Despite this, indoor air temperature dynamics in varied levels were seen during the daytime in both buildings.

Results show that an increase of indoor air temperature up to 31 to 32 degrees C across all 3 m × 3 m zones along sections—A-A, B-B, and C-C—in Building "B" (see Table 7) while temperature readings in the internal zone close to west façade and east facades along D-D remained at 34 and 31, respectively, at midday (Figure 5) when the sun was just above the building and no direct radiation could penetrate through the sides. Solar axis to buildings in Colombo between the 23rd and 27th April is normally over the head, therefore there is no way of entering solar access from sides. However, a middle floor was investigated in this study so that effect of heat from vehicular traffic was considered minimal. This thermal behaviour explicitly suggests the presence of heat stress on the air conditioning system from the façade. Office equipment inside the three floors remained switched off from the morning to assist the research, so that indoor heat generation was limited to a minimum. As mentioned in a previous section, the BEI of this building was approximately 120 kWh/m^2/a, but heat stress of 10.7 °C indoor air temperature above the set-point is an issue to be addressed.

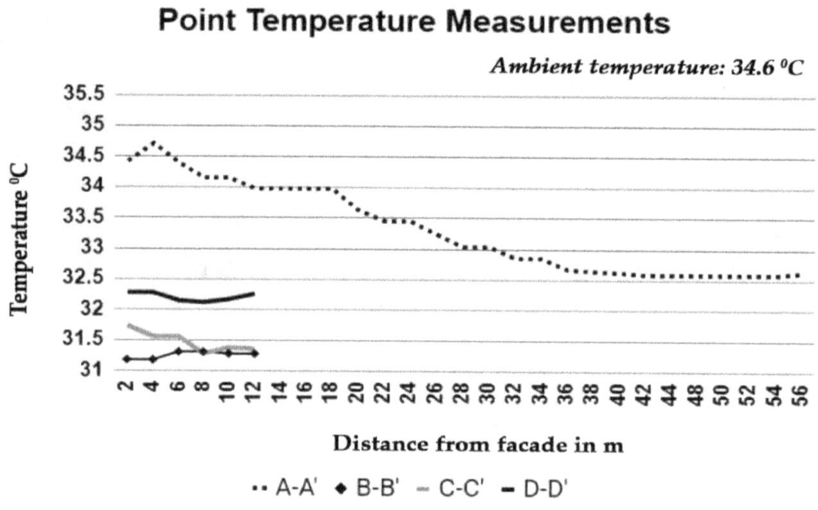

Figure 5. Multizone air temperature behavior in this air conditioned Building "B"- move between 31.5 and 34.7 °C despite set point temperature at 24 °C; BEI is around 120 k/Whm^2/a, but high risk of heat gain is visible.

The set-point temperature of both buildings was at 24 °C. Despite similarities in set point temperatures, Building "A" was assessed with lesser elevation of indoor air temperature ranging between 24.4 and 25.1 °C (Figure 6). It has a highly dense AC diffuser system than Building "B" and results show that it is capable of maintaining indoor air temperature closer to the set point temperature, but at a much higher energy cost at a BEI of 320 kWh/m^2/a. Results suggest the criticality of shallow plan forms on end user energy demand of the building stock.

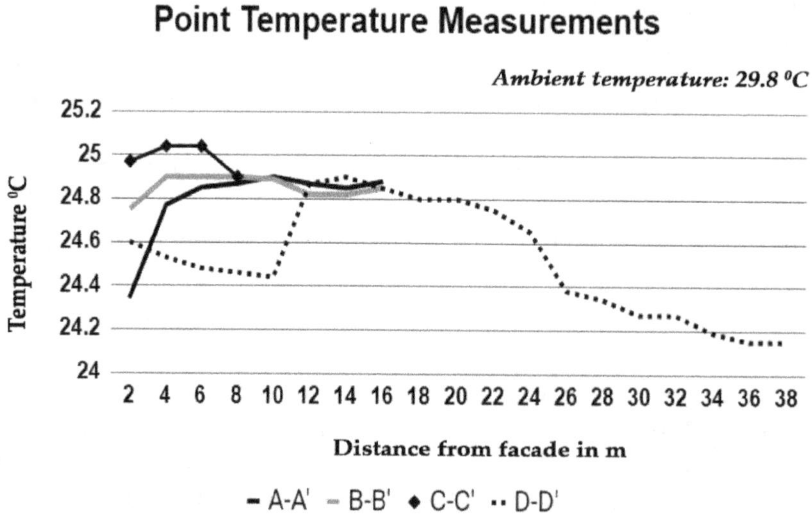

Figure 6. Multizone air temperature behavior of this air conditioned Building "A" moves just 1.1 °C above set point temperature of 24 °C but at a very high energy cost of nearly 340 kWh/m^2/a.

Shallow plan depths of Building "B" and "A" are 12 m and 16 m, respectively. Length of the floor plates of Buildings "B' and A" are 50 m and 38 m, respectively. Multizone air temperature distribution across the depths and lengths of floor plates are shown in the two graphs above (Figures 5 and 6). Considerable dynamics of air temperatures well above set point temperature in all zones across these depths and lengths demonstrate increased demand on cooling energy due to high levels of external gains through facades.

Performance behavior of Buildings "B" and "A" in Phase III are shown in Figures 7 and 8 respectively. During this period maximum ambient moved around 28 °C, an ambient climatic situation which is not extreme and also lower than the ambient levels during Phase II, (which fluctuated between 31 and 34 degree C). Air just outside the building, measured at the 8th floor level revealed a heat built up situation just outside the façade, suggesting solar gain from outside.

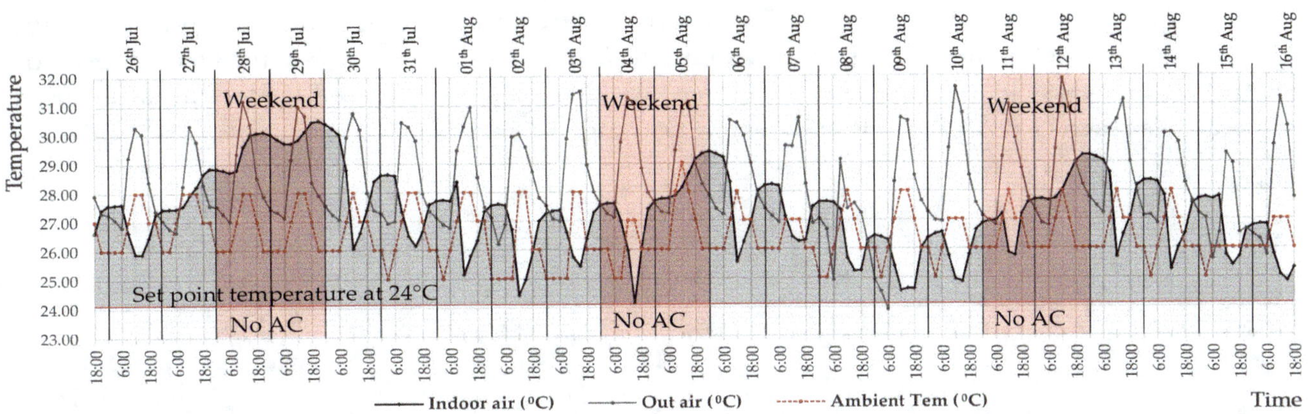

Figure 7. Building "B": Hourly severe heat stress behavior in terms of air temperature on the 8th floor over a period of 20 days in July–August 2018; indoor air moved well above set point temperature.

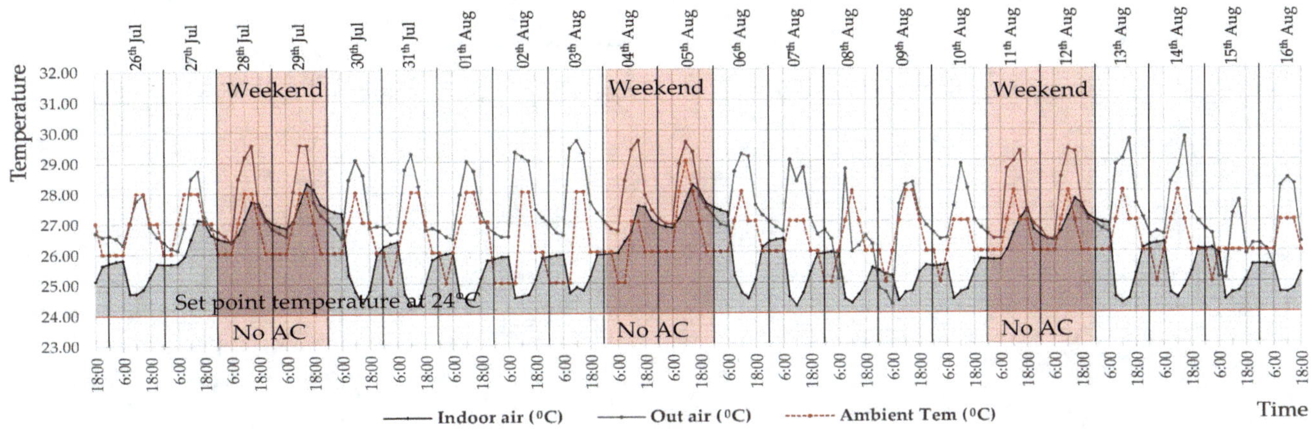

Figure 8. Building "A": Hourly moderate heat stress behavior in terms of air temperature in the 8th floor over a period of 20 days in July–August 2018. Indoor air moved closer to set point but at a much higher energy cost of BEI.

Cooling system of the building has been able to control the indoor air to keep its temperature at approximately 26–26.5 °C during midday, but results show an overheating situation in nights and during weekends. Overheating situation in the night indicates a level of heat stress from the heat released from the envelope. Overheated situations in the weekends show a greater level of heat stress from the facades.

Thermal behavior of Building "A" followed a similar pattern to the Building "B" but showed relatively a smaller dynamics from the set point temperature at a much higher energy cost at a BEI of 320 kWh/m^2/a (see Figure 8), similar to Phase II outcome. Situations in weekends showed a severe heat stress due to environmental gain on the indoor air. The results indicate the following.

- Overheated microclimates, well above the ambient, outside the buildings at high elevations
- Severe heat stress and indoor overheating during weekends (nonfunctional hours)

- Higher indoor air temperature levels than the set point temperature through daytime AND elevation of indoor air above ambient in early morning hours

Wall surface temperature behavior was quantified during Phase III. Hourly values of internal wall surface temperatures of four exterior facades were averaged to a single wall surface value in hourly terms and then compared with corresponding indoor air. Results show a close interaction of indoor air with internal wall surface temperatures reflecting a heat stress impact from facades (Figures 9 and 10). Despite air conditioning, internal wall surfaces reach higher values up to 28.5–29.5 °C until noon indicating heat gain from outside. Internal wall surface starts to drop after 3 pm when direct exposure of facades to outside radiation starts diminishing indicating a release of stored heat to indoors from walls. Heat sink capacity of thermal mass in the envelope was not observed. Indoor air and surface temperatures reach close to each other by early morning at ~6 am.

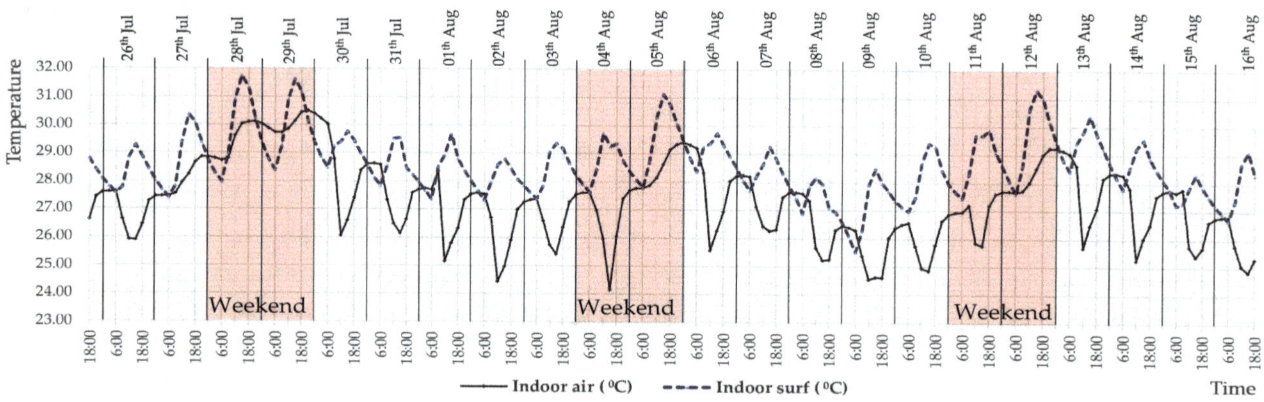

Figure 9. Wall surface and indoor air temperature behavior in Building "B".

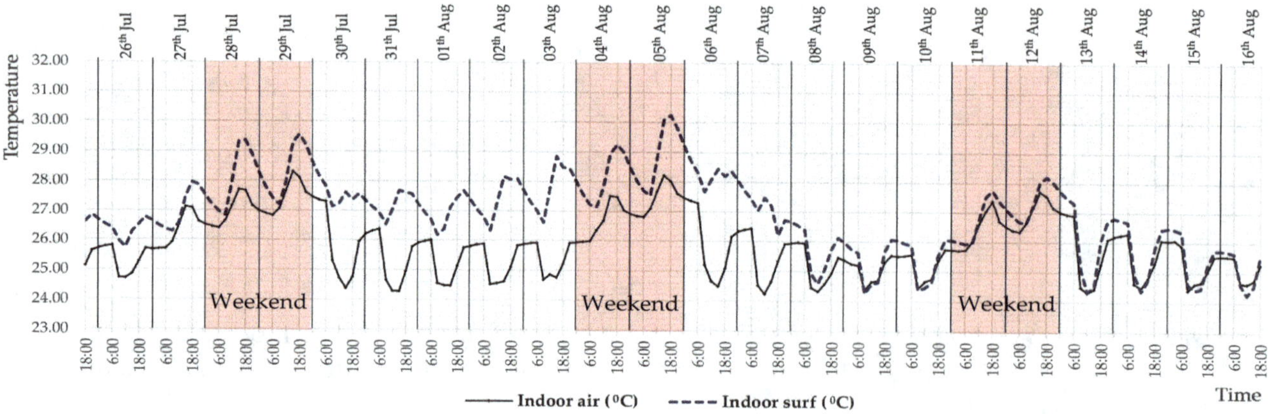

Figure 10. Wall surface and indoor air temperature behavior in Building "A".

7. Conclusions—Generalized Findings

Results suggest a problem associated with shallow plan form buildings. Although the work is mainly focused on two specific buildings, they represent most of the thermophysical characteristics of the larger building population in Colombo. Inward heat transfer across the plan depth up to 16 m in Building "A" and 12 m in Building "B" from the façade is a critical problem that may be addressed through appropriate façade design. Findings suggest that presence of heat stress in terms of higher air temperature in air conditioned indoors along the length of the plan form is visible.

Elevation of indoor air temperature in air conditioned environments by 10.5 °C above the set point temperature is considered critical in respect to energy demand. The two buildings investigated fall within the following two extremes of thermal performance scenarios:

1. Air conditioned buildings with shallow plan forms may be able to maintain indoor air temperatures close to set point temperatures at 24 °C but at a very high energy cost.

2. Air conditioned buildings with shallow plan forms can maintain a building energy index of ~120 kWh/m^2/a, but maintaining indoor air temperature close to set point temperature becomes problematic and sometime an elevation of indoor air temperature by even 10.5 °C above the set point temperature could be visible.

More research is underway to further strengthen these findings for a wider section of buildings in warm humid climates. The survey on plan form characteristics and orientation of a multi-level office building population in Colombo showed that nearly 80 per cent of buildings do not have appropriate orientation for solar defense. Their main building facades are facing either east or west orientations. A sound repositioning effort that focuses on minimizing environmental heat stress on building facades and thus exploring energy efficiency in multi-level commercial office building sector for the life of both exiting stock and new buildings is critical. This provides a level of opportunity for the building stakeholders to improve their asset values for the life of buildings and contribute to emission reduction targets.

While it is important to recognize that the results from the field investigations are not fully generalized due to limitations, the outcome on the heat stress behavior on conditioned office environments serves as a detailed comprehension for energy sustainability of office buildings. The better that architects and engineers understand the indoor thermal environment inside conditioned office buildings, the better design research will be able to link technological characteristics of envelope and plan form with energy consumption levels from heat stress on envelope towards a deeper comprehension of how best to reduce environmental heat stress on facades in the face of warming climates.

Author Contributions: U.R., Conceptualization, Methodology, Formal Analysis, Investigation, Draft Preparation, Final Writing, Project administration, U.R.

Acknowledgments: The author acknowledges Grant No 109-13 of National Research Council of Sri Lanka for funding of equipment, as well as Indrika Rajapaksha and Waruni Jayasinghe for assisting the project administration; Ruksala Ishani, and Supun Rodrigo who have provided assistance in data collection.

References

1. IPCC. Contribution of Working Groups I, II and III to the Fifth Assessment Report of the Intergovernmental Panel on Climate Change. In *Climate Change 2014: Synthesis Report*; Core Writing Team, Pachauri, R.K., Meyer, L.A., Eds.; IPCC: Geneva, Switzerland, 2014.
2. Seo, S.; Mendelsohn, R.; Munasinghe, M. Climate change and agriculture in Sri Lanka: A Ricardian valuation. *Environ. Dev. Econ.* **2005**, *10*, 581–596. [CrossRef]
3. Garnaut., R. *The Garnaut Climate Change Review*; Final Report; The Cambridge University Press: Cambridge, UK, 2008. Available online: http://www.garnautreview.org.au (accessed on 12 November 2017).
4. Coley, D.; Kershaw, T. Changes in internal temperatures within the built environment as a response to a climate change. In *Building and Environment*; Elsevier Science: Amsterdam, The Netherlands, 2010; Volume 45, pp. 89–93.
5. Torcellini, P.; Pless, S.; Deru, M.; Crawley, D. *Zero Energy Buildings: A Critical Look at the Definition*; Conference Paper NREL/CP-550-39833; U.S. Department of Energy: Washington, DC, USA, June 2006.
6. Guan, L. Preparation of future weather data to study the impact of climate change on buildings. *Build. Environ.* **2009**, *44*, 793–800. [CrossRef]

7. Howden, S.M.; Crimp, S. Effect of Climate and Climate Change on Electricity Demand in Australia. In Proceedings of the Integrating Models for Natural Resources Management Across Discipline, Issue and Scales, Canberra, Australia, 10–13 December 2001; Ghassemi, F., Whetton, P., Little, R., Littleboy, M., Eds.; MSSANZ Inc.: Canberra, ACT, Australia, 2001; pp. 655–660.
8. Lee, W.L.; Yik, F.W.H. Regulatory and voluntary approaches for enhancing building energy efficiency. *Prog. Energy Combust. Sci.* **2004**, *30*, 377–499. [CrossRef]
9. GBCA (Green Building Council of Australia. The Value of Green Star—A Dacade of Environmental Benefits, May 2013. Available online: http://www.gbca.org.au (accessed on 22 January 2019).
10. Hyde, R.; Rajapaksha, I.; Groenhout, N.; Barram, F.; Rajapaksha, U.; Shahriar, A.N.M.; Candido, C. Towards a methodology for retrofitting commercial buildings using bioclimatic principles. In Proceedings of the 43rd ANZAScA Conference, The University of Tasmania, Hobart, Australia, 25–27 November 2009. Available online: http://trove.nla.gov.au/version/170149721 (accessed on 20 January 2019).
11. De Wilde, P.; Coley, D. The implications of a changing climate for buildings. *Build. Environ.* **2012**, *55*, 1–7. [CrossRef]
12. Levin, M.D.; Price, L.; Martin, N. Mitigation options for carbon dioxide emissions from buildings. *Energy Policy* **1996**, *24*, 937–949. [CrossRef]
13. Burton, S.; Sala, M. *Energy Efficient Office Refurbishment*; Earthscan: London, UK, 2001.
14. Jenerette, G.D.; Harlan, S.L.; Brazel, A.; Jones, N.L.; Stefanov, W.L. Regional relationships between surface temperature, vegetation, and human settlement in a rapidly urbanizing ecosystem. *Landsc. Ecol.* **2007**, *22*, 353–365. [CrossRef]
15. McEvoy, D. Climate change and cities. In *Built Environment*; Alexandrine Press: Marcham, UK, 2007; Volume 33, No. 01; pp. 5–9.
16. Manjula, R.; Estoque, R.C.; Murayama, Y. An Urban Heat Island Study of the Colombo Metropolitan Area, Sri Lanka, Based on Landsat Data (1997–2017). *ISPRS Int. J. Geo-Inf.* **2017**, *6*, 189.
17. SLSEA. Energy Consumption in Buildings, Website of Sri Lanka Sustainable Energy Authority. 2017. Available online: www.energy.gov.lk (accessed on 27 September 2018).
18. Wijayatunga, P.D.C.; Fernando, W.J.L.S.; Shrestha, R.M. Greenhouse gas emission mitigation in the Sri Lanka power sector supply side and demand side options. *Energy Convers. Manag.* **2003**, *44*, 3247–3265. [CrossRef]
19. De Dear, R. The Adaptive Model of Thermal Comfort: Macquarie University's ASHRAE RP-884 Project. 2004. Available online: http://aws.mq.edu.au/rp-884/ashrae_rp884_home.html (accessed on 28 August 2018).
20. Kwok Alison, G.; Rajkovich Nicholas, B. Addressing climate change in comfort standards. *Build. Environ.* **2010**, *45*, 18–22. [CrossRef]
21. LaSalle. Commercial Buildings Going Green. 2004. Available online: www.joneslasalle (accessed on 20 November 2017).
22. Olgyay, V. *Design with Climate: Bioclimatic Approach to Architectural Regionalism*; Princeton University Press: Princeton, NJ, USA, 1963.
23. Hyde, R. *Climate Responsive Design: A Study of Buildings in Moderate and Hot Humid Climates*; E and FN Spon: London, UK, 2000.
24. Hawkes, D. The theoretical basis of comfort in the selective control of environment. *Energy Environ.* **1982**, *5*, 127–134. [CrossRef]
25. Guan, L. Implication of global warming on Air- Conditioned office buildings in Australia. *Build Res. Inf.* **2009**, *37*, 43–54. [CrossRef]
26. Dascalaki, E.G.; Droutsa, K.; Gaglia, A.G.; Kontoyiannidis, S.; Balaras, C.A. Data collection and analysis of the building stock and its energy consumption. *Energy Build.* **2010**, *42*, 1231–1237. [CrossRef]
27. Lomas, K.J. Architectural design of an advanced naturally ventilated building form. *Energy Build.* **2007**, *39*, 166–181. [CrossRef]
28. Szokolay, S.V. *An Introduction to Architectural Science: The Basis of Sustainable Design*, 3rd ed.; Routledge, Taylor and Francis Group: New York, NY, USA, 2014.
29. Peterkin, N. Rewards for passive solar design in the Building Code of Australia. *Renew. Energy* **2009**, *34*, 440–443. [CrossRef]
30. Kabre, C. Trends in solar control in contemporary buildings. In Proceedings of the Thermal Performance and Comfort—ANzAScA 97 International Conference Session 2, The University of Queensland, Australia, 29 September–3 Octomber 1997; pp. 19–26.

31. Haase, M.; da Silva, F.M.; Amato, A. Simulation of ventilated facades in hot and humid climates. *Energy Build.* **2009**, *41*, 361–373. [CrossRef]
32. Eskin, N.; Turkmen, H. Analysis of annual heating and cooling energy requirements for office buildings in different climates in Turkey. *Energy Build.* **2008**, *40*, 763–773. [CrossRef]
33. Napier, J. Climate Based Façade Design for Business Buildings with Examples from Central London. *Buildings* **2015**, *5*, 16–38. [CrossRef]
34. Trubiano, F. Performance Based Envelopes: A Theory of Spatialized Skins and the Emergence of the Integrated Design Professional. *Buildings* **2013**, *3*, 689–712. [CrossRef]
35. Parsons, K. Heat stress standards ISO 7243 and its global application. *Ind. Health* **2006**, *44*, 368–379. [CrossRef]
36. Chowdhury, S.; Hamada, Y.; Ahmed, K.S. Prediction and comparison of monthly indoor heat stress (WBGT and PHS) for RMG production spaces in Dhaka, Bangladesh. *Sustain. Cities Soc.* **2017**, *29*, 41–57. [CrossRef]
37. Lam, J.C. Energy analysis of commercial buildings in subtropical climates. *Build. Environ.* **2000**, *35*, 19–26. [CrossRef]
38. Baker, N.V. *Passive and Low Energy Building Design for Tropical Island Climates*; Commonwealth Secretariate: London, UK, 1987.
39. Feng, G.; Chi, D.; Xu, X.; Dou, B.; Sun, Y.; Fu, Y. Study on the Influence of Window-wall Ratio on the Energy Consumption of Nearly Zero Energy Buildings. *Procedia Eng.* **2017**, *205*, 730–737. Available online: www.sciencedirect.com (accessed on 18 November 2018). [CrossRef]
40. McKeen, P.; Fung, A.S. The Effect of Building Aspect Ratio on Energy Efficiency: A Case Study for Multi-Unit Residential Buildings in Canada Philip. *Buildings* **2014**, *4*, 336–354. [CrossRef]
41. Rajapaksha, I.; Rajapaksha, U. Criticality of Building Morphology on End Use Energy Demand: Evidence based assessment of urban office stock in Colombo Metropolitan Region. In Proceedings of the Symposium of Sri Lanka Sustainable Energy Authority, Colombo, Sri Lanka, 6–7 December 2017.
42. Lee, W.V.; Steemers, K. Exposure duration in overheating assessment: A retrofit modeling study. *Build. Res. Inf.* **2017**, *45*, 6–12.
43. Jayatilake, K.; Rajapaksha, U. Interactive architecture and contextual adaptability: Issues of energy sustainability in contemporary tall office buildings in Colombo. In *Building the Future—Sustainable and Resilient Built Environments, Proceedings of 9th FARU International Research Conference, Colombo, Sri Lanka, 9–10 September 2016*; Rajapaksha, U., Ed.; University of Moratuwa: Moratuwa, Sri Lanka, 2016; pp. 354–366.
44. Yau, Y.H.; Hasbi, S. A review of climate change impacts on commercial buildings and their technical services in the tropics. *Renew. Sustain. Energy Rev.* **2013**, *18*, 430–441. [CrossRef]
45. Goulding, J.R.; Lewis, J.O.; Steemers, T.C. (Eds.) *Energy Conscious Design*; BT. Batsford for the Commission of the European Communities: London, UK, 1992.
46. UN DESA. *2018 Revision of World Urbanization Prospects*; United Nations—Department of Economic and Social Affairs, Multimedia Library Publications, 16 May 2018; Population Division of UN. Available online: www.un.org (accessed on 20 January 2018).
47. UNFPA. *UNFPA Report on 'State of the World's Population 2007*; UNFPA: New York, NY, USA, 2007.
48. Ukkwatta, N.L.; Dayawansa, N.D.K. Urban Heat Islands and the Energy Deamn: An Analysis for Colombo City of Sri Lanka Using Thermal Remote sensing data. *Soc. Soc. Manag.* **2012**, *1*, 124–131.
49. Mortimer, N.D.; Ashley, A.; Elsayed, M.; Kelly, M.D.; Rix, J.H.R. Developing a database of energy use in the UK non-domestic building stock. *Energy Policy* **1999**, *27*, 451–468. [CrossRef]
50. Choudhary, R.; Tiran, W. Influence of district features on energy consumption in non-domestic buildings. *Build. Res. Inf.* **2014**, *42*, 32–46. [CrossRef]
51. Depecker, P.; Menezo, C.; Virgone, J.; Lepers, S. Design of building shapes and energetic consumption. *Build. Environ.* **2001**, *36*, 627–635. [CrossRef]
52. Catalina, T.; Virgone, J.; Iordach, V. Study on the impact of building form on the energy consumption. In Proceedings of the Building Simulation, 12th Conference of International Building Performance Simulation Association, Sydney, Australia, 14–16 November 2011; pp. 1726–1729.
53. Tsikaloudaki, K.; Laskos, K.; Bikas, D. On the establishment of climatic zones in Europe with reagrd to the energy performance of buildings. *Energies* **2012**, *5*, 32–44. [CrossRef]
54. Wijayatunga, P.; Attalage, R. Analysis of rural households energy supplies in Sri Lanka: Energy efficiency, fuel switching and barriers to expansion. *Energy Convers. Manag.* **2003**, *44*, 1123–1130. [CrossRef]

55. Ratnaweera, C.; Hestnes, A.G. Enhanced cooling in typical Srilankan dwellings. *Energy Build.* **1996**, *23*, 183–190. [CrossRef]
56. ECG19. Energy Consumption Guide: Energy Use in Offices. 2003. Available online: http:www.cibse.org (accessed on 9 September 2018).

11

Photocatalytic Functionalized Aggregate: Enhanced Concrete Performance in Environmental Remediation

Amer Hakki [1,2,*], Lu Yang [1,3], Fazhou Wang [3], Ammar Elhoweris [2], Yousef Alhorr [2] and Donald E. Macphee [1,*]

1 Department of Chemistry, University of Aberdeen, Meston Building, Meston Walk, Aberdeen AB24 3UE, UK; yanglu@whut.edu.cn
2 Gulf Organisation of Research and Development, Qatar Science and Technology Park, Doha 210162, Qatar; a.elhoweris@gord.qa (A.E.); alhorr@gord.qa (Y.A.)
3 State Key Laboratory of Silicate Materials for Architectures, Wuhan University of Technology, 122# Luoshi Road, Wuhan 430070, China; fzhwang@whut.edu.cn
* Correspondence: a.hakki@gord.qa (A.H.); d.e.macphee@abdn.ac.uk (D.E.M.)

Abstract: Engineering of effective photocatalytically active structures is of great importance as it introduces a solution for some existing air pollution problems. This can be practically achieved through the bonding of particulate photocatalysts to the surface of construction materials, such as aggregates, with a suitable stable binding agent. However, the accessibility of the photocatalytically active materials to both the air pollutants and sunlight is an essential issue which must be carefully considered when engineering such structures. Herein, different amounts of commercial TiO_2 were supported on the surface of quartz sand, as an example of aggregates, with a layer of silica gel acting as a binder between the photocatalyst and the support. The thus prepared photocatalytically active aggregates were then supported on the surface of mortars to measure their performance for NOx removal. The obtained materials were characterized by electron microscopy (SEM and TEM), Fourier Transform Infrared Spectroscopy (FTIR), X-ray Diffraction (XRD), and UV-vis Absorption Spectroscopy. Very good coverage of the support's surface with the photocatalyst was successfully achieved as the electron microscopic images showed. FTIR spectroscopy confirmed the chemical bonding, i.e., interfacial Ti–O–Si bonds, between the photocatalyst and the silica layer. The photocatalytic activities of the obtained composites were tested for photocatalytic removal of nitrogen oxides, according to the ISO standard method (ISO 22197-1). The obtained aggregate-exposed mortars have shown up to ca. four times higher photocatalytic performance towards NO removal compared to the sample in which the photocatalyst is mixed with cement, however, the nitrate selectivity can be affected by Ti–O–Si bonding.

Keywords: environmental remediation; air pollution; photocatalytic construction materials; nitric oxides; functionalized aggregate

---

1. Introduction

Air pollution is a serious problem which directly affects everyone who lives in populated cities around the world [1]. Incontrovertibly, fossil fuel combustion for energy production and in the transportation sector, together with metallurgical industries, i.e., cementitious and construction industries, are the major sources of most anthropogenically produced pollutants [2]. Common air pollutants that pose risks to human health include nitrogen oxides (NOx; mainly NO and NO_2), carbon monoxide (CO), sulfur oxides (SOx), ozone (O_3), volatile organic compounds (VOCs), and particulate matter (PM) [3]. Among them, NOx gases have attracted a lot of concern since they cause or worsen serious health hazards for humans, including diseases such as emphysema and bronchitis, and aggravate existing heart disorders. Moreover, NOx gases are responsible for acid rains and smog and contribute to greenhouse warming [4].

Under solar irradiation, some photocatalysts such as TiO_2 have the potential to reduce ambient concentrations of NOx from areas in which the concentrations of these pollutants exceed national exposure level guidelines [5]. The light-induced oxidation reactions occur on the surface of irradiated TiO_2 in the presence of molecular oxygen and convert the toxic NO gas into nitrate through subsequent oxidation steps, with the intermediate formation of nitrogen dioxide: $NO \rightarrow NO_2 \rightarrow NO_3^-$ [6–9]. The thus formed nitrate (NO_3^-) on the TiO_2 surface can be rinsed and the catalyst can be recovered for further photocatalytic cycles.

Because of their high surface area and their direct contact with urban atmospheres, construction materials are excellent carriers to which photocatalytic materials can be applied for air purification. Recently, photocatalytically active surfaces have established an important commercial as well as technological position in modern construction technologies [10]. Some trials on the application of TiO_2 photocatalysts in several concrete applications have already been conducted. Richard Meier's Dives in Misericordia Church project in Rome [11], the New Road Construction Concepts (NR2C), the Air Quality Innovation Programme (IPL) in the Netherlands, and Photocatalytic Innovative Coverings Applications for Depollution (PICADA) are examples of such trials and projects [5]. These approaches have typically focused on the addition of the photocatalytic materials to the binder phase, where their performance can be strongly influenced by poor dispersibility or occlusion of the added photocatalytic powder [12–14]. Moreover, some reported studies address the impact of photocatalyst additions on the practical characteristics of concrete, e.g., workability and durability [15]. Despite our level of understanding of these mechanisms, there have been few innovative solutions to maintain the cement-free photocatalyst performance level over time.

Therefore, engineering photocatalytically active structures in a way that ensures high performance in air purification is of great importance from both environmental and economical perspectives [16–18]. The approach suggested in this work is supportive of TiO_2-based photocatalysts on aggregate materials such that the photocatalyst is positioned above the surface of the concrete. Placement of such aggregates on the surface of the concrete structure during construction ensures that the photocatalyst will have free access to atmospheric pollutants (particularly NOx), rainwater and illuminating radiation, and be free of agglomeration issues. However, although this approach has been previously reported, efficient binding of photocatalytic material to the support surface is still a challenging issue.

2. Materials and Methods

2.1. Synthesis of TiO_2–SiO_2–Quartz Composites

Standard quartz sand (Q) was first sieved to obtain a sample with the desired particle size distribution (1–3 mm) before it was washed with deionized water several times and dried at 95 °C. A commercial photocatalyst (CristalACTiV™ PC105, Cristal, Stallingborough, UK) was supported on the obtained Q using a silica-based binder prepared from the precursor tetraethyl orthosilicate (TEOS, Sigma-Aldrich, Schnelldorf, Germany). A TEOS mother solution was prepared by adding the exact amount of TEOS into an ethanol:water:HCl mixture at a TEOS:ethanol:water:HCl molar ratio of $1:3:4:4 \times 10^{-3}$, which was stirred for 10 days at room temperature in a sealed vessel [19]. Then the required volumes of the obtained solution were added to 200 mL of ethanol, in which the required amounts of TiO_2 were suspended, to get 0.5, 1, 2.5, 5, and 10 wt % TiO_2:Quartz in the final product. These values were chosen because many reports identified the efficacy of TiO_2 photocatalysts conventionally mixed into cementitious materials (with TiO_2 loadings from 1 to 10 wt % as a fraction of the cement content) [20–22]. The molar ratio of TiO_2:TEOS was always kept at 1:1. The suspension was then kept gently stirring at room temperature overnight. It was then added dropwise to 40 g of Q with continuous stirring at 80 °C under reduced pressure. The obtained materials were dried at 90 °C overnight followed by heat treatment at 200 °C for 4 h. The thus obtained modified Q samples were then cooled in air, washed with deionized water, dried at 90 °C, and sieved again to collect particles larger than 1 mm in order to separate modified quartz from loosely or non-connected TiO_2-binder. The resulting composite samples are denoted QTx where x represents expected TiO_2 content. In order to simplify the characterisation, fine powdered Quartz (particles sizes in the range 20–50 μm obtained by ball milling of commercial quartz (Aldrich) and sieving) was also modified with 10 wt % TiO_2 following a similar procedure; the obtained sample is denoted FQT.

2.2. Characterization

X-ray diffraction (XRD) patterns were obtained using a PANalytical diffractometer (X'Pert[3] Powder, Malvern Panalytical, UK) equipped with a CuKa1 1.54 Å X-ray source. FTIR spectra were recorded using a PerkinElmer Spectrum Two equipped with UATR (Single Reflection Diamond, PerkinElmer, Inc, City, State, USA). UV-vis diffuse reflectance spectra of the samples were recorded using a Cary 60 UV-vis spectrophotometer (Agilent Technology, City, State, USA) equipped with a fibre optic coupler. Barium sulphate was used as a reference in the range of 250 to 600 nm. The resulting reflectance spectra were transformed into apparent absorption spectra by using the Kubelka−Munk function $F(R\infty) = (1 - R\infty)^2/2R\infty$. The amount of TiO_2 loaded on the surface of Q was analyzed by X-ray fluorescence spectroscopy (XRF) using Rigaku NexQC. For this purpose, a calibration series of TiO_2–SiO_2 was prepared by mixing the required amount of TiO_2 (PC105) and SiO_2 (ball milled commercial quartz) to get a series from 0.1 to 20 wt % TiO_2. Morphologies of samples were observed using a scanning electron microscope (SEM, Zeiss EVO MA10, Zeiss, City, State, USA) equipped with an energy dispersive X-ray spectrometer (EDS, Oxford INCA) for elemental composition analyses. The transmission electron microscopy (TEM) was performed on a JEOL-JEM-2000EX microscope operated with an accelerating voltage of 200 kV; images were captured with a Gatan Erlangshen ES500W camera.

2.3. Photocatalytic Performance Test

The removal of nitrogen oxides (NOx), as examples of air pollutants, using the obtained quartz sand modified with the photocatalyst was tested to check their efficiency in purification of the polluted air [23]. Figure 1 illustrates the air-purification test set-up which was established for this purpose. The set-up consists of the following (see Figure 1): gas suppliers; i.e., synthetic air and NO in nitrogen (BOC, Guildford, UK), gas flow controllers (Bronkhorst, Newmarket Suffolk, UK) (1), a humidity supplier unit (2), a photocatalytic reactor according to ISO standard design (ISO 22197-1) (3), a UV(A) irradiation source (4), and a NOx analyzer (5). The gas supplies were NO (100 ppm) in N_2 and synthetic air (BOC, UK). The required concentration (1000 ppbv), flow rates (5×10^{-5} $m^3 \cdot s^{-1}$), and humidity (ca. 40%, confirmed by a Rotronic hygropalm) were achieved by controlling the flow of each gas by the gas flow controllers (1). The photoreactor was constructed from PMMA (Poly(methyl methacrylate)), covered with borosilicate glass, and was positioned below the output from an SS0.5 kW, 500 W fully reflective solar simulator equipped with a 1.5 AM filter (Sciencetech, London, Ontario, Canada). The irradiation was adjusted to ensure that the test sample (6) received a light intensity of 10 Wm^{-2} at $\lambda < 420$ nm. The light intensity was measured using a broadband thermopile detector (Gentec-EO-XLP12-3S-H2-D0) located exactly where the test sample should be placed. A Thermo Scientific Model 42i-HL High Level NO-NO_2-NOx Analyzer (Air Monitors Ltd., Gloucestershire, UK) was used to monitor the concentrations of NO, NO_2, and total NOx in the outlet gas flow. Aggregate exposed mortars were prepared by supporting 10 g of bare or TiO_2-modified quartz sand on the top of mortars (with top surface area of 64 cm^2 and thickness of 15 mm) in order to examine their efficiencies in NOx removal. For comparison, a reference sample was prepared according to the conventional mixing method (CMM) in which TiO_2 was first mixed with cement at a 10 wt % ratio. Once prepared, a 2 mm thick layer of the mortar mixture was applied to the top of the pre-made mortar similar to those used to prepare the aggregate exposed mortars.

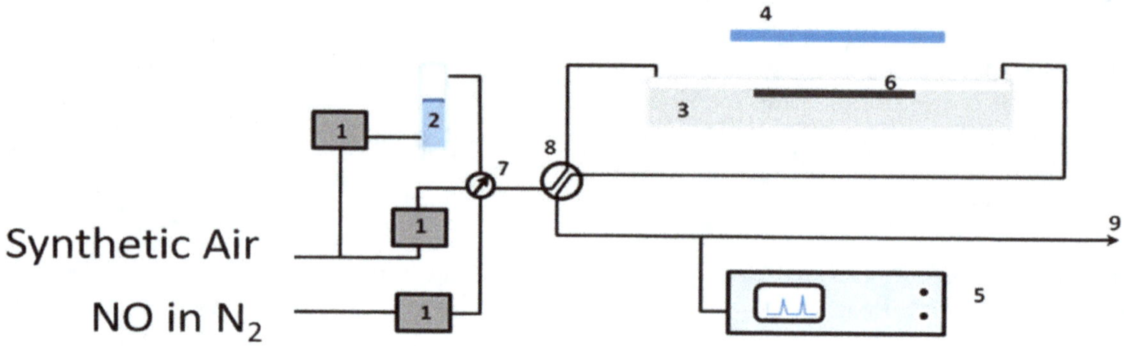

Figure 1. The photocatalytic testing apparatus: (1) mass flow controllers, (2) humidity supplier, (3) photocatalytic reactor, (4) UV(A) irradiation source, (5) NOx analyzer, (6) test sample, (7) and (8) valves, and (9) gas stream outlet.

3. Results and Discussion

XRD patterns of uncoated quartz sand (Q) and the prepared TiO_2–SiO_2–Quartz composite (FQT) are shown in Figure 2. The appearance of a clear XRD pattern for anatase TiO_2 in FQT sample confirms the successful support of TiO_2 on the quartz in FQT.

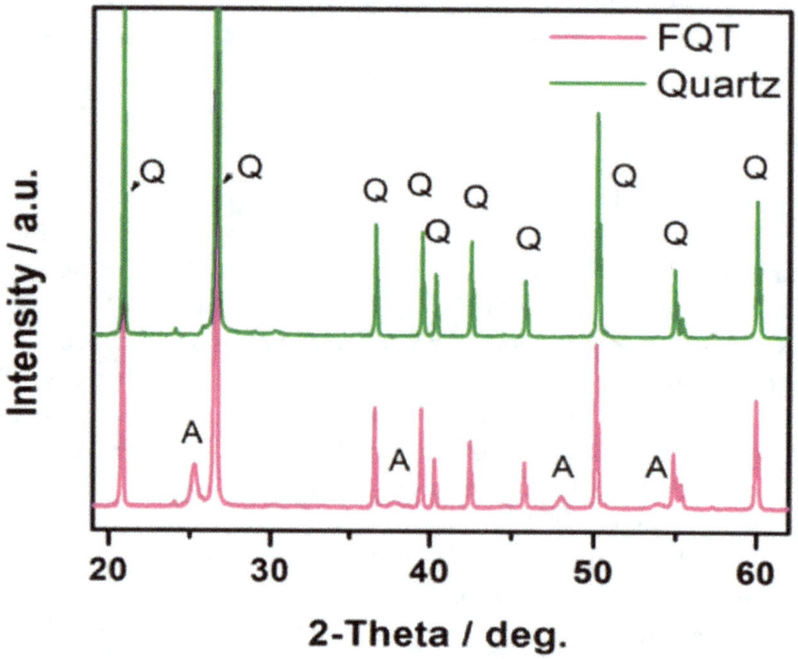

Figure 2. XRD patterns of bare quartz and TiO_2–SiO_2–Quartz composite (FQT), Q and A refer to quartz and anatase patterns, respectively.

Evidence for the formation of a chemical bond between the TiO_2 and SiO_2 binding layers can clearly be observed from the FTIR spectrum in the range between 900–960 cm^{-1} (Figure 3), the absorption being assigned to the stretching vibrational mode of the Ti–O–Si bond. No similar absorption peak due to this mode was observed for SiO_2 in the bare Q sample.

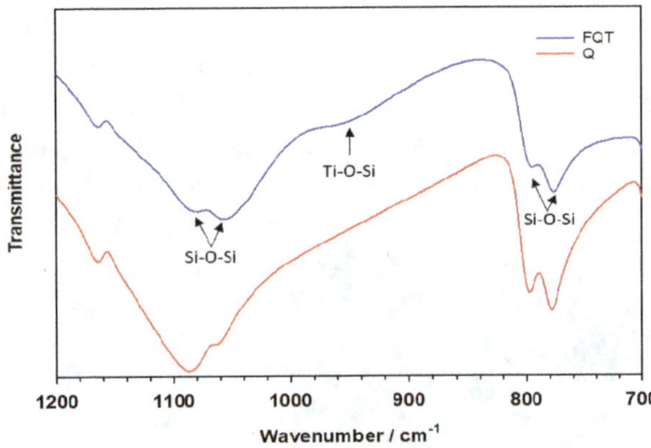

Figure 3. FTIR spectra of bare quartz (Q) and the TiO_2–SiO_2–Quartz composite (FQT).

Figure 4 shows the diffuse reflectance UV-vis absorption spectra of free TiO_2 and FQT composite samples. Considering TiO_2 as an indirect semiconductor, the modified Kubelka–Munk function $[F(R\infty)h\nu]^{1/2}$ was plotted as a function of the incident photon energy which allowed the determination of the photocatalyst's band gap. The spectra are additional evidence for the presence of TiO_2 in the FQT sample and show that the loading of TiO_2, via the herein employed method, on the surface of quartz through a SiO_2 binding layer has negligible effect on its band gap.

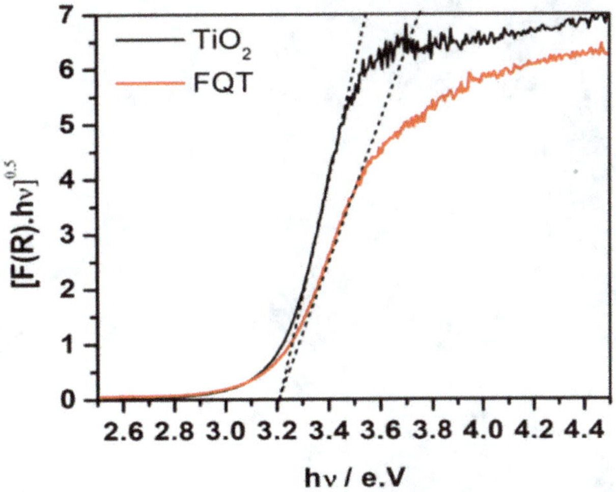

Figure 4. UV-vis absorption spectra presented as the modified Kubelka–Munk function of free TiO_2 and TiO_2–SiO_2–Quartz composite (FQT).

The amounts of TiO_2 loaded on the surface of Q, as recorded by XRF analyses, are summarized in Table 1. As can be seen from this table, the actual amount of TiO_2 loaded is lower than the calculated values for all the prepared samples, meaning that some loss of TiO_2 occurred during the preparation procedure. However, the lower the loaded amount the lower the TiO_2 loss. This indicates a limitation of the modifying layer thickness over which all the extra added amount of TiO_2-binder is lost.

Table 1. Calculated and analyzed loaded amount of TiO_2 on quartz via silica gel binder.

Sample ID	TiO_2 (wt %)		TiO_2 Loss (%)
	Calculated	Analyzed	
QT0.5	0.5	0.35	30
QT1	1.0	0.63	37
QT2.5	2.4	0.72	70
QT5	4.7	2.9	38
QT10	9.1	2.67	71

The level of coverage of the support (Q) with TiO_2 was examined using the SEM-EDS technique. Figure 5 compares the SEM-EDS of bare quartz coated with a silicate layer derived from TEOS in which commercial TiO_2 (PC105) is dispersed. The effectiveness of a silicate-based layer on quartz (FQT) for the efficient support of TiO_2 can be clearly seen from these SEM images. Very good coverage of TiO_2 over the sample is achieved as a result of coating a silicate-based film over the quartz substrate.

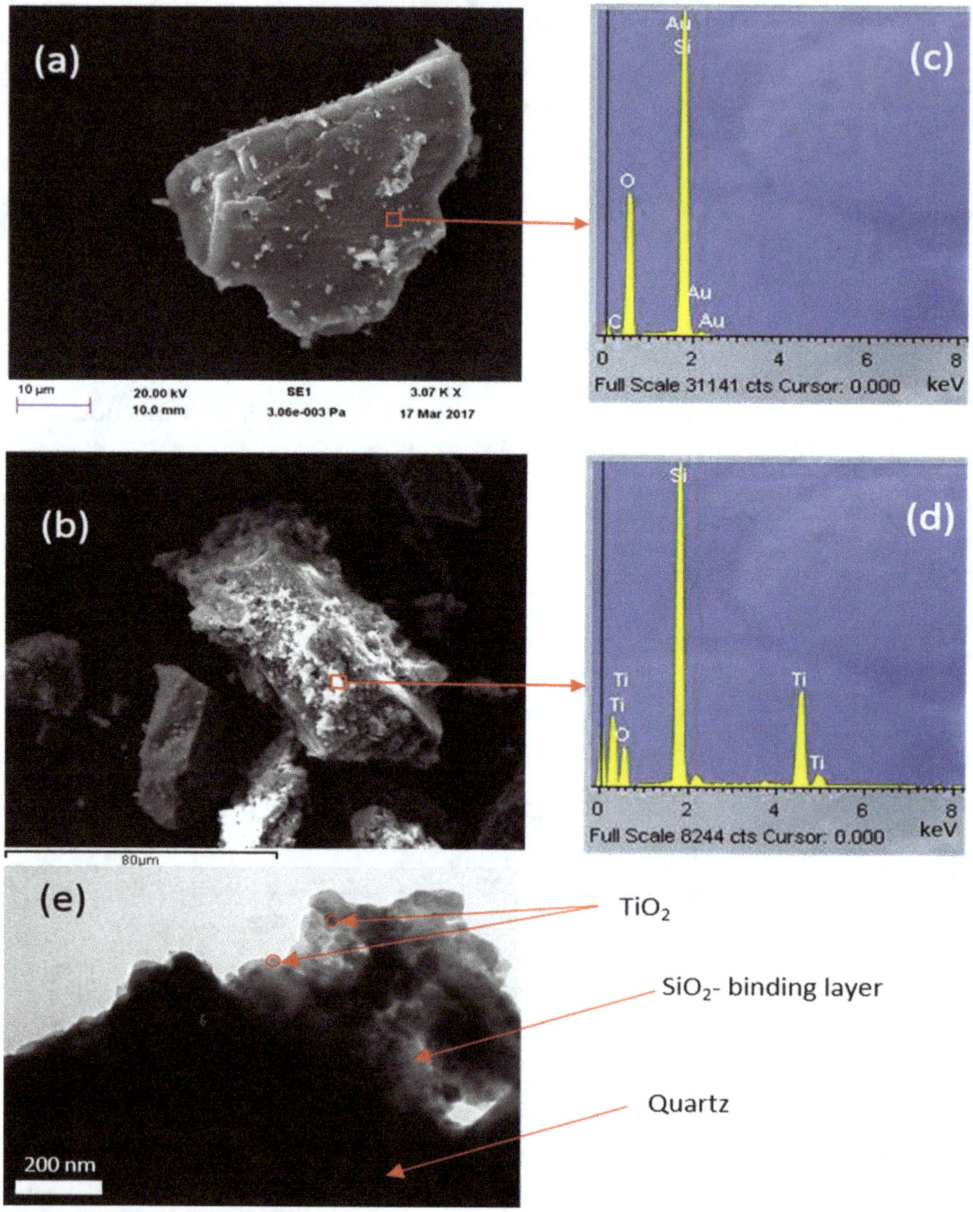

Figure 5. SEM images for (**a**) uncoated quartz and (**b**) TiO_2 (commercial PC105) immobilized on quartz via a SiO_2 binder formed from tetraethyl orthosilicate (TEOS). EDS of these samples are presented in (**c**,**d**), and the TEM image for the same samples in (**e**).

Although a few areas might be clear of silicate coating, Figure 5 shows that the silicate layer has generally been immobilized nicely on the grains. Consequently, in this case, TiO_2, associated with the silicate-based gel phase, is also similarly distributed and is not bonded directly to the quartz surface. This is consistent with the TEM image in Figure 5e. Comparing the EDS analyses (Figure 5c,d) indicates the spreading of TiO_2 with the silicate layer.

In order to examine the effectiveness of the obtained modified quartz sand samples for the removal of nitrogen oxides from polluted air, aggregate exposed mortars (with top surface areas of

64 cm^2) were prepared, on top of which 10 g of bare or TiO_2-modified quartz sand were supported. An image of the obtained mortars is presented in Figure 6 which shows a clear change in the color of the quartz sand to a whiter one by increasing the amounts of loaded TiO_2.

Figure 6. Images of the prepared aggregate exposed mortars. On the top of each sample, 10 g of bare (Q) or TiO_2-modified quartz sand (QTx) were supported.

Figure 7 shows an example of the changes in the concentrations of nitrogen oxides, i.e., NO, NOx, and NO_2, in the gas stream flows over the QT10 sample in the dark and under irradiation. The concentration of NO is constant at ca. 1000 ppb as long as the light is off. When the light is switched on, the initial NO concentration drops with a simultaneous increase in NO_2 concentration, as NO_2 is one of the NO oxidation products. Consequently, the concentration of NOx, which reflects the total oxidation of NO to HNO_3, is reduced during the illumination time.

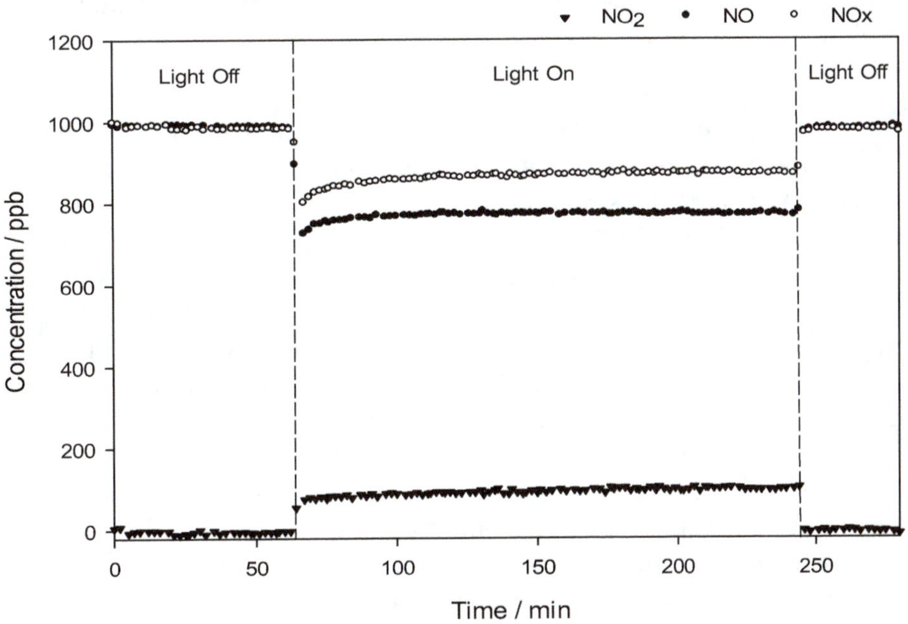

Figure 7. Changes in the concentrations of nitrogen oxides, i.e., NO, NO_2, and NOx, as a function of time in the presence of the QT10 sample.

The activities of the obtained samples for De-NOx remediation were determined by measuring the photonic efficiencies (ξ), which is defined as the ratio of the reaction rate (NO and NOx removal as well as NO_2 formation) and the incident photon flux [24]. Equation (1) was used to calculate ξ [25], where

$\dot{V}$ is the volumetric flow rate, c_d the concentration of each of the nitrogen oxide gases recorded under dark conditions, c_i the concentration of the same gases recorded under illumination, p the pressure, N_A the Avogadro constant, h the Plank constant, c the speed of light, I the incident irradiation intensity, λ the employed wavelength assuming monochromatic light (365 nm), A the irradiated area, R the gas constant, and T the absolute temperature. The obtained results are illustrated in Figure 8a. The results of nitrate selectivity, which was calculated according to Equation (2) employing the obtained ξ NO and ξ NO_2, both measured on the same sample, are shown in Figure 8b.

$$\xi = \frac{\dot{V}(c_d - c_i)\, pN_A hc}{I\lambda ART} \quad (1)$$

$$S = \frac{\xi_{NOx}}{\xi_{NO}} \quad (2)$$

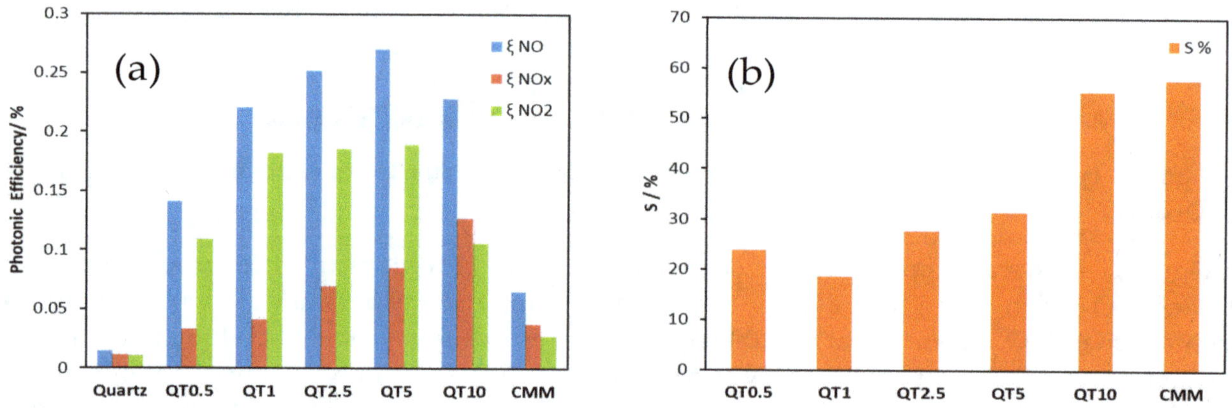

Figure 8. Photonic efficiencies of the aggregate exposed mortars prepared from bare quartz or TiO_2-modified quartz for NO and NOx removal and for NO_2 formation (**a**), and nitrate selectivity recorded for the same samples (**b**).

As can be seen from Figure 8a, the De-NOx performance (0.14–0.23% ξ NO) is clearly enhanced by supporting the photocatalyst on quartz sand when compared to its performance (0.06% ξ NO) when mixed with cement (CMM). This means that supporting TiO_2 on the quartz surface via a silica gel binder ensures a much higher effective surface area which is translated into higher photocatalytic efficiency. It is worth noting that the herein recorded photonic efficiency for NO oxidation measured for TiO_2 supported on quartz (the QT5 sample for example) is about 58% of that reported in literature for the highly active bare anatase TiO_2 (Hombikat UV100) [25]. Bearing in mind that QT5 had only 2.9% of the TiO_2 loaded, a significantly efficient utilization of the valuable photocatalyst is consequently achieved by supported systems. It is also worth mentioning that although the De-NOx performance improves by increasing the amount of photocatalyst loaded from QT1 to QT5, QT1 might be preferred from an economical viewpoint as this sample showed a ξ NO of more than 80% of that measured for QT5 although QT1 had ca. five times less TiO_2 compared to QT5. On the other hand, nitrate selectivity (Figure 8b) was also enhanced by increasing the amount of photocatalyst loaded but did not exceed that of CMM. The high selectivity is surely a valuable opportunity and improved by a higher loading amount. This observation can be explained by the fact that the porous silica gel coating presents a support for TiO_2 both as TiO_2 bonded to the silica gel pore walls (which usually negatively affects the selectivity) but also as TiO_2 trapped in the pores. This non-bonded TiO_2 offers increased activity and selectivity. Factors which control nitrate selectivity appear to be complex. Relevant variables include TiO_2 polymorphism, defect state, availability of water, etc., but the role of substrate binding must also be considered.

4. Conclusions

TiO_2 has been successfully loaded on the surface of aggregate, with quartz used as an example, utilizing a silicate-based binder. Aggregate exposed mortars with these photocatalytically active aggregates on the surface have shown an excellent performance at the photocatalytic removal of NO from polluted air. The herein developed aggregate exposed mortars have shown up to ca. four times higher efficiency for NO removal compared with the sample prepared according to the conventional mixing method (CMM). Thus, the herein prepared supported TiO_2 aggregates offer much greater utilization efficiency for the valuable photocatalyst and thus correspond to a considerably lower cost of use in photocatalytic concrete.

Author Contributions: Conceptualization, A.H. and D.E.M.; methodology, A.H. and L.Y.; validation, A.H., L.Y. and D.E.M.; formal analysis, A.H.; investigation, A.H.; resources, D.E.M. and F.W.; data curation, A.H.; writing—original draft preparation, A.H.; writing—review and editing, D.E.M., A.E. and Y.A.; visualization, A.H., D.E.M. and A.E.; supervision, D.E.M. and F.W.; project administration, D.E.M.; funding acquisition, D.E.M. and F.W.

Acknowledgments: The authors gratefully acknowledge funding from the UK Engineering and Physical Sciences Research Council (Grant Ref: EP/M003299/1) and the Natural Science Foundation of China (No. 51461135005) International Joint Research Project (EPSRC-NSFC). L. Zheng and M. R. Jones from the Division of Civil Engineering, University of Dundee, Dundee, UK are gratefully acknowledged for the preparation of the mortars.

References

1. West, J.J.; Cohen, A.; Dentener, F.; Brunekreef, B.; Zhu, T.; Armstrong, B.; Bell, M.L.; Brauer, M.; Carmichael, G.; Costa, D.L.; et al. What We Breathe Impacts Our Health: Improving Understanding of the Link between Air Pollution and Health. *Environ. Sci. Technol.* **2016**, *50*, 4895–4904. [CrossRef] [PubMed]
2. Walton, H.; Dajnak, D.; Beevers, S.; Williams, M.; Watkiss, P.; Hunt, A. Understanding the Health Impacts of Air Pollution in London. *King's College London*, 15 July 2015.
3. WHO. *Ambient (Outdoor) Air Quality and Health*; WHO: Geneva, Switzerland, 2014.
4. Abdul-Wahab, S.A.; Chin Fah En, S.; Elkamel, A.; Ahmadi, L.; Yetilmezsoy, K. A review of standards and guidelines set by international bodies for the parameters of indoor air quality. *Atmos. Pollut. Res.* **2015**, *6*, 751–757. [CrossRef]
5. Serpone, N. Heterogeneous Photocatalysis and Prospects of TiO_2-Based Photocatalytic DeNOxing the Atmospheric Environment. *Catalysts* **2018**, *8*, 553. [CrossRef]
6. Engel, A.; Große, J.; Dillert, R.; Bahnemann, D.W. The Influence of Irradiance and Humidity on the Photocatalytic Conversion of Nitrogen(II) Oxide. *J. Adv. Oxid. Technol.* **2015**. [CrossRef]
7. Devahasdin, S.; Fan, C.; Li, K.; Chen, D.H. TiO_2photocatalytic oxidation of nitric oxide: Transient behavior and reaction kinetics. *J. Photochem. Photobiol. A Chem.* **2003**, *156*, 161–170. [CrossRef]
8. Maggos, T.; Bartzis, J.G.; Liakou, M.; Gobin, C. Photocatalytic degradation of NOx gases using TiO_2-containing paint: A real scale study. *J. Hazard. Mater.* **2007**, *146*, 668–673. [CrossRef] [PubMed]
9. Hüsken, G.; Hunger, M.; Brouwers, H.J.H. Experimental study of photocatalytic concrete products for air purification. *Build. Environ.* **2009**, *44*, 2463–2474. [CrossRef]
10. Ballari, M.M.; Brouwers, H.J.H. Full scale demonstration of air-purifying pavement. *J. Hazard. Mater.* **2013**, *254–255*, 406–414. [CrossRef] [PubMed]
11. Cassar, L. Photocatalysis of cementitious materials: Clean buildings and clean air. *MRS Bull.* 2004. [CrossRef]
12. Folli, A.; Pochard, I.; Nonat, A.; Jakobsen, U.H.; Shepherd, A.M.; Macphee, D.E. Engineering photocatalytic Cements: Understanding TiO_2surface chemistry to control and modulate photocatalytic performances. *J. Am. Ceram. Soc.* **2010**, *939*, 3360–3369. [CrossRef]
13. MacPhee, D.E.; Folli, A. Photocatalytic concretes—The interface between photocatalysis and cement chemistry. *Cem. Concr. Res.* **2016**, *85*, 48–54. [CrossRef]
14. Folli, A.; Pade, C.; Hansen, T.B.; De Marco, T.; MacPhee, D.E. TiO_2 photocatalysis in cementitious systems: Insights into self-cleaning and depollution chemistry. *Cem. Concr. Res.* **2012**, *42*, 539–548. [CrossRef]
15. Zhang, R.; Cheng, X.; Hou, P.; Ye, Z. Influences of nano-TiO_2 on the properties of cement-based materials: [illegible] drying shrinkage. *Constr. Build. Mater.* **2015**, *81*, 35–41. [CrossRef]

16. Yang, L.; Wang, F.; Hakki, A.; Macphee, D.E.; Liu, P.; Hu, S. The influence of zeolites fly ash bead/TiO_2 composite material surface morphologies on their adsorption and photocatalytic performance. *Appl. Surf. Sci.* **2017**, *392*, 687–696. [CrossRef]
17. Hakki, A.; Yang, L.; Wang, F.; Macphee, D.E. The Effect of Interfacial Chemical Bonding in TiO_2 Composites on Their Photocatalytic NOx Abatement Performance. *J. Vis. Exp.* **2017**, *4*. [CrossRef]
18. Yang, L.; Hakki, A.; Wang, F.; Macphee, D.E. Different Roles of Water in Photocatalytic DeNOx Mechanisms on TiO_2: Basis for Engineering Nitrate Selectivity? *ACS Appl. Mater. Interfaces* **2017**, *9*, 17034–17041. [CrossRef]
19. Fateh, R.; Dillert, R.; Bahnemann, D. Preparation and characterization of transparent hydrophilic photocatalytic TiO_2/SiO_2 thin films on polycarbonate. *Langmuir* **2013**, *29*, 3730–3739. [CrossRef]
20. Hanus, M.J.; Harris, A.T. Nanotechnology innovations for the construction industry. *Prog. Mater. Sci.* **2013**, *58*, 1056–1102. [CrossRef]
21. Lucas, S.S.; Ferreira, V.M.; De Aguiar, J.L.B. Incorporation of titanium dioxide nanoparticles in mortars—Influence of microstructure in the hardened state properties and photocatalytic activity. *Cem. Concr. Res.* **2013**, *43*, 112–120. [CrossRef]
22. Ballari, M.M.; Hunger, M.; Hüsken, G.; Brouwers, H.J.H. NOx photocatalytic degradation employing concrete pavement containing titanium dioxide. *Appl. Catal. B Environ.* **2010**, *95*, 245–254. [CrossRef]
23. ISO. *ISO 22197-1: Fine Ceramics (Advanced Ceramics, Advanced Technical Ceramics)—Test Method for air-purification Performance of Semiconducting Photocatalytic Materials—Part 1: Removal of Nitric Oxide*; ISO: Geneva, Switzerland, 2007.
24. Kisch, H.; Bahnemann, D. Best Practice in Photocatalysis: Comparing Rates or Apparent Quantum Yields? *J. Phys. Chem. Lett.* **2015**, *6*, 1907–1910. [CrossRef] [PubMed]
25. Freitag, J.; Domínguez, A.; Niehaus, T.A.; Hülsewig, A.; Dillert, R.; Frauenheim, T.; Bahnemann, D.W. Nitrogen(II) oxide charge transfer complexes on TiO_2: A new source for visible-light activity. *J. Phys. Chem. C* **2015**, *119*, 4488–4501. [CrossRef]

Permissions

 Every chapter published in this book has been scrutinized by our experts. Their significance has been extensively debated. The topics covered herein carry significant findings which will fuel the growth of the discipline. They may even be implemented as practical applications or may be referred to as a beginning point for another development.

The contributors of this book come from diverse backgrounds, making this book a truly international effort. This book will bring forth new frontiers with its revolutionizing research information and detailed analysis of the nascent developments around the world.

We would like to thank all the contributing authors for lending their expertise to make the book truly unique. They have played a crucial role in the development of this book. Without their invaluable contributions this book wouldn't have been possible. They have made vital efforts to compile up to date information on the varied aspects of this subject to make this book a valuable addition to the collection of many professionals and students.

This book was conceptualized with the vision of imparting up-to-date information and advanced data in this field. To ensure the same, a matchless editorial board was set up. Every individual on the board went through rigorous rounds of assessment to prove their worth. After which they invested a large part of their time researching and compiling the most relevant data for our readers.

The editorial board has been involved in producing this book since its inception. They have spent rigorous hours researching and exploring the diverse topics which have resulted in the successful publishing of this book. They have passed on their knowledge of decades through this book. To expedite this challenging task, the publisher supported the team at every step. A small team of assistant editors was also appointed to further simplify the editing procedure and attain best results for the readers.

Apart from the editorial board, the designing team has also invested a significant amount of their time in understanding the subject and creating the most relevant covers. They scrutinized every image to scout for the most suitable representation of the subject and create an appropriate cover for the book.

The publishing team has been an ardent support to the editorial, designing and production team. Their endless efforts to recruit the best for this project, has resulted in the accomplishment of this book. They are a veteran in the field of academics and their pool of knowledge is as vast as their experience in printing. Their expertise and guidance has proved useful at every step. Their uncompromising quality standards have made this book an exceptional effort. Their encouragement from time to time has been an inspiration for everyone.

The publisher and the editorial board hope that this book will prove to be a valuable piece of knowledge for researchers, students, practitioners and scholars across the globe.

List of Contributors

Hiba Najini, Mutasim Nour and Fadi Ghaith
School of Engineering and Physical Sciences, Heriot-Watt University, Edinburgh EH14 4AS, UK

Sulaiman Al-Zuhair
Chemical Engineering Department, UAE University, Al-Ain 15551, UAE

Uthpala Rathnayake and Cheuk Lun Chow
Department of Architecture and Civil Engineering, City University of Hong Kong, Hong Kong, China

Denvid Lau
Department of Architecture and Civil Engineering, City University of Hong Kong, Hong Kong, China
Department of Civil and Environmental Engineering, Massachusetts Institute of Technology, Cambridge, MA 02139, USA

Wael Alattyih and Husnain Haider
Department of Civil Engineering, College of Engineering, Qassim University, Buraydah, Qassim 51452, Saudi Arabia

Halim Boussabaine
Faculty of Business and Law, The British University in Dubai, Dubai 345015, UAE

Camille Pajot, Benoit Delinchant and Yves Maréchal
University Grenoble Alpes, CNRS, Grenoble INP, G2Elab, 38000 Grenoble, France

Damien Frésier
Gaz et Electricité de Grenoble, 57 Rue Pierre Semard, 38000 Grenoble, France

Shahana Y. Janjua and Prabir K. Sarker
School of Civil and Mechanical Engineering, Curtin University, Perth 6102, Australia

Wahidul K. Biswas
Sustainable Engineering Group, Curtin University, Perth 6102, Australia

Faiz Shaikh
School of Civil and Mechanical Engineering, Curtin University, Perth, WA 6120, Australia

Stephen Y. C. Yim
Development and Construction Division, Housing Department, Housing Authority Headquarters, Ho Man Tin, Hong Kong, China

S. Thomas Ng and M. U. Hossain
Department of Civil Engineering, The University of Hong Kong, Pokfulam, Hong Kong, China

James M.W. Wong
Research & Development, Construction Industry Council, 38/F, COS Centre, 56 Tsun Yip Street, Kwun Tong, Kowloon, Hong Kong, China

Jeeyoung Park and Stephen Wittkopf
School of Engineering and Architecture, Lucerne University of Applied Sciences and Arts, Technikumstrasse 21, CH-6048 Horw, Switzerland

Dirk Hengevoss
Institute for Ecopreneurship, School for Life Sciences, University of Applied Sciences Northwestern Switzerland, Hofackerstrasse 30, 4132 Muttenz, Switzerland

Umair Hasan and Andrew Whyte
School of Civil and Mechanical Engineering, Curtin University, Perth, WA 6845, Australia

Hamad Al Jassmi
Department of Civil and Environmental Engineering, United Arab Emirates University, Al Ain 17666, UAE

Upendra Rajapaksha
Department Architecture, Faculty of Architecture, University of Moratuwa, Moratuwa 10400, Sri Lanka

Donald E. Macphee
Department of Chemistry, University of Aberdeen, Meston Building, Meston Walk, Aberdeen AB24 3UE, UK

Ammar Elhoweris and Yousef Alhorr
Gulf Organisation of Research and Development, Qatar Science and Technology Park, Doha 210162, Qatar

Amer Hakki
Department of Chemistry, University of Aberdeen, Meston Building, Meston Walk, Aberdeen AB24 3UE, UK
Gulf Organisation of Research and Development, Qatar Science and Technology Park, Doha 210162, Qatar

Fazhou Wang
State Key Laboratory of Silicate Materials for Architectures,Wuhan University of Technology, 122# Luoshi Road, Wuhan 430070, China

Lu Yang
Department of Chemistry, University of Aberdeen, Meston Building, Meston Walk, Aberdeen AB24 3UE, UK
State Key Laboratory of Silicate Materials for Architectures,Wuhan University of Technology, 122# Luoshi Road, Wuhan 430070, China

Index

A

Air Conditioned Buildings, 206, 208, 223

B

Building Components, 52, 96-101, 103, 105-107, 109-111, 114, 116, 145, 173

Building Envelope, 2, 21, 24, 39, 47, 87-88, 95, 115, 149-150, 207-208

Building Integrated Photovoltaic, 19, 46, 150, 153

Building Sector, 1-2, 5, 21, 48, 50, 72-73, 115, 131-132, 146-147, 206-207, 223

Building Stock, 86, 210-216, 218, 220, 224-225

C

Chiller Load, 8-9, 11, 15

Concrete Block Wall, 103, 106-107, 110

Condensate Water, 7, 11

Construction Equipment, 50, 107-108

Construction Projects, 1, 3, 51, 60, 74, 79-80, 195

Cooling Load, 2, 8-9, 11, 14, 30, 152, 206, 214

Cross-code Analysis, 1, 8, 10-11

D

Demolition Waste, 100, 106-108, 147, 171, 201-202

Design Parameters, 2-3, 8-11, 14, 19, 52, 71, 210

E

Economic Analysis, 1, 7, 9, 12, 14-16, 80, 202

Economic Cost, 39, 41, 43-44

Electricity Consumption, 5, 8, 137, 140, 156-158, 162, 165

Electricity Production, 154, 162-163, 165

Energy Demand, 1, 3, 5, 20-21, 23-24, 46, 96-97, 100, 103, 106, 111, 150, 154, 159, 164, 200, 202, 210, 213, 215, 220, 222, 225

Energy Distribution, 5

Energy Efficiency, 1-2, 19-20, 23-24, 30, 36, 38, 42-45, 48, 54, 71-72, 120, 202, 207-208, 210, 223-225

Energy Intensity, 2-3, 5, 132

Energy Performance, 1, 3-5, 8-9, 12, 14-15, 19-21, 30, 33, 36, 41-42, 45, 47, 49, 117, 152-153, 167, 208, 225

Energy Utility, 5, 8, 15, 210

Environmental Impact, 2, 7-8, 16, 24, 44, 69, 103, 111, 140, 145-147, 149, 152-153, 155, 159, 162-166, 170-171, 173, 175, 195

Environmental Load, 170-171, 174, 194-195, 219

Environmental Performance, 96-97, 99-101, 106, 111, 115, 132, 134, 145-146, 148-149, 200-201

Environmental Risks, 52, 54, 56, 70, 72-73

F

Fire Performance, 23-25, 36-37, 42, 44, 46, 48

Fire Safety, 23-26, 37, 42-43, 45

Footing Systems, 101, 103

Full-time Workers, 182, 187, 191

Functional Risks, 52, 56, 71-72

G

Glass Façade, 23, 30, 36-37, 45, 161

Green Building, 1-5, 7, 11-12, 14, 18-24, 38-39, 41-44, 48, 50-55, 58, 60-64, 67, 70, 72-75, 79-80, 119-120, 128, 130, 170, 224

Green Star Concrete, 119-120, 123-126, 128-130

Greenhouse Gas Emissions, 40, 96, 115-116, 147-149, 171, 173-174, 202

H

Heat Energy, 26-27, 29

Heat Stress, 206, 208, 211, 219-223, 225

High-rise Buildings, 2-3, 5, 132, 147

I

Indoor Overheating, 206, 208, 221

L

Life Cycle Assessment, 46, 96, 99-100, 103, 111, 115-118, 129, 131-132, 146-150, 152, 154-155, 167-168, 202, 204

M

Management Risk, 50, 70, 74

N

Non-structural Components, 96-97, 100-101, 103, 107

O

Occupancy Density, 3, 5, 7, 10-11, 14

Operational Energy, 97, 206-207

P

Peak Shaving, 82-84, 86, 89-90, 92

Photovoltaic Laminate, 151, 154, 159

Post-use Stage, 105-111, 114

Power Density, 5, 14

Project Management, 50, 66, 69-70

Project Value, 50, 52, 60, 62, 64, 70-72, 74-75

Public Transport Services, 173, 186, 195

R
Rain Cladding, 150, 161, 163, 166
Reference Building, 4-5, 7, 10-11, 14-15, 96, 101, 103, 107, 110-111
Replacement Intervals, 96-97, 101
Residential Building, 3, 20, 103, 115-117, 131-132, 134, 144-146, 173, 182, 187, 196
Risk Factors, 50-53, 55-56, 58, 60-64, 67, 69-75, 80
Road Networks, 170, 172, 195, 200
Roof System, 100-101, 103

S
Sensitivity Index, 5, 8
Service Life, 52, 96-101, 103, 105-109, 111, 114-117, 137-138, 144-145
Shallow Plan Forms, 206, 211, 214, 216, 218, 220, 223
Solar Heat Gain, 23, 33, 47, 208, 215
Structural Components, 96-97, 100-101, 103, 107, 110-111
Syntactic Constraints, 169, 179

T
Thermal Mass, 27-28, 39, 42, 45, 48, 94, 146, 208-209, 222
Thermal Model, 82, 87-88
Thermal Performance, 20, 31-33, 36, 45-47, 206, 210-211, 213, 218, 222, 224
Timber Roof Frame, 100, 103
Transportation System, 170-172, 175, 182, 194

V
Value Creation, 50-56, 60, 64, 67, 69-70, 72, 74, 79
Value Engineering, 50, 52, 60, 72-74, 80-81

W
Wash Water, 119-126, 129
Water Consumption, 1, 4, 7, 10-12, 14, 96, 100-101, 103, 109-110, 117, 158
Water Layer, 23-24, 28-29, 38-39
Water Performance, 1, 14-15
Water Utility, 7, 14-15
Water Wall, 23-24, 27-33, 36, 38-39, 41-43, 45-46

www.ingramcontent.com/pod-product-compliance
Lightning Source LLC
LaVergne TN
LVHW081027210926
848485LV00004B/964